Statistical Regression and Classification

From Linear Models to Machine Learning

T0321030

CHAPMAN & HALL/CRC
Texts in Statistical Science Series

Series Editors
Joseph K. Blitzstein, *Harvard University, USA*
Julian J. Faraway, *University of Bath, UK*
Martin Tanner, *Northwestern University, USA*
Jim Zidek, *University of British Columbia, Canada*

Statistical Process Control: Theory and Practice, Third Edition
G.B. Wetherill and D.W. Brown

Generalized Additive Models: An Introduction with R, Second Edition
S. Wood

Epidemiology: Study Design and Data Analysis, Third Edition
M. Woodward

Practical Data Analysis for Designed Experiments
B.S. Yandell

Texts in Statistical Science

Statistical Regression and Classification
From Linear Models to Machine Learning

Norman Matloff

University of California, Davis, USA

CRC Press
Taylor & Francis Group
Boca Raton   London   New York

CRC Press is an imprint of the
Taylor & Francis Group an **informa** business

A CHAPMAN & HALL BOOK

Contents

Preface

Why write yet another regression book? There is a plethora of books out there already, written by authors whom I greatly admire, and whose work I myself have found useful. I might cite the books by Harrell [61] and Fox [50], among many, many excellent examples. Note that I am indeed referring to general books on regression analysis, as opposed to more specialized work such as [64] and [76], which belong to a different genre. My book here is intended for a traditional (though definitely modernized) regression course, rather than one on statistical learning.

Yet, I felt there is an urgent need for a different kind of book. So, why is this regression book different from all other regression books? First, **it modernizes the standard treatment of regression methods**. In particular:

- The book supplements classical regression models with introductory material on machine learning methods.

- Recognizing that these days, classification is the focus of many applications, the book covers this topic in detail, especially the multiclass case.

- In view of the voluminous nature of many modern datasets, there is a chapter on Big Data.

- There is much more hands-on involvement of computer usage.

Other major senses in which this book differs from others are:

- Though presenting the material in a mathematically precise manner, the book aims to provide much needed applied insight for the practicing analyst, remedying the "too many equations, too few explanations" problem.

For instance, the book not only shows how the math works for transformations of variables, but also raises points on why one might refrain from applying transformations.

- The book features a recurring interplay between parametric and nonparametric methods. For instance, in an example involving currency data, the book finds that the fitted linear model predicts substantially more poorly than a k-nearest neighbor fit, suggesting deficiencies in the linear model. Nonparametric analysis is then used to further investigate, providing parametric model assessment in a manner that is arguably more insightful than classical residual plots.

- For those interested in computing issues, many of the book's chapters include optional sections titled Computational Complements, on topics such as data wrangling, development of package source code, parallel computing and so on.

  Also, many of the exercises are code-oriented. In particular, in some such exercises the reader is asked to write "mini-CRAN" functions,[1] short but useful library functions that can be applied to practical regression analysis. Here is an example exercise of this kind:

  > Write a function stepAR2() that works similarly to stepAIC(), except that this new function uses adjusted R^2 as its criterion for adding or deleting a predictor. The call form will be

  ```
  stepAR2(lmobj,direction='fwd',
      nsteps=ncol(lmobj$model)-1)
  ```

  > where the arguments are...

- For those who wish to go into more depth on mathematical topics, there are Mathematical Complements sections at the end of most chapters, and math-oriented exercises. The material ranges from straightforward computation of mean squared error to esoteric topics such as a proof of the Tower Property, $E\left[E(V|U_1, U_2) \mid U_1\right] = E(V \mid U_1)$, a result that is used in the text.

As mentioned, **this is still a book on traditional regression analysis.** In contrast to [64], this book is aimed at a traditional regression course. Except for Chapters 10 and 11, the primary methodology used is linear and generalized linear parametric models, covering both the Description and Prediction goals of regression methods. We are just as interested in

---

[1]CRAN is the online repository of user-contributed R code.

Description applications of regression, such as measuring the gender wage gap in Silicon Valley, as we are in forecasting tomorrow's demand for bike rentals. An entire chapter is devoted to measuring such effects, including discussion of Simpson's Paradox, multiple inference, and causation issues. The book's examples are split approximately equally in terms of Description and Prediction goals. Issues of model fit play a major role.

The book includes more than 75 full examples, using real data. But concerning the above comment regarding "too many equations, too few explanations,", merely including examples with real data is not enough to truly tell the story in a way that will be useful in practice. Rather few books go much beyond presenting the formulas and techniques, thus leaving the hapless practitioner to his own devices. Too little is said in terms of what the equations really mean in a practical sense, what can be done with regard to the inevitable imperfections of our models, which techniques are too much the subject of "hype," and so on.

As a nonstatistician, baseball great Yogi Berra, put it in his inimitable style, "In theory there is no difference between theory and practice. In practice there is." This book aims to remedy this gaping deficit. It develops the material in a manner that is *mathematically precise* yet always maintains as its top priority — borrowing from a book title of the late Leo Breiman — *a view toward applications.*

In other words:

> The philosophy of this book is to not only prepare the analyst to know *how* to do something, but also to understand *what* she is doing. For successful application of data science techniques, the latter is just as important as the former.

Some further examples of how this book differs from the other regression books:

Intended audience and chapter coverage:

This book is aimed at both practicing professionals and use in the classroom. It aims to be both accessible and valuable to this diversity of readership.

In terms of classroom use, with proper choice of chapters and appendices, the book could be used as a text tailored to various discipline-specific audiences and various levels, undergraduate or graduate. I would recommend that the core of any course consist of most sections of Chapters 1-4 (excluding the Math and Computational Complements sections), with coverage of at least introductory sections of Chapters 5, 6, 7, 8 and 9 for all audiences. Beyond that, different types of disciplines might warrant different choices of further material. For example:

- **Statistics students:** Depending on level, at least some of the Mathematical Complements and math-oriented exercises should be involved. There might be more emphasis on Chapters 6, 7 and 9.

- **Computer science students:** Here one would cover more of classification, machine learning and Big Data material, Chapters 5, 8, 10, 11 and 12. Also, one should cover the Computational Complements sections and associated "mini-CRAN" code exercises.

- **Economics/social science students:** Here there would be heavy emphasis on the Description side, Chapters 6 and 7, with special emphasis on topics such as Instrumental Variables and Propensity Matching in Chapter 7. Material on generalized linear models and logistic regression, in Chapter 4 and parts of Chapter 5, might also be given emphasis.

- **Student class level:** The core of the book could easily be used in an undergraduate regression course, but aimed at students with background in calculus and matrix algebra, such as majors in statistics, math or computer science. A graduate course would cover more of the chapters on advanced topics, and would likely cover more of the Mathematical Complements sections.

- **Level of mathematical sophistication:** In the main body of the text, i.e., excluding the Mathematical Complements sections, basic matrix algebra is used throughout, but use of calculus is minimal. As noted, for those instructors who want the mathematical content, it is there in the Mathematical Complements sections, but the main body of the text requires only the matrix algebra and a little calculus.

The reader must of course be familiar with terms like *confidence interval*, *significance test* and *normal distribution*. Many readers will have had at least some prior exposure to regression analysis, but this is not assumed, and the subject is developed from the beginning.

The reader is assumed to have some prior experience with R, but at a minimal level: familiarity with function arguments, loops, if-else and vector/matrix operations and so on. For those without such background, there are many gentle tutorials on the Web, as well as a leisurely introduction in a statistical context in [21]. Those with programming experience can also read the quick introduction in the appendix of [99]. My book [100] gives a detailed treatment of R as a programming language, but that level of sophistication is certainly not needed for the present book.

A comment on the field of machine learning:

Mention should be made of the fact that this book's title includes both the word *regression* and the phrase *machine learning*. The latter phrase is included to reflect that the book includes some introductory material on machine learning, in a regression context.

Much has been written on a perceived gap between the statistics and machine learning communities [23]. This gap is indeed real, but work has been done to reconcile them [16], and in any event, the gap is actually not as wide as people think.

My own view is that machine learning (ML) consists of the development of regression models with the Prediction goal. Typically nonparametric (or what I call semi-parameteric) methods are used. Classification models are more common than those for predicting continuous variables, and it is common that more than two classes are involved, sometimes a great many classes. All in all, though, it's still regression analysis, involving the conditional mean of Y given X (reducing to $P(Y = 1|X)$ in the classification context).

One often-claimed distinction between statistics and ML is that the former is based on the notion of a sample from a population whereas the latter is concerned only with the content of the data itself. But this difference is more perceived than real. The idea of cross-validation is central to ML methods, and since that approach is intended to measure how well one's model generalizes beyond our own data, it is clear that ML people do think in terms of samples after all. Similar comments apply to ML's citing the variance-vs.-bias tradeoff, overfitting and so on

So, at the end of the day, we all are doing regression analysis, and this book takes this viewpoint.

Code and software:

The book also makes use of some of my research results and associated software. The latter is in my package **regtools**, available from CRAN [97]. A number of other packages from CRAN are used. Note that typically we use only the default values for the myriad arguments available in many functions; otherwise we could fill an entire book devoted to each package! Cross-validation is suggested for selection of tuning parameters, but with a warning that it too can be problematic.

In some cases, the **regtools** source code is also displayed within the text, so as to make clear exactly what the algorithms are doing. Similarly, data wrangling/data cleaning code is shown, not only for the purpose of "hands-on" learning, but also to highlight the importance of those topics.

Thanks:

Conversations with a number of people have enhanced the quality of this book, some via direct comments on the presentation and others in discussions not directly related to the book. Among them are Charles Abromaitis, Stuart Ambler, Doug Bates, Oleksiy Budilovsky, Yongtao Cao, Tony Corke, Tal Galili, Frank Harrell, Harlan Harris, Benjamin Hofner, Jiming Jiang, Hyunseung Kang, Martin Mächler, Erin McGinnis, John Mount, Richard Olshen, Pooja Rajkumar, Ariel Shin, Chuck Stone, Jessica Tsoi, Yu Wu, Yihui Xie, Yingkang Xie, Achim Zeileis and Jiaping Zhang.

A seminar presentation by Art Owen introduced me to the application of random effects models in recommender systems, a provocative blend of old and new. This led to the MovieLens examples and other similar examples in the book, as well as a vigorous new research interest for me. Art also led me to two Stanford statistics PhD students, Alex Chin and Jing Miao, who each read two of the chapters in great detail. Special thanks also go to Nello Cristianini, Hui Lin, Ira Sharenow and my old friend Gail Gong for their detailed feedback.

Thanks go to my editor, John Kimmel, for his encouragement and much-appreciated patience, and to the internal reviewers, David Giles, Robert Gramacy and Christopher Schmidt. Of course, I cannot put into words how much I owe to my wonderful wife Gamis and our daughter Laura, both of whom inspire all that I do, including this book project.

Website:

Code, errata, extra examples and so on are available at

http://heather.cs.ucdavis.edu/regclass.html.

A final comment:

My career has evolved quite a bit over the years. I wrote my dissertation in abstract probability theory [105], but turned my attention to applied statistics soon afterward. I was one of the founders of the Department of Statistics at UC Davis. Though a few years later I transferred into the new Computer Science Department, I am still a statistician, and much of my CS research has been statistical, e.g., [98]. Most important, my interest in regression has remained strong throughout those decades.

I published my first research papers on regression methodology way back in the 1980s, and the subject has captivated me ever since. My long-held wish has been to write a regression book, and thus one can say this work is 30 years in the making. I hope you find its goals both worthy and attained. Above all, I simply hope you find it an interesting read.

List of Symbols and Abbreviations

Y: the response variable
X: vector of predictor variables
$\widetilde{X}$: X with a 1 prepended
$X^{(j)}$: the jth predictor variable
n: number of observations
p: number of predictors
Y_i: value of the response variable in observation i
X_i: vector of predictors in observation i
$X_i^{(j)}$: value of the jth predictor variable in observation i
A: $n \times (p+1)$ matrix of the predictor data in a linear model
D: length-n vector of the response data in a linear model
H: the *hat matrix*, $A(A'A)^{-1}A'$
$\mu(t)$: the regression function $E(Y|X=t)$
$\sigma^2(t)$: $Var(Y|X=t)$
$\widehat{\mu}(t)$: estimated value of $\mu(t)$
β: vector of coefficients in a linear/generalized linear model
$\widehat{\beta}$: estimated value of β
$\prime$: matrix transpose
I: multiplicative identity matrix
k-NN: k-Nearest Neighbor method
MSE: Mean Squared (Estimation) Error
MSPE: Mean Squared Prediction Error
CART: Classification and Regression Trees
SVM: Support Vector Machine
NN: neural network
PCA: Principal Components Analysis
NMF: Nonnegative Matrix Factorization
OVA: One vs. All classification

AVA: All vs. All classification
LDA: Linear Discriminant Analysis

Chapter 1

Setting the Stage

This chapter will set the stage for the book, previewing many of the major concepts to be presented in later chapters. The material here will be referenced repeatedly throughout the book.

1.1 Example: Predicting Bike-Sharing Activity

Let's start with a well-known dataset, **Bike Sharing**, from the Machine Learning Repository at the University of California, Irvine.[1] Here we have daily/hourly data on the number of riders, weather conditions, day-of-week, month and so on. Regression analysis, which relates the mean of one variable to the values of one or more other variables, may turn out to be useful to us in at least two ways:

- **Prediction:**

  The managers of the bike-sharing system may wish to predict ridership, say for the following question:

  > Tomorrow, Sunday, is expected to be sunny and cool, say 62 degrees Fahrenheit. We may wish to predict the number of riders, so that we can get some idea as to how many bikes will need repair. We may try to predict ridership, given the

---

[1] Available at *https://archive.ics.uci.edu/ml/datasets/Bike+Sharing+Dataset*.

1

weather conditions, day of the week, time of year and so on.

Some bike-sharing services actually have trucks to move large numbers of bikes to locations that are expected to have high demand. Prediction would be even more useful here.

- **Description:**

  We may be interested in determining what factors affect ridership. How much effect, for instance, does wind speed have in influencing whether people wish to borrow a bike?

These twin goals, Prediction and Description, will arise frequently in this book. Choice of methodology will often depend on the goal in the given application.

1.2 Example of the Prediction Goal: Body Fat

Prediction is difficult, especially about the future — baseball great, Yogi Berra

The great baseball player Yogi Berra was often given to malapropisms, one of which supposedly was the quote above. But there is more than a grain of truth to this, because indeed we may wish to "predict" the present or even the past.

For example, consiser the **bodyfat** data set, available in the R package, **mfp**, available on CRAN [5]. (See Section 1.20.1 for information on CRAN packages, a number of which will be used in this book.) Direct measurment of body fat is expensive and unwieldy, as it involves underwater weighing. Thus it would be highly desirable to "predict" that quantity from easily measurable variables such as height, age, weight, abdomen circumference and so on.

In scientific studies of ancient times, there may be similar situations in which we "predict" past unknown quantities from present known ones.

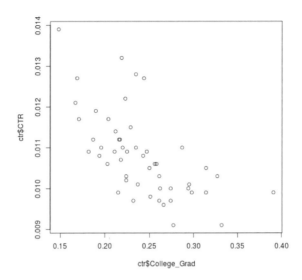

Figure 1.1: Click rate vs. college rate

1.3 Example: Who Clicks Web Ads?

One of the most common applications of machine learning methods is in marketing. Sellers wish to learn which types of people might be interested in a given product. The reader is probably familiar with Amazon's *recommender system*, in which the viewer who indicates interest in a given book, say, is shown a list of similar books.[2]

We will discuss recommender systems at several points in this book, beginning with Section 3.2.4. A more general issue is the *click-through rate* (CTR), meaning the proportion of viewers of a Web page who click on a particular ad on the page. A simple but very engaging example was discussed online [53]. The data consist of one observation per state of the U.S.[3] There was one predictor, the proportion of college graduates in the state, and a response variable, the CTR.

---

[2]As a consumer, I used to ignore these, but now with the sharp decline in the number of bricks-and-mortar bookstores which I could browse, I now often find Amazon's suggestions useful.

[3]We use the classical statistical term **observation** here, meaning a single data point, in this case data for a single state. In the machine learning community, it is common to use the term **case**.

A plot of the data, click rate vs. college rate, is in Figure 1.1. There definitely seems to be something happening here, with a visible downward trend to the points. But how do we quantify that? One approach to learning what relation, if any, educational level has to CTR would be to use regression analysis. We will see how to do so in Section 1.8.

1.4 Approach to Prediction

Even without any knowledge of statistics, many people would find it reasonable to predict via subpopulation means. In the above bike-sharing example, say, this would work as follows.

Think of the "population" of all days, past, present and future, and their associated values of number of riders, weather variables and so on.[4] Our data set is considered a sample from this population. Now consider the subpopulation consisting of all days with the given conditions: Sundays, sunny skies and 62-degree temperatures.

It is intuitive that:

> A reasonable prediction for tomorrow's ridership would be the mean ridership among all days in the subpopulation of Sundays with sunny skies and 62-degree temperatures.

In fact, such a strategy is optimal, in the sense that it minimizes our expected squared prediction error, as discussed in Section 1.19.3 of the Mathematical Complements section at the end of this chapter. But what is important for now is to note that in the above prediction rule, we are dealing with a *conditional* mean: Mean ridership, *given* day of the week is Sunday, skies are sunny, and temperature is 62.

Note too that we can only calculate an *estimated* conditional mean. We wish we had the true population value, but since our data is only a sample, we must always keep in mind that we are just working with estimates.

---

[4]This is a somewhat slippery notion, because there may be systemic differences from the present and the distant past and distant future, but let's suppose we've resolved that by limiting our time range.

1.5 A Note about E(), Samples and Populations

To make this more mathematically precise, note carefully that in this book, as with many other books, the *expected value* functional $E()$ refers to the population mean. Say we are studying personal income, I, for some population, and we choose a person at random from that population. Then $E(I)$ is not only the mean of that random variable, but much more importantly, it is the mean income of all people in that population.

Similarly, we can define conditional means, i.e., means of subpopulations. Say G is gender. Then the conditional expected value, $E(I \mid G = \text{male})$ is the mean income of all men in the population.

To illustrate this in the bike-sharing context, let's define some variables:

- R, the number of riders

- W, the day of the week

- S, the sky conditions, e.g., sunny

- T, the temperature

We would like our prediction Q to be the conditional mean,

$$Q = E(R \mid W = \text{Sunday}, S = \text{sunny}, T = 62) \tag{1.1}$$

There is one major problem, though: We don't know the value of the right-hand side of (1.1). All we know is what is in our sample data, whereas the right-side of (1.1) is a population value, and thus unknown.

The difference between sample and population is of course at the very core of statistics. In an election opinion survey, for instance, we wish to know p, the proportion of people in the population who plan to vote for Candidate Jones. But typically only 1200 people are sampled, and we calculate the proportion of Jones supporters among them, $\widehat{p}$, using that as our estimate of p. (Note that the "hat" notation ^ is the traditional one for "estimate of.") This is why the news reports on these polls always include the *margin of error*.[5]

---

[5] This is actually the radius of a 95% confidence interval for p.

Similarly, though we would like to know the value of E(R | W = Sunday, S = sunny, T = 62), **it is an unknown population value, and thus must be estimated from our sample data**, which we'll do later in this chapter.

Readers will greatly profit from constantly keeping in mind this distinction between populations and samples.

Another point is that in statistics, the populations are often rather conceptual in nature. On the one hand, in the election poll example above, there is a concrete population involved, the population of all voters. On the other hand, consider the bike rider data in Section 1.1. Here we can think of our data as being a sample from the population of all bikeshare users, past, present and future.

Before going on, a bit of terminology, again to be used throughout the book: We will refer to the quantity to be predicted, e.g., R above, as the *response variable*, and the quantities used in prediction, e.g., W, S and T above, as the *predictor variables*. Other popular terms are *dependent variable* for the response and *independent variables* or *regressors* for the predictors. The machine learning community uses the term *features* rather than *predictors*.

1.6   Example of the Description Goal: Do Baseball Players Gain Weight As They Age?

Nothing in life is to be feared, it is only to be understood. Now is the time to understand more, so that we may fear less — Marie Curie

Though the bike-sharing data set is the main example in this chapter, it is rather sophisticated for introductory material. Thus we will set it aside temporarily, and bring in a simpler data set for now. We'll return to the bike-sharing example in Section 1.15.

This new dataset involves 1015 major league baseball players, courtesy of the UCLA Statistics Department. You can obtain the data as the data set **mlb** in **freqparcoord**, a CRAN package authored by Yingkang Xie and myself [104].[6] The variables of interest to us here are player weight W, height H and age A, especially the first two.

---

[6]We use the latter version of the dataset here, in which we have removed the Designated Hitters.

Here are the first few records:

```
> library(freqparcoord)
> data(mlb)
> head(mlb)
           Name  Team       Position  Height
1   Adam_Donachie  BAL        Catcher     74
2      Paul_Bako  BAL        Catcher     74
3 Ramon_Hernandez  BAL        Catcher     72
4   Kevin_Millar  BAL  First_Baseman     72
5    Chris_Gomez  BAL  First_Baseman     73
6  Brian_Roberts  BAL Second_Baseman     69
   Weight    Age PosCategory
1     180  22.99     Catcher
2     215  34.69     Catcher
3     210  30.78     Catcher
4     210  35.43   Infielder
5     188  35.71   Infielder
6     176  29.39   Infielder
```

1.6.1 Prediction vs. Description

Recall the Prediction and Description goals of regression analysis, discussed in Section 1.1. With the baseball player data, we may be more interested in the Description goal, such as:

> Ahtletes strive to keep physically fit. Yet even they may gain weight over time, as do people in the general population. To what degree does this occur with the baseball players? This question can be answered by performing a regression analysis of weight against height and age, which we'll do in Section 1.9.1.2.[7]

On the other hand, there doesn't seem to be much of a Prediction goal here. It is hard to imagine much need to predict a player's weight. One example of this, though, is working with missing data, in which we wish to predict any value that might be unavailable.

However, for the purposes of explaining the concepts, we will often phrase things in a Prediction context. In the baseball player example, it will turn

---

[7]The phrasing here, "regression analysis of ... against ...," is commonly used in this field. The quantity before "against" is the response variable, and the ones following are the predictors.

out that by trying to predict weight, we can deduce effects of height and age. In particular, we can answer the question posed above concerning weight gain over time.

So, suppose we will have a continuing stream of players for whom we only know height (we'll bring in the age variable later), and need to predict their weights. Again, we will use the conditional mean to do so. For a player of height 72 inches, for example, our prediction might be

$$\widehat{W} = E(W \mid H = 72) \tag{1.2}$$

Again, though, this is a population value, and all we have is sample data. How will we estimate $E(W \mid H = 72)$ from that data?

First, some important notation: Recalling that μ is the traditional Greek letter to use for a population mean, let's now use it to denote a function that gives us subpopulation means:

For any height t, define

$$\mu(t) = E(W \mid H = t) \tag{1.3}$$

which is the mean weight of all people in the population who are of height t.

Since we can vary t, this is indeed a function, and it is known as *the regression function of W on H.*

So, $\mu(72.12)$ is the mean population weight of all players of height 72.12, $\mu(73.88)$ is the mean population weight of all players of height 73.88, and so on. These means are population values and thus unknown, but they do exist.

So, to predict the weight of a 71.6-inch-tall player, we would use $\mu(71.6)$ — if we knew that value, which we don't, since once again this is a population value while we only have sample data. So, we need to estimate that value from the (height, weight) pairs in our sample data, which we will denote by $(H_1, W_1), ...(H_{1015}, W_{1015})$. How might we do that? In the next two sections, we will explore ways to form our estimate, $\widehat{\mu}(t)$. (Keep in mind that for now, we are simply exploring, especially in the first of the following two sections.)

1.6.2   A First Estimator

Our height data is only measured to the nearest inch, so instead of estimating values like $\mu(71.6)$, we'll settle for $\mu(72)$ and so on. A very natural estimate for $\mu(72)$, again using the "hat" symbol to indicate "estimate of," is the mean weight among all players in our sample for whom height is 72, i.e.

$$\widehat{\mu}(72) = \text{ mean of all } W_i \text{ such that } H_i = 72 \tag{1.4}$$

R's **tapply()** can give us all the $\widehat{\mu}(t)$ at once:

```
> library(freqparcoord)
> data(mlb)
> muhats <- tapply(mlb$Weight, mlb$Height, mean)
> muhats
        67        68        69        70        71        72
 172.5000  173.8571  179.9474  183.0980  190.3596  192.5600
        73        74        75        76        77        78
 196.7716  202.4566  208.7161  214.1386  216.7273  220.4444
        79        80        81        82        83
 218.0714  237.4000  245.0000  240.5000  260.0000
```

In case you are not familiar with **tapply()**, here is what just happened. We asked R to partition the Weight variable into groups according to values of the Height variable, and then compute the mean weight in each group. So, the mean weight of people of height 72 in our sample was 192.5600. In other words, we would set $\widehat{\mu}(72) = 192.5600$, $\widehat{\mu}(74) = 202.4566$, and so on. (More detail on **tapply()** is given in the Computational Complements section at the end of this chapter.)

Since we are simply performing the elementary statistics operation of estimating population means from samples, we can form confidence intervals (CIs). For this, we'll need the "n" and sample standard deviation for each height group:

```
> tapply(mlb$Weight, mlb$Height, length)
 67  68  69  70  71  72  73  74  75  76  77  78
  2   7  19  51  89 150 162 173 155 101  55  27
 79  80  81  82  83
 14   5   2   2   1
> tapply(mlb$Weight, mlb$Height, sd)
        67        68        69        70        71        72
 10.60660  22.08641  15.32055  13.54143  16.43461  17.56349
```

73	74	75	76	77	78
16.41249	18.10418	18.27451	19.98151	18.48669	14.44974

79	80	81	82	83
28.17108	10.89954	21.21320	13.43503	NA

Here is how that first call to **tapply()** worked. Recall that this function partitions the data by the Height variables, resulting in a weight vector for each height value. We need to specify a function to apply to each of the resulting vectors, which in this case we choose to be R's **length()** function. The latter then gives us the count of weights for each height value, the "n" that we need to form a CI. By the way, the NA value is due to there being only one player with height 83, which is makes life impossible for **sd()**, as it divides from "n-1."

An approximate 95% CI for $\mu(72)$, for example, is then

$$190.3596 \pm 1.96 \ \frac{17.56349}{\sqrt{150}} \tag{1.5}$$

or about (187.6,193.2).

The above analysis takes what is called a *nonparametric* approach. To see why, let's proceed to a parametric one, in the next section.

1.6.3   A Possibly Better Estimator, Using a Linear Model

All models are wrong, but some are useful — famed statistician George Box

[In spite of] innumerable twists and turns, the Yellow River flows east — Confucious

So far, we have assumed nothing about the shape that $\mu(t)$ would have, if it were plotted on a graph. Again, it is unknown, but the function does exist, and thus it does correspond to *some* curve. But we might consider making an assumption on the shape of this unknown curve. That might seem odd, but you'll see below that this is a very powerful, intuitively reasonable idea.

Toward this end, let's plot those values of $\widehat{\mu}(t)$ we found above. We run

```
> plot(67:83, muhats)
```

producing Figure 1.2.

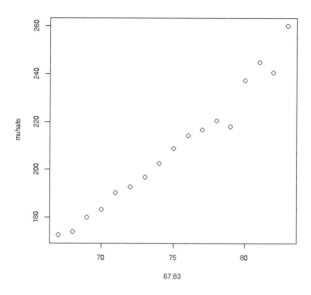

Figure 1.2: Plotted $\widehat{\mu}(t)$

Interestingly, the points in this plot seem to be near a straight line. Just like the quote of Confucious above concerning the Yellow River, visually we see something like a linear trend, in spite of the "twists and turns" of the data in the plot. This suggests that our unknown function $\widehat{\mu}(t)$ has a linear form, i.e., that

$$\mu(t) = c + dt \tag{1.6}$$

for some constants c and d, over the range of t appropriate to human heights. Or, in English,

$$\text{mean weight} = c + d \times \text{ height} \tag{1.7}$$

Don't forget the word *mean* here! We are assuming that the *mean* weights in the various height subpopulations have the form (1.6), NOT that weight itself is this function of height, which can't be true.

This is called a *parametric* model for $\mu(t)$, with parameters c and d. We will use this below to estimate $\mu(t)$. Our earlier estimation approach, in

Section 1.6.2, is called *nonparametric*. It is also called *assumption-free* or *model-free*, since it made no assumption at all about the shape of the $\mu(t)$ curve.

Note the following carefully:

- Figure 1.2 suggests that our straight-line model for $\mu(t)$ may be less accurate at very small and very large values of t. This is hard to say, though, since we have rather few data points in those two regions, as seen in our earlier R calculations; there is only one person of height 83, for instance.

  But again, in this chapter we are simply exploring, so let's assume for now that the straight-line model for $\widehat{\mu}(t)$ is reasonably accurate. We will discuss in Chapter 6 how to assess the validity of this model.

- Since $\mu(t)$ is a population function, the constants c and d are population values, thus unknown. However, we can estimate them from our sample data. We do so using R's **lm()** ("linear model") function:[8]

```
> lmout <- lm(mlb$Weight ~ mlb$Height)
> lmout
Call:
lm(formula = mlb$Weight ~ mlb$Height)

Coefficients:
(Intercept)     mlb$Height
  -151.133          4.783
```

This gives $\widehat{c} = -151.133$ and $\widehat{d} = 4.783$. We can superimpose the fitted line to Figure 1.2, using R's **abline()** function, which adds a line with specified slope and intercept to the currently-displayed plot:

```
> abline(coef=coef(lmout))
```

The result is shown in Figure 1.3.

Note carefully that we do not expect the line to fit the points exactly. On the contrary, the line is only an estimate of $\mu(t)$, the condiitonal *mean* of weight given height, not weight itself.

---

[8]Details on how the estimation is done will be given in Chapter 2.

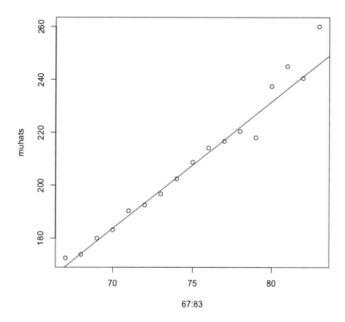

Figure 1.3: Turkish student evaluations

We would then set, for instance (using the "check" instead of the hat, so as to distinguish from our previous estimator)

$$\check{\mu}(72) = -151.133 + 4.783 \times 72 = 193.2666 \tag{1.8}$$

So, using this model, we would predict a slightly heavier weight than our earlier prediction.

By the way, we need not type the above expression into R by hand. Here is why: Writing the expression in matrix-multiply form, it is

$$(-151.133, 4.783) \begin{pmatrix} 1 \\ 72 \end{pmatrix} \tag{1.9}$$

Be sure to see the need for that 1 in the second factor; it is used to pick up the -151.133. Now let's use that matrix form to show how we can conveniently compute that value in R:[9]

The key is that we can exploit the fact that R's **coef()** function fetches the coefficients c and d for us:

```
> coef(lmout)
(Intercept)    mlb$Height
-151.133291      4.783332
```

Recalling that the matrix-times-matrix operation in R is specified via the %*% operator, we can now obtain our estimated value of $\mu(72)$ as

```
> coef(lmout) %*% c(1,72)
              [,1]
[1,]   193.2666
```

We can form a confidence interval from this too, which for the 95% level will be

$$\check{\mu}(72) \pm 1.96 \text{ s.e.}[(\check{\mu}(72)] \tag{1.10}$$

where $s.e.[]$ signifies *standard error*, the estimated standard deviation of an estimator. Here $\check{\mu}(72)$, being based on our random sample data, is itself random, i.e., it will vary from sample to sample. It thus has a standard deviation, which we call the standard error. We will see later that $s.e.[(\check{\mu}(72)]$ is obtainable using the R **vcov()** function:

```
> tmp <- c(1,72)
> sqrt(tmp %*% vcov(lmout) %*% tmp)
              [,1]
[1,]   0.6859655
> 193.2666 + 1.96 * 0.6859655
[1]   194.6111
> 193.2666 - 1.96 * 0.6859655
[1]   191.9221
```

(More detail on **vcov()** and **coef()** as R functions is presented in Section 1.20.4 in the Computational Complements section at the end of this chapter.)

---

[9]In order to gain a solid understanding of the concepts, we will refrain from using R's **predict()** function for now. It will be introduced later, though, in Section 1.10.3.

So, an approximate 95% CI for $\mu(72)$ under this model would be about (191.9,194.6).

1.7 Parametric vs. Nonparametric Models

Now here is a major point: The CI we obtained from our linear model, (191.9,194.6), was narrower than what the nonparametric approach gave us, (187.6,193.2); the former has width of about 2.7, while the latter's is 5.6. In other words:

> A parametric model is — if it is (approximately) valid — more powerful than a nonparametric one, yielding estimates of a regression function that tend to be more accurate than what the nonparametric approach gives us. This should translate to more accurate prediction as well.

Why should the linear model be more effective? Here is some intuition, say for estimating $\mu(72)$: As will be seen in Chapter 2, the **lm()** function uses *all* of the data to estimate the regression coefficients. In our case here, all 1015 data points played a role in the computation of $\tilde{\mu}(72)$, whereas only 150 of our observations were used in calculating our nonparametric estimate $\widehat{\mu}(72)$. The former, being based on much more data, should tend to be more accurate.[10]

On the other hand, in some settings it may be difficult to find a valid parametric model, in which case a nonparametric approach may be much more effective. *This interplay between parametric and nonparametric models will be a recurring theme in this book.*

1.8 Example: Click-Through Rate

Let's try a linear regression model on the CTR data in Section 1.3. The file can be downloaded from the link in [53].

```
> ctr <- read.table('State_CTR_Date.txt',
    header=TRUE, sep='\t')
```

---

[10]Note the phrase *tend to* here. As you know, in statistics one usually cannot say that one estimator is always better than another, because anomalous samples do have some nonzero probability of occurring.

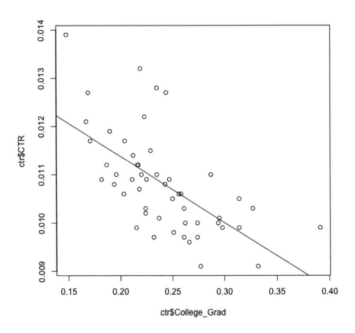

Figure 1.4: CTR data and fitted line

```
> lmout <- lm( ctr $CTR ~ ctr $ College_Grad )
> lmout
...
Coefficients :
       (Intercept)    ctr $College_Grad
          0.01412              −0.01373
...
```

A scatter plot of the data, with the fitted line superimposed, is shown in
Figure 1.4. It was generated by the code

```
> plot ( ctr $ College_Grad, ctr $CTR)
> abline ( coef=coef ( lmout ))
```

The relation between education and CTR is interesting, but let's put this in perspective, by considering the standard deviation of **College_Grad**:

```
> sd(ctr$College_Grad)
[1]  0.04749804
```

So, a "typical" difference between one state and another is something like 0.05. Multiplying by the -0.01373 figure above, this translates to a difference in click-through rate from state to state of about 0.0005. This is certainly not enough to have any practical meaning.

So, putting aside such issues as whether our data constitute a sample from some "population" of potential states, the data suggest that there is really no substantial relation between educational level and CTR. The original blog post on this data, noting the negative value of $\hat{d}$, cautioned that though this seems to indicate that the more-educated people click less, "correlation is not causation." Good advice, but it's equally important to note here that even if the effect is causal, it is tiny.

1.9   Several Predictor Variables

Now let's predict weight from height and age. We first need some notation.

Say we are predicting a response variable Y from variables $X^{(1)}, ..., X^{(k)}$. The regression function is now defined to be

$$\mu(t_1, ..., t_k) = E(Y \mid X^{(1)} = t_1, ..., X^{(k)} = t_k) \tag{1.11}$$

In other words, $\mu(t_1, ..., t_k)$ is the mean Y among all units (people, cars, whatever) in the population for which $X^{(1)} = t_1, ..., X^{(k)} = t_k$.

In our baseball data, Y, $X^{(1)}$ and $X^{(2)}$ might be weight, height and age, respectively. Then $\mu(72, 25)$ would be the population mean weight among all players of height 72 and age 25.

We will often use a vector notation

$$\mu(t) = E(Y \mid X = t) \tag{1.12}$$

with $t = (t_1, ..., t_k)'$ and $X = (X^{(1)}, ..., X^{(k)})'$, where $'$ denotes matrix transpose.[11]

---

[11]Our vectors in this book are column vectors. However, since they occupy a lot of

1.9.1   Multipredictor Linear Models

Let's consider a parametric model for the baseball data,

$$\text{mean weight} = c + d \times \text{ height} + e \times \text{ age} \qquad (1.14)$$

1.9.1.1   Estimation of Coefficients

We can again use **lm()** to obtain sample estimates of c, d and e:

```
> lm(mlb$Weight ~ mlb$Height + mlb$Age)
...
Coefficients:
(Intercept)      mlb$Height         mlb$Age
  -187.6382          4.9236          0.9115
```

Note that the notation mlb$Weight ~mlb$Height + mlb$Age simply means "predict weight from height and age." The variable to be predicted is specified to the left of the tilde, and the predictor variables are written to the right of it. The + does not mean addition.

A shorter formulation is

```
> lm(Weight ~ Height + Age, data=mlb)
```

You can see that if we have many predictors, this notation is more compact and convenient.

And, shorter still, we could write

```
> lm(Weight ~ ., data=mlb[,4:6])
```

Here the period means "all the other variables." Since we are restricting the data to be columns 4 and 6 of **mlb**, Height and Age, the period means those two variables.

---

space on a page, we will often show them as transposes of rows. For instance, we will often write $(5, 12, 13)'$ instead of

$$\begin{pmatrix} 5 \\ 12 \\ 13 \end{pmatrix} \qquad (1.13)$$

So, the output shows us the estimated coefficientsis, e.g., $\widehat{d} = 4.9236$. Our estimated regression function is

$$\widehat{\mu}(t_1, t_2) = -187.6382 + 4.9236\ t_1 + 0.9115\ t_2 \qquad (1.15)$$

where t_1 and t_2 are height and age, respectively.

Setting $t_1 = 72$ and $t_2 = 25$, we find that

$$\widehat{\mu}(72, 25) = 189.6485 \qquad (1.16)$$

and we would predict the weight of a 72-inch tall, age 25 player to be about 190 pounds.

1.9.1.2   The Description Goal

It was mentioned in Section 1.1 that regression analysis generally has one or both of two goals, Prediction and Description. In light of the latter, some brief comments on the magnitudes of the estimated coefficientsis would be useful at this point:

- We estimate that, on average (a key qualifier), each extra inch in height corresponds to almost 5 pounds of additional weight.

- We estimate that, on average, each extra year of age corresponds to almost a pound in extra weight.

That second item is an example of the Description goal in regression analysis. We may be interested in whether baseball players gain weight as they age, like "normal" people do. Athletes generally make great efforts to stay fit, but we may ask how well they succeed in this. The data here seem to indicate that baseball players indeed are prone to some degree of "weight creep" over time.

1.9.2   Nonparametric Regression Estimation: k-NN

Now let's drop the linear model assumption (1.14), and estimate our regression function "from scratch." So this will be a model-free approach, thus termed *nonparametric* as explained earlier.

Our analysis in Section 1.6.2 was model-free. But here we will need to broaden our approach, as follows.

1.9.2.1   Looking at Nearby Points

Again say we wish to estimate, using our data, the value of $\mu(72, 25)$. A potential problem is that there likely will not be any data points in our sample that exactly match those numbers, quite unlike the situation in (1.4), where $\widehat{\mu}(72)$ was based on 150 data points. Let's check:

```
> z <- mlb[mlb$Height == 72 & mlb$Age == 25,]
> z
[1]  Name          Team          Position
[4]  Height        Weight        Age
[7]  PosCategory
<0 rows> (or 0-length row.names)
```

(Recall that in R, we use a single ampersand when "and-ing" vector quantities, but use a double one for ordinary logical expressions.)

So, indeed there were no data points matching the 72 and 25 numbers. Since the ages are recorded to the nearest 0.01 year, this result is not surprising. But at any rate we thus cannot set $\widehat{\mu}(72, 25)$ to be the mean weight among our sample data points satisfying those conditions, as we did in Section 1.6.2. And even if we had had a few data points of that nature, that would not have been enough to obtain an accurate estimate $\widehat{\mu}(72, 25)$.

Instead, we use data points that are *close* to the desired prediction point. Again taking the weight/height/age case as a first example, this means that we would estimate $\mu(72, 25)$ by the average weight in our sample data among those data points for which height is *near* 72 and age is *near* 25.

1.9.2.2   Measures of Nearness

Nearness is generally defined as *Euclidean distance*:

$$\text{distance}[(s_1, s_2, ..., s_k), (t_1, t_2, ..., t_k)] = \sqrt{((s_1 - t_1)^2 + ... + (s_k - t_k)^2}$$

(1.17)

For instance, the distance from a player in our sample of height 72.5 and age 24.2 to the point (72,25) would be

$$\sqrt{(72.5 - 72)^2 + (24.2 - 25)^2} = 0.9434 \qquad (1.18)$$

Note that the Euclidean distance between $s = (s_1, ..., s_k)$ and $t = (t_1, ..., t_k)$ is simply the Euclidean norm of the difference $s - t$ (Section A.1).

1.9.2.3 The k-NN Method, and Tuning Parameters

The *k-Nearest Neighbor* (k-NN) method for estimating regression functions is simple: Find the k data points in our sample that are closest to the desired prediction point, and average their values of the response variable Y.

A question arises as to how to choose the value of k. Too large a value means we are including "nonrepresentative" data points, but too small a value gives us too few points to average for a good estimate. We will return to this question later, but will note that due to this nature of k, we will call k a *tuning* parameter. Various tuning parameters will come up in this book.

1.9.2.4 Nearest-Neighbor Analysis in the regtools Package

We will use the k-NN functions in my **regtools** package, available on CRAN [97]. The main computation is performed by **knnest()**, with preparatory nearest-neighbor computation done by **preprocessx()**. The call forms are

```
preprocessx(x,kmax,xval=FALSE)
knnest(y,xdata,k,nearf=meany)
```

In the first, **x** is our predictor variable data, one column per predictor. The argument **kmax** specifies the maximum value of k we wish to use (we might try several), and **xval** refers to cross-validation, a concept to be introduced later in this chapter. The essence of **preprocessx()** is to find the **kmax** nearest neighbors of each observation in our dataset, i.e., row of **x**.

The arguments of **knnest()** are as follows. The vector **y** is our response variable data; **xdata** is the output of **preprocessx()**; **k** is the number of nearest neighbors we wish to use. The argument **nearf** specifies the function we wish to be applied to the Y values of the neighbors; the default is the mean, but instead we could for instance specify the median. (This flexibility will be useful in other ways as well.)

The return value from **knnest()** is an object of class **'knn'**.

1.9.2.5   Example: Baseball Player Data

There is also a **predict** function associated with **knnest()**, with call form

predict (kout , predpts , needtoscale)

Here **kout** is the return value of a call to **knnest()**, and each row of **regestpts** is a point at which we wish to estimate the regression function. Also, if the points to be predicted are not in our original data, we need to set **needtoscale** to TRUE.

For example, let's estimate $\mu(72, 25)$, based on the 20 nearest neighbors at each point.

```
> data(mlb)
> library(regtools)
> xd <- preprocessx(mlb[,c(4,6)],20)
> kout <- knnest(mlb[,5],xd,20)
> predict(kout,c(72,25),TRUE)
187.4
```

So we would predict the weight of a 72-inches tall, age 25 player to be about 187 pounds, not much different — in this instance — from what we obtained earlier with the linear model.

1.10   After Fitting a Model, How Do We Use It for Prediction?

As noted, our goal in regression analysis could be either Prediction or Description (or both). How specifically does the former case work?

1.10.1   Parametric Settings

The parametric case is the simpler one. We fit our data, write down the result, and then use that result in the future whenever we are called upon to do a prediction.

Recall Section 1.9.1.1. It was mentioned there that in that setting, we probably are not interested in the Prediction goal, but just as an illustration, suppose we do wish to predict. We fit our model to our data — called our *training data* — resulting in our estimated regression function, (1.15). From now on, whenever we need to predict a player's weight, given his height and age, we simply plug those values into (1.15).

1.10.2 Nonparametric Settings

The nonparametric case is a little more involved, because we have no explicit equation like (1.15). Nevertheless, we use our training data in the same way. For instance, say we need to predict the weight of a player whose height and age are 73.2 and 26.5, respectively. Our predicted value will then be $\hat{\mu}(73.2, 26.5)$. To obtain that, we go back to our training data, find the k nearest points to (73.2,26.5), and average the weights of those k players. We would go through this process each time we are called upon to perform a prediction.

A variation:

A slightly different approach, which is used in **regtools**, is as follows. Denote our training set data as $(X_1, Y_1), ..., (X_n, Y_n)$, where again the X_i are typically vectors, e.g., (height,age). We estimate our regression function at each of the points X_i, forming $\hat{\mu}(X_i), i = 1, ..., n$. Then, when faced with a new case (X, Y) for which Y is unknown, we find the *single* closest X_i to X, and guess Y to be 1 or 0, depending on whether $\hat{\mu}(X_i) > 0.5$. Since $\hat{\mu}(X_i)$ already incorporates the neighborhood-averaging operation, doing so for our new point would be largely redundant. Using only the single closest point saves both computation time and storage space.

1.10.3 The Generic predict() Function

Consider this code:

```
> lmout <- lm(Weight ~ Height + Age, data=mlb)
> predict(lmout, data.frame(Height = 72, Age = 25))
       1
189.6493
```

We fit the model as in Section 1.9.1.1, and then predicted the weight of a player who is 72 inches tall and age 25. We use $\hat{\mu}(72, 25)$ for this, which of course we could obtain as

```
> coef(lmout) %*% c(1,72,25)
         [,1]
```

[1 ,]   189.6493

But the **predict()** function is simpler and more explicitly reflects what we want to accomplish.

By the way, **predict** is a *generic* function. This means that R will *dispatch* a call to **predict()** to a function specific to the given class. In this case, **lmout** above is of class 'lm', so the function ultimately executed above is **predict.lm'**. Similarly, in Section 1.9.2.5, the call to **predict()** goes to **predict.knn()**. More details are in Section 1.20.4.

IMPORTANT NOTE: To use **predict()** with **lm()**, the latter must be called in the **data** = form shown above, and the new data to be predicted must be a data frame with the same column names.

1.11   Overfitting, and the Variance-Bias Tradeoff

One major concern in model development is *overfitting*, meaning to fit such an elaborate model that it "captures the noise rather than the signal." This description is often heard these days, but it is vague and potentially misleading. We will discuss it in detail in Chapter 9, but it is of such importance that we introduce it here in this prologue chapter.

The point is that, after fitting our model, we are concerned that it may fit our training data well but not predict well on new data in the future.[12] Let's look into this further:

1.11.1   Intuition

To see how overfitting may occur, consider the famous *bias-variance trade-off*, illustrated in the following example. Again, keep in mind that the treatment will at this point just be intuitive, not mathematical.

Long ago, when I was just finishing my doctoral study, I had my first experience with statistical consulting. A chain of hospitals was interested in comparing the levels of quality of care given to heart attack patients at its various locations. A problem was noticed by the chain regarding straight comparison of raw survival rates: One of the locations served a

---

[12]Note that this assumes that nothing changes in the system under study between the time we collect our training data and the time we do future predictions.

largely elderly population, and since this demographic presumably has more difficulty surviving a heart attack, this particular hospital may misleadingly appear to be giving inferior care.

An analyst who may not realize the age issue here would thus be biasing the results. The term "bias" here doesn't mean deliberate distortion of the analysis, just that the model has a systemic bias, i.e., it is "skewed," in the common vernacular. And it is permanent bias, in the sense that it won't disappear, no matter how large a sample we take.

Such a situation, in which an important variable is not included in the analysis, is said to be *underfitted*. By adding more predictor variables in a regression model, in this case age, we are reducing bias.

Or, suppose we use a regression model that is linear in our predictors, but the true regression function is nonlinear. This is bias too, and again it won't go away even if we make the sample size huge. This is often called *model bias* by statisticians; the economists call the model *misspecified*.

On the other hand, we must keep in mind that our data is a sample from a population. In the hospital example, for instance, the patients on which we have data can be considered a sample from the (somewhat conceptual) population of all patients at this hospital, past, present and future. A different sample would produce different regression coefficient estimates. In other words, there is variability in those coefficients from one sample to another, i.e., variance. We hope that that variance is small, which gives us confidence that the sample we have is representative.

But the more predictor variables we have, the more collective variability there is in the inputs to our regression calculations, and thus the larger the variances of the estimated coefficients.[13] If those variances are large enough, the bias-reducing benefit of using a lot of predictors may be overwhelmed by the increased variability of the results. This is called *overfitting*.

In other words:

> In deciding how many (and which) predictors to use, we have a tradeoff. The richer our model, the less bias, but the higher the variance.

In Section 1.19.2 it is shown that for any statistical estimator $\widehat{\theta}$ (that has finite variance),

$$\text{mean squred error} = \text{squared bias} + \text{variance}$$

---

[13]I wish to thank Ariel Shin for this interpretation.

Our estimator here is $\widehat{\mu}(t)$. This shows the tradeoff: Adding variables, such as age in the hospital example, reduces squared bias but increases variance. Or, equivalently, removing variables reduces variance but exacerbates bias. It may, for example, be beneficial to accept a little bias in exchange for a sizable reduction in variance, which we may achieve by removing some predictors from our model.

The trick is to somehow find a "happy medium," easier said than done. Chapter 9 will cover this in depth, but for now, we introduce a common method for approaching the problem:

1.11.2   Example: Student Evaluations of Instructors

In Section 9.9.5 we will analyze a dataset consisting of student evaluations of instructors. Let's defer the technical details until then, but here is a sneak preview.

The main variables here consist of 28 questions on the instructor, such as "The quizzes, assignments, projects and exams contributed to helping the learning." The student gives a rating of 1 to 5 on each question.

Figure 1.5 describes this data, plotting frequency of occurrence against the questions (the 28, plus 4 others at the beginning). Again, don't worry about the details now, but it basically shows there are 3 kinds of instructors: one kind gets very righ ratings on the 28 questions, across the board; one kind gets consistently medium-high ratings; and the third kind gets low ratings across all the questions.

This indicates that we might reduce those 28 questions to just one, in fact any one of the 28.

1.12   Cross-Validation

The proof of the pudding is in the eating — old English saying

Toward that end, i.e., proof via "eating," it is common to artificially create a set of "new" data and try things out there. Instead of using all of our collected data as our training set, we set aside part of it to serve as simulated "new" data. This is called the *validation set* or *test set*. The remainder will be our actual training data. In other words, we randomly partition our original data, taking one part as our training set and the other part to

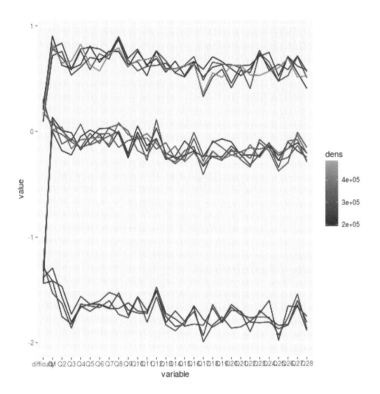

Figure 1.5: Turkish student evaluations (see color insert)

play the role of new data. We fit our model, or models, to the training set, then do prediction on the test set, pretending its response variable values are unknown. We then compare to the real values. This will give us an idea of how well our models will predict in the future. The method is called *cross-validation*.

The above description is a little vague, and since there is nothing like code to clarify the meaning of an algorithm, let's develop some. Here first is code to do the random partitioning of **data**, with a proportion **p** to go to the training set:

```
xvalpart <- function(data,p) {
   n <- nrow(data)
   ntrain <- round(p*n)
```

```
    trainidxs <- sample(1:n,ntrain,replace=FALSE)
    list(train=data[trainidxs,],
        valid=data[-trainidxs,])
}
```

R uses - in array indices for exclusion, e.g.,

```
> x <- c(5,12,13)
> x[-2]
[1]    5 13
```

Thus, using the expression **-trainidxs** above gives us the validation cases.

Now to perform cross-validation, we'll consider the parametric and non-parametric cases separately, in the next two sections.

1.12.1   Linear Model Case

To do cross-validation for linear models, we could use this code.[14]

1.12.1.1   The Code

```
# arguments:
#
#    data:  full data
#    ycol:  column number of resp. var.
#    predvars:  column numbers of predictors
#    p:  prop. for training set
#    meanabs:  see 'value' below

# value:  if meanabs is TRUE, the mean absolute
#         prediction error; otherwise, an R list
#         containing pred., real Y

xvallm <- function(data,ycol,predvars,p,meanabs=TRUE){
    tmp <- xvalpart(data,p)
    train <- tmp$train
    valid <- tmp$valid
    # fit model to training data
```

---

[14]There are sophisticated packages on CRAN for this, such as **cvTools** [4]. But to keep things simple, and to better understand the concepts, we will write our own code. Similarly, as mentioned, we will not use R's **predict()** function for the time being.

```
trainy <- train[,ycol]
trainpreds <- train[,predvars]
# using matrix form in lm() call
trainpreds <- as.matrix(trainpreds)
lmout <- lm(trainy ~ trainpreds)
# apply fitted model to validation data; note
# that %*% works only on matrices, not data frames
validpreds <- as.matrix(valid[,predvars])
predy <- cbind(1,validpreds)%*% coef(lmout)
realy <- valid[,ycol]
if (meanabs) return(mean(abs(predy - realy)))
list(predy = predy, realy = realy)
}
```

1.12.1.2 Applying the Code

Let's try cross-validdtion on the weight/height/age data, using mean absolute prediction error as our criterion for prediction accuracy:

```
library(freqparcoord)
data(mlb)
xvallm(mlb,5,c(4,6),2/3)

> xvallm(mlb,5,c(4,6),2/3)
[1] 13.38045
```

So, on average we would be off by about 13 pounds. We might improve upon this by using the data's Position variable, but we'll leave that for later.

Keep in mind the randomness, though. We randomly split the data, and would get a different result if we were to run the code again. This point is explored in Exercise 1 at the end of this chapter. Also, we will later discuss an extension, i*r-fold cross-validation*, in Section 2.9.6.

1.12.2 k-NN Case

Here is the code for performing cross-validation for k-NN:

```
# arguments:
#
#    data:  full data
#    ycol:  column number of resp. var.
```

```
#      k:    number  of  nearest  neighbors
#      p:    prop.  for  training  set
#      meanabs:   see  'value'  below

# value:   if  meanabs  is  TRUE,  the  mean  absolute
#              prediction  error;  otherwise,  an  R  list
#              containing  pred.,  real  Y

xvalknn <-
      function(data, ycol, predvars, k, p, meanabs=TRUE){
      # cull  out  just  Y  and  the  Xs
      data <- data[, c(predvars, ycol)]
      ycol <- length(predvars) + 1
      tmp <- xvalpart(data, p)
      train <- tmp$train
      valid <- tmp$valid
      valid <- as.matrix(valid)
      xd <- preprocessx(train[, -ycol], k)
      kout <- knnest(train[, ycol], xd, k)
      predy <- predict(kout, valid[, -ycol], TRUE)
      realy <- valid[, ycol]
      if (meanabs) return(mean(abs(predy - realy)))
      list(predy = predy, realy = realy)
}
```

So, how well does k-NN predict?

```
> library(regtools)
> set.seed(9999)
> xvalknn(mlb, 5, c(4,6), 25, 2/3)
[1]  14.32817
```

The two methods gave similar results. However, not only must we keep
in mind the randomness of the partitioning of the data, but we also must
recognize that this output above depended on choosing a value of 25 for **k**,
the number of nearest neighbors. We could have tried other values of **k**,
and in fact could have used cross-validation to choose the "best" value.

1.12.3   Choosing the Partition Sizes

One other problem, of course, is that we did have a random partition of
our data. A different one might have given substantially different results.

In addition, there is the matter of choosing the sizes of the training and validation sets (e.g., via the argument **p** in **xvalpart()**). We have a classical tradeoff at work here: Let k be the size of our training set. If we make k too large, the validation set will be too small for an accurate measure of prediction accuracy. We won't have that problem if we set k to a smaller size, but then we are measuring the predictive ability of only k observations, whereas in the end we will be using all n observations for predicting new data.

The *Leaving One-Out Method* and its generalizations solves this problem, albeit at the expense of much more computation. It will be presented in Section 2.9.5.

1.13 Important Note on Tuning Parameters

Recall how k-NN works: To predict a new case for which $X = t$ but Y is unknown, we look at the our existing data. We find the k closest neighbors to t, then average their Y values. That average becomes our predicted value for the new case.

We refer to k as a tuning parameter, to be chosen by the user. Many methods have multiple tuning parameters, making the choice a challenge. One can of course choose their values using cross validation, and in fact the **caret** package includes methods to automate the process, simultaneously optimizing over many tuning parameters.

But cross-validation can have its own overfitting problems (Section 9.3.2). One should not be lulled into a false sense of security.

The late Leo Breiman was suspicious of tuning parameters, and famously praised one regression method (*boosting*), as "the best off-the-shelf method" available — meaning that the method works well without tweaking tuning parameters. His statement may have been overinterpreted regarding the boosting method, but the key point here is that Breiman was not a fan of tuning parameters.

A nice description of Breiman's view was given in an obituary by Michael Jordan, who noted [78],

> Another preferred piece of Breimanesque terminology was "off-the-shelf," again a rather physical metaphor. Leo tended to be suspicious of "free parameters;" procedures should work with little or no "tuning."

Breiman's concerns about tuning parameters extended to choosing those parameters via cross-validation. The latter is an important tool, but should be used with a healthy dose of skepticism.

This issue will come up often, since many commonly-used methods do have various tuning parameters; some have multiple, complex tuning parameters.

Again, this point must be kept in mind:

> Optimizing for a tuning parameter is **inherently prone to overfitting**, as we are optimizing for particular data. If we have multiple tuning parameters, the potential for overfitting is compounded.

1.14   Rough Rule of Thumb

The issue of how many predictors to use to simultaneously avoid overfitting and still produce a good model is nuanced, and in fact this is still not fully resolved. Chapter 9 will be devoted to this complex matter.

Until then, though it is worth using the following:[15]

> **Rough Rule of Thumb (Tukey):** For a data set consisting of n observations, use fewer than $\sqrt{(n)}$ predictors.

1.15   Example: Bike-Sharing Data

We now return to the bike-sharing data (Section 1.1). Our little excursion to the simpler data set, involving baseball player weights and heights, helped introduce the concepts in a less complex setting. The bike-sharing data set is more complicated in several ways:

- **Complication (a):** It has more potential predictor variables.

- **Complication (b):** It includes some *nominal* (or *categorical*) variables, such as Day of Week. The latter is technically numeric, 0 through 6, but those codes are just names. Hence the term *nominal*. In R, by the way, the formal term for such variables is *factors*.

---

[15]Unfortunately, reference unknown.

The problem is that there is no reason, for instance, that Sunday, Thursday and Friday should have an ordinal relation in terms of ridership just because, say, $0 < 4 < 5$.

- **Complication (c):** It has some potentially nonlinear relations. For instance, people don't like to ride bikes in freezing weather, but they are not keen on riding on really hot days either. Thus we might suspect that the relation of ridership to temperature rises at first, eventually reaching a peak, but declines somewhat as the temperature increases further.

Now that we know some of the basic issues from analyzing the baseball data, we can treat this more complicated data set.

Let's read in the bike-sharing data. We'll look at one of the files in that dataset, **day.csv**. We'll restrict attention to the first year,[16] and since we will focus on the registered riders, let's shorten the name for convenience:

```
> shar <- read.csv("day.csv",header=TRUE)
> shar <- shar[1:365,]
> names(shar)[15] <- "reg"
```

1.15.1   Linear Modeling of $\mu(t)$

In view of Complication (c) above, the inclusion of the word *linear* in the title of our current section might seem contradictory. But one must look carefully at *what* is linear or not, and we will see shortly that, yes, we can use linear models to analyze nonlinear relations.

Let's first check whether the ridership/temperature relation seems nonlinear, as we have speculated:

```
plot(shar$temp,shar$reg)
```

The result is shown in Figure 1.6.

There seem to be some interesting groupings among the data, likely due to the other variables, but putting those aside for now, the plot does seem to suggest that ridership is slightly associated with temperature in the "first rising, then later falling" form as we had guessed.

---

[16]There appears to have been some systemic change in the second year, and while this could be modeled, we'll keep things simple by considering only the first year.

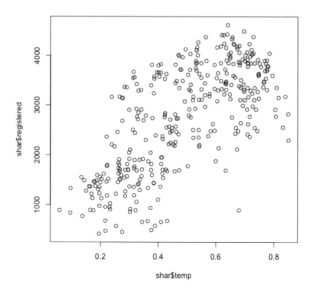

Figure 1.6: Ridership vs. temperature

Thus a linear model of the form

$$\text{mean ridership} = c + d \times \text{ temperature} \qquad (1.19)$$

would seem inappropriate. But don't give up so quickly! A model like

$$\text{mean ridership} = c + d \times \text{ temperature} + e \times \text{ temperature}^2 \qquad (1.20)$$

i.e., with a temperature-squared term added, might work fine. A negative value for e would give us the "first up, then later down" behavior we want our model to have.

And there is good news — the model (1.20) is actually linear! We say that the expression is *linear in the parameters*, even though it is nonlinear with respect to the temperature variable. This means that if we multiply each of c, d and e by, say, 8, then the values of the left and right sides of the equation both increase eightfold.

Anotber way to see this is that in calling **lm()**, we can simply regard squared temperature as a new variable:

```
> shar$temp2 <- shar$temp^2
> lm(shar$reg ~ shar$temp + shar$temp2)
```

Call:
lm(**formula** = shar$reg ~ shar$temp + shar$temp2)

```
Coefficients:
(Intercept)      shar$temp     shar$temp2
     -378.9          9841.8        -6169.8
```

And note that, sure enough, the coefficient of the squared term, $\widehat{e} = -6169.8$, did indeed turn out to be negative.

Of course, we want to predict from many variables, not just temperature, so let's now turn to Complication (b) cited earlier, the presence of nominal data. This is not much of a problem either.

Such situations are generally handled by setting up what are called *indicator variables* or *dummy variables*. The former term alludes to the fact that our variable will *indicate* whether a certain condition holds or not, with 1 coding the yes case and 0 indicating no.

We could, for instance, set up such a variable for Tuesday data:

```
> shar$tues <- as.integer(shar$weekday == 2)
```

Indeed, we could define six variables like this, one for each of the days Monday through Saturday. Note that Sunday would then be indicated indirectly, via the other variables all having the value 0. A direct Sunday variable would be redundant, and in fact would present mathematical problems, as we'll see in Chapter 8. (Actually, R's **lm()** function can deal with factor variables directly, as shown in Section 9.7.5.1. But we take the more basic route here, in order to make sure the underlying principles are clear.)

However, let's opt for a simpler analysis, in which we distinguish only between weekend days and weekdays, i.e. define a dummy variable that is 1 for Monday through Friday, and 0 for the other days. Actually, those who assembled the data set already defined such a variable, which they named **workingday**.[17]

---

[17]More specifically, a value of 1 for this variable indicates that the day is in the Monday-Friday range *and* it is not a holiday.

We incorporate this into our linear model:

$$\text{mean reg} = c + d \times \text{temp} + e \times \text{temp}^2 + f \text{ workingday} \qquad (1.21)$$

There are several other dummy variables that we could add to our model, but for this introductory example let's define just one more:

```
> shar$clearday <- as.integer(shar$weathersit == 1)
```

So, our regression model will be

$$
\begin{aligned}
\text{mean reg} \quad = \quad & \beta_0 + \beta_1 \text{ temp} + \beta_2 \text{ temp}^2 \\
+ \quad & \beta_3 \text{ workingday} + \beta_4 \text{ clearday} \qquad (1.22)
\end{aligned}
$$

As is traditional, here we have used subscripted versions of the Greek letter β to denote our equation coefficients, rather than c, d and so on.

So, let's run this through **lm()**:

```
> lmout <- lm(reg ~ temp+temp2+workingday+clearday,
      data = shar)
```

The return value of **lm()**, assigned here to **lmout**, is a very complicated R object, of class 'lm'. We shouldn't inspect it in detail now, but let's at least print the object, which in R's interactive mode can be done simply by typing the name, which automatically calls **print()** on the object:[18]

```
> lmout
...
...
Coefficients:
(Intercept)            temp            temp2    workingday
    -1362.6         11059.2          -7636.4         686.0
   clearday
      518.9
```

Remember, the population function $\mu(t)$ is unnown, so the β_i are unknown. The above coefficients are merely sample-based estimates. For example, using our usual "hat" notation to mean "estimate of," we have that

$$\widehat{\beta_3} = 686.0 \qquad (1.23)$$

---

[18]See more detail on this in Section 1.20.4.

The estimated regression function is then

$$\hat{\mu}(t_1, t_2, t_3, t_4) = -1362.6 + 11059.2t_1 - 7636.4t_2 + 686.0t_3 + 518.9t_4 \quad (1.24)$$

where $t_2 = t_1^2$.

So, what should we predict for the number of riders on the type of day described at the outset of this chapter — Sunday, sunny, 62 degrees Fahrenheit? First, note that the designers of the data set have scaled the **temp** variable to [0,1], as

$$\frac{\text{Celsius temperature} - \text{minimum}}{\text{maximum} - \text{minimum}} \quad (1.25)$$

where the minimum and maximum here were -8 and 39, respectively. This form may be easier to understand, as it is expressed in terms of where the given temperature fits on the normal range of temperatures. A Fahrenheit temperature of 62 degrees corresponds to a scaled value of 0.525. So, our predicted number of riders is

```
> coef(lmout) %*% c(1,0.525,0.525^2,0,1)
            [,1]
[1,]  2857.677
```

So, our predicted number of riders for sunny, 62-degree Sundays will be about 2858. How does that compare to the average day?

```
> mean(shar$reg)
[1]  2728.359
```

So, we would predict a somewhat above-average level of ridership.

As noted earlier, one can also form confidence intervals and perform significance tests on the β_i. We'll go into this in Chapter 2, but some brief comments on the magnitudes and signs of the $\hat{\beta}_i$ are useful at this point:

- As noted, the estimated coefficient of **temp2** is negative, consistent with our intuition. Note, though, that it is actually less negative than when we predicted **reg** from only temperature and its square. This change is typical, and will be discussed in detail in Chapter 7.

- The estimated coefficient for **workingday** is positive. This too matches our intuition, as presumably many of the registered riders use the bikes to commute to work. The value of the estimate here, 686.0, indicates that, for fixed temperature and weather conditions, weekdays tend to have close to 700 more registered riders than weekends.

- Similarly, the coefficient of **clearday** suggests that for fixed temperature and day of the week, there are about 519 more riders on clear days than on other days.

1.15.2   Nonparametric Analysis

Let's see what k-NN gives us as our predicted value for sunny, 62-degree Sundays, say with $k = 20$:

```
> shar1 <-
      shar[,c('workingday','temp','reg','clearday')]
> xd <- preprocessx(shar1[,-3],20)
> kout <- knnest(shar1$reg,xd,20)
> predict(kout,c(0,0.525,1),TRUE)
2881.8
```

This is again similar to what the linear model gave us. This probably means that the linear model was pretty good, but we will discuss this in detail in Chapter 6.

1.16   Interaction Terms, Including Quadratics

Let's take another look at (1.22), specifically the term involving the variable **workingday**, a dummy indicating a nonholiday Monday through Friday. Our estimate for β_3 turned out to be 686.0, meaning that, holding temperature and the other variables fixed, there is a mean increase of about 686.0 riders on working days.

But look at our model, (1.22). The (estimated) values of the right-hand side will differ by 686.0 for working vs. nonworking days, no matter what the temperature is. In other words, the working day effect is the same on low-temprerature days as on warmer days. For a broader model that does not make this assumption, we could add an *interaction term*, consisting of a product of **workingday** and **temp**:

$$\begin{aligned}
\text{mean reg} \;=\;& \beta_0 + \beta_1 \text{ temp} + \beta_2 \text{ temp}^2 \\
+\;& \beta_3 \text{ workingday} + \beta_4 \text{ clearday} \quad\quad (1.26) \\
+\;& \beta_5 \text{ temp} \times \text{workingday} \quad\quad\quad\quad (1.27)
\end{aligned}$$

Note that the temp^2 term is also an interaction term, the interaction of the **temp** variable with itself.

How does this model work? Let's illustrate it with a new data set.

1.16.1 Example: Salaries of Female Programmers and Engineers

This data is from the 2000 U.S. Census, consisting of 20,090 programmers and engineers in the Silicon Valley area. The data set is included in the **freqparcoord** package on CRAN [104]. Suppose we are working toward a Description goal, specifically the effects of gender on wage income.

As with our bike-sharing data, we'll add a quadratic term, in this case on the age variable, reflecting the fact that many older programmers and engineers encounter trouble finding work [108]. Let's restrict our analysis to workers having at least a Bachelor's degree, and look at the variables **age**, **age2**, **sex** (coded 1 for male, 2 for female), **wkswrked** (number of weeks worked), **ms**, **phd** and **wageinc** (wage income). Other than an age^2 term, we'll start out with no interaction terms.

```
> library(freqparcoord)
> data(prgeng)
> prgeng$age2 <- prgeng$age^2
> edu <- prgeng$educ
> prgeng$ms <- as.integer(edu == 14)
> prgeng$phd <- as.integer(edu == 16)
> prgeng$fem <- prgeng$sex - 1
> tmp <- prgeng[edu >= 13,]
> pe <- tmp[,c(1,12,9,13,14,15,8)]
> pe <- as.matrix(pe)
```

Our model is

$$\begin{aligned}
\text{mean wageinc} \quad = \quad & \beta_0 + \beta_1 \text{ age} + \beta_2 \text{ age}^2 + \beta_3 \text{ wkswrkd} \\
+ \quad & \beta_4 \text{ ms} + \beta_5 \text{ phd} \\
+ \quad & \beta_6 \text{ fem}
\end{aligned} \qquad (1.28)$$

We find the following:

```
> lm( wageinc ~
     age+age2+wkswrkd+ms+phd+fem , data=prgeng )
...
Coefficients:
(Intercept)             age              age2           wkswrkd
 -81136.70          3900.35           -40.33           1196.39
         ms             phd               fem
   15431.07         23183.97        -11484.49
```

The model probably could use some refining, for example variables we have omitted, such as occupation. But as a preliminary statement, the results are striking in terms of gender: With age, education and so on held constant, women are estimated to have incomes about $11,484 lower than comparable men.

But this analysis implicitly assumes that the female wage deficit is, for instance, uniform across educational levels. To see this, consider (1.28). Being female makes a β_6 difference, no matter what the values of **ms** and **phd** are. (For that matter, this is true of **age** too, though we won't model that here for simplicity.) To generalize our model in this regard, let's define two interaction variables, the product of **ms** and **fem**, and the product of **phd** and **fem**.

Our model is now

$$\begin{aligned}
\text{mean wageinc} \quad = \quad & \beta_0 + \beta_1 \text{ age} + \beta_2 \text{ age}^2 + \beta_3 \text{ wkswrkd} \\
+ \quad & \beta_4 \text{ ms} + \beta_5 \text{ phd} \\
+ \quad & \beta_6 \text{ fem} + \beta_7 \text{ msfem} + \beta_8 \text{ phdfem}
\end{aligned} \qquad (1.29)$$

So, now instead of there being a single number for the "female effect," β_6, we how have three:

- Female effect for holders of a Bachelor's degree: β_6

- Female effect for Master's degree holders: $\beta_6 + \beta_7$

- Female effect for PhD degree holders $\beta_6 + \beta_8$

So, let's rerun the regression analysis:

```
> prgeng$msfem <- prgeng$ms * prgeng$fem
> prgeng$phdfem <- prgeng$phd * prgeng$fem
> lm(wageinc ~
      age+age2+wkswrkd+ms+phd+fem+msfem+phdfem,
      data=prgeng)
...
Coefficients:
(Intercept)            age           age2        wkswrkd
  -81216.78        3894.32         -40.29        1195.31
         ms            phd            fem          msfem
   16433.67       25325.31      -10276.80       -4157.25
     phdfem
  -14061.64
```

Let's compute the estimated values of the female effects, first for a worker with less than a graduate degree. This is -10276.80. For the Master's case, the mean female effect is estimated to be -10276.80 - 4157.25 = -14434.05. For a PhD, the figure is -10276.80 - 14861.64 = -25138.44. In other words, Once one factors in educational level, the gender gap is seen to be even worse than before.

Thus we still have many questions to answer, especially since we haven't considered other types of interactions yet. This story is not over yet, and will be pursued in detail in Chapter 7.

Rather than creating the interaction terms "manually" as is done here, one can use R colon operator, e.g., **ms:fem**, which automates the process. This was not done above, so as to ensure that the reader fully understands the meaning of interaction terms. But this is how it would go:

```
> lm(wageinc ~ age+age2+wkswrkd+ms+phd+fem+
      ms:fem+phd:fem, data=prgeng)
...
Coefficients:
(Intercept)            age           age2
  -81216.78        3894.32         -40.29
    wkswrkd             ms            phd
    1195.31       16433.67       25325.31
```

fem	ms : fem	phd : fem
-10276.80	-4157.25	-14061.64

For information on the colon and related operators, type **?formula** at the
R prompt.

1.16.2   Fitting Separate Models

Suppose we have a model that includes a dummy predictor D, and we form
interaction terms between D and other predictors. In essence, this is the
same as fitting two regression models without interaction terms, one for the
subpopulation $D = 1$ and the other for $D = 0$. To see this, consider again
the census data above.

To keep things simple, let's just one other predictor, the age variable, and
take D to be the dummy variable for female:

```
> data(prgeng)
> prgeng$fem <- prgeng$sex - 1
> fm <- which(prgeng$fem == 1)
> male <- prgeng[-fm,]    # data from male subpop
> female <- prgeng[fm,]   # data from female subpop
> lm(wageinc ~ age, data=male)
Coefficients:
(Intercept)              age
    44313.2            486.2
> lm(wageinc ~ age, data=female)
Coefficients:
(Intercept)              age
      30551              503
> lm(wageinc ~ age+fem+age*fem, data=prgeng)
Coefficients:
(Intercept)         age         fem      age:fem
    44313.2       486.2    -13761.7         16.8
```

Look at that last result. For a female worker, **fem** and **age:fem** would be
equal to 1 and **age**, respectively. That means the coefficent for **age** would
be $486.2 + 16.8 = 503$, which matches the 503 value obtained from running
lm() with **data = female**. For a male worker, **fem** and **age:fem** would
both be 0, and the **age** coefficent is then 486.2, matching the **lm()** results
for the **male** data. The intercept terms match similarly.

The reader may be surprised that the estimated age coefficient is higher for the women than the men. The problem is that the intercept term is much lower for women, and the line for men is above that for women for all reasonable values of age. At age 50, for instance, the estimated mean for men is $44313.2 + 486.2 \times 50 = 68623.2$, while for women it is $30551 + 503 \times 50 = 55701$.

If our goal is Description, running separate regression models like this may be much easier to interpret. This is highly encouraged. However, things become unwieldy if we have multiple dummies; if there are d of them, we must fit 2^d separate models.

1.16.3 Saving Your Work

Readers who are running the book's examples on their computers may find it convenient to use R's **save()** and **load()** functions. Our **pe** data above will be used again at various points in the book, so it is worthwhile to save it:

```
> save(pe, file='pe.save')
```

Later — days, weeks, whatever — you can reload it by simply typing

```
load('pe.save')
```

Your old **pe** object will now be back in memory. This is a lot easier than re-loading the original **prgeng** data, adding the **fem**, **ms** and **phd** variables, etc.

1.16.4 Higher-Order Polynomial Models

Theoretically, we need not stop with quadratic terms. We could add cubic terms, quartic terms and so on. Indeed, the famous Stone-Weierstrass Theorem [123] says that any continuous function can be approximated to any desired accuracy by some high-order polynomial.

But this is not practical. In addition to the problem of overfitting there are numerical issues. In other words, roundoff errors in the computation would render it meaningless at some point, and indeed **lm()** will refuse to compute if it senses a situation like this. See Exercise 1 in Chapter 8.

1.17   Classification Techniques

Recall the hospital example in Section 1.11.1. There the response variable
is nominal, represented by a dummy variable taking the values 1 and 0,
depending on whether the patient survives or not. This is referred to as
a *classification problem*, because we are trying to predict which class the
population unit belongs to — in this case, whether the patient will belong
to the survival or nonsurvival class. We could set up dummy variables
for each of the hospital branches, and use these to assess whether some
were doing a better job than others, while correcting for variations in age
distribution from one branch to another. (Thus our goal here is Description
rather than directly Prediction itself.)

The point is that we are predicting a 1-0 variable. In a marketing context, we might be predicting which customers are more likely to purchase
a certain product. In a computer vision context, we may want to predict
whether an image contains a certain object. In the future, if we are for-
tunate enough to develop relevant data, we might even try our hand at
predicting earthquakes.

Classification applications are extremely common. And in many cases there
are more than two classes, such as in identifying many different printed
characters in computer vision.

In a number of applications, it is desirable to actually convert a problem
with a numeric response variable into a classification problem. For instance,
there may be some legal or contractual aspect that comes into play when our
variable V is above a certain level $\mathbf{c}$, and we are only interested in whether
the requirement is satisfied. We could replace V with a new variable

$$Y = \begin{cases} 1, & \text{if } V > c \\ 0, & \text{if } V \leq c \end{cases} \tag{1.30}$$

Classification methods will play a major role in this book.

1.17.1   It's a Regression Problem!

Recall that the regression function is the conditional mean:

$$\mu(t) = E(Y \mid X = t) \tag{1.31}$$

(As usual, X and t may be vector-valued.) In the classification case, Y is an indicator variable, which implies that we know its mean is the probability that $Y = 1$ (Section 1.19.1). In other words,

$$\mu(t) = P(Y = 1 \mid X = t) \tag{1.32}$$

The great implication of this is that *the extensive knowledge about regression analysis developed over the years can be applied to the classification problem.*

One intuitive strategy would be to guess that $Y = 1$ if the conditional probability of 1 is greater than 0.5, and guess 0 otherwise. In other words,

$$\text{guess for } Y = \begin{cases} 1, & \text{if } \mu(X) > 0.5 \\ 0, & \text{if } \mu(X) \leq 0.5 \end{cases} \tag{1.33}$$

It turns out that this strategy is optimal, in that it minimizes the overall misclassification error rate (see Section 1.19.4 in the Mathematical Complements portion of this chapter). However, it should be noted that this is not the only possible criterion that might be used. We'll return to this issue in Chapter 5.

As before, note that (1.32) is a population quantity. We'll need to estimate it from our sample data.

1.17.2 Example: Bike-Sharing Data

Let's take as our example the situation in which ridership is above 3500 bikes, which we will call HighUsage:

```
> shar$highuse <- as.integer(shar$reg > 3500)
```

We'll try to predict that variable. Let's again use our earlier example, of a Sunday, clear weather, 62 degrees. Should we guess that this will be a High Usage day?

We can use our k-NN approach just as before. Indeed, we don't need to re-run **preprocessx()**.

```
> kout <- knnest(as.integer(shar1$reg > 3500),xd,20)
> predict(kout,c(0,0.525,1),TRUE)
0.1
```

We estimate that there is a 10% chance of that day having HighUsage.

The parametric case is a little more involved. A model like

$$
\begin{aligned}
\text{probability of HighUsage} \; = \; & \beta_0 + \beta_1 \text{ temp} + \beta_2 \text{ temp}^2 \\
+ \; & \beta_3 \text{ workingday } + \beta_4 \text{ clearday} \quad (1.34)
\end{aligned}
$$

could be used, but would not be very satisfying. The left-hand side of (1.34), as a probability, should be in [0,1], but the right-hand side could in principle fall far outside that range.

Instead, the most common model for conditional probability is *logistic regression*:

$$
\begin{aligned}
\text{probability of HighUsage} \; = \; & \ell(\beta_0 + \beta_1 \text{ temp} + \beta_2 \text{ temp}^2 \\
+ \; & \beta_3 \text{ workingday } + \beta_4 \text{ clearday}) \quad (1.35)
\end{aligned}
$$

where $\ell(s)$ is the *logistic function*,

$$
\ell(s) = \frac{1}{1 + e^{-s}} \tag{1.36}
$$

Our model, then, is

$$
\mu(t_1, t_2, t_3, t_4) = \frac{1}{1 + e^{-(\beta_0 + \beta_1 t_1 + \beta_2 t_2 + \beta_3 t_3 + \beta_4 t_4)}} \tag{1.37}
$$

where t_1 is temperature, t_2 is the square of temperature, and so on. We wish to estimate $\mu(62, 62^2, 0, 1)$.

Note the form of the curve, shown in Figure 1.7 The appeal of this model is clear at a glance: First, the logistic function produces a value in [0,1], as appropriate for modeling a probability. Second, it is a monotone increasing function in each of the variables in (1.35), just as was the case in (1.22) for predicting our numeric variable, **reg**. Other motivations for using the logistic model will be discussed in Chapter 4.

R provides the **glm()** ("generalized linear model") function for several non-linear model families, including the logistic,[19] which is designated via **family = binomial**:

---

[19]Often called "logit," by the way.

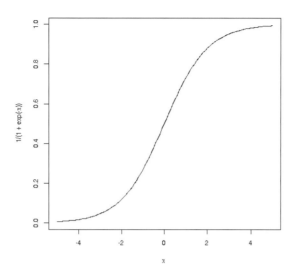

Figure 1.7: Logistic function

```
> shar$highuse <- as.integer(shar$reg > 3500)
> glmout <- glm(highuse ~
      temp+temp2+workingday+clearday ,
    data=shar , family=binomial)
> tmp <- coef(glmout) %*% c(1,0.525,0.525^2,0,1)
> 1/(1+exp(-tmp))
          [,1]
[1,]  0.1010449
```

So, our parametric model gives an almost identical result here to the one arising from k-NN, about a 10% probability of HighUsage.

1.18 Crucial Advice: Don't Automate, Participate!

Data science should not be a "spectator sport"; the methodology is effective only if the users *participate*. Avoid ceding the decision making to the computer output. For example:

- Statistical significance does not imply practical importance, and conversely.

- A model is just that — just an approximation to reality, hopefully useful but never exact.

- Don't rely solely on variable selection algorithms to choose your model (Chapter 9).

- "Read directions before use" — make sure you understand what a method really does before employing it.

1.19   Mathematical Complements

1.19.1   Indicator Random Variables

A random variable W is an indicator variable, if it is equal to 1 or 0, depending on whether a certain event Q occurs or not. Two simple properties are very useful:

- $EW = P(Q)$

  This follows from

$$EW = 1 \cdot P(Q) + 0 \cdot P(\text{not } Q) = P(Q) \qquad (1.38)$$

- $Var(W) = P(Q) \cdot [1 - P(Q)]$

  True because

$$Var(W) = E(W^2) - (EW)^2 = E(W) - E(W^2) = EW(1 - EW) \qquad (1.39)$$

  where the second equality stems from $W^2 = W$ (remember, W is either 1 or 0). Then use the first bullet above!

1.19.2   Mean Squared Error of an Estimator

Say we are estimating some unknown population value θ, using an estimator $\widehat{\theta}$ based on our sample data. Then a natural measure of the accuracy of our estimator is the *Mean Squared Error* (MSE),

$$E[(\widehat{\theta} - \theta)^2] \qquad (1.40)$$

This is the squared distance from our estimator to the true value, averaged over all possible samples.

Let's rewrite the quantity on which we are taking the expected value:

$$\left(\widehat{\theta} - \theta\right)^2 = \left(\widehat{\theta} - E\widehat{\theta} + E\widehat{\theta} - \theta\right)^2 = (\widehat{\theta} - E\widehat{\theta})^2 + (E\widehat{\theta} - \theta)^2 + 2(\widehat{\theta} - E\widehat{\theta})(E\widehat{\theta} - \theta) \qquad (1.41)$$

Look at the three terms on the far right of (1.41). The expected value of the first is $Var(\widehat{\theta})$, by definition of variance.

As to the second term, $E\widehat{\theta} - \theta$ is the *bias* of $\widehat{\theta}$, the tendency of $\widehat{\theta}$ to over- or underestimate θ over all possible samples.

What about the third term? Note first that $E\widehat{\theta} - \theta$ is a constant, thus factoring out of the expectation. But for what remains,

$$E(\widehat{\theta} - E\widehat{\theta}) = 0 \qquad (1.42)$$

Taking the expected value of both sides of (1.41), and taking the above remarks into account, we have

$$\begin{aligned} \text{MSE}(\widehat{\theta}) &= Var(\widehat{\theta}) + (E\widehat{\theta} - \theta)^2 & (1.43) \\ &= \text{variance} + \text{bias}^2 & (1.44) \end{aligned}$$

In other words:

> The MSE of $\widehat{\theta}$ is equal to the variance of $\widehat{\theta}$ plus squared bias of $\widehat{\theta}$.

1.19.3 $\mu(t)$ Minimizes Mean Squared Prediction Error

Claim: *Consider all the functions f() with which we might predict Y from X, i.e., $\widehat{Y} = f(X)$. The one that minimizes mean squared prediction error, $E[(Y - f(X))^2]$, is the regression function, $\mu(t) = E(Y \mid X = t)$.*

(Note that the above involves population quantities, not samples. Consider the quantity $E[(Y - f(X))^2]$, for instance. It is the mean squared prediction error (MSPE) over all (X, Y) pairs in the population.)

To derive this, first ask, for any (finite-variance) random variable W, what number c minimizes the quantity $E[(W - c)^2]$? The answer is $c = EW$. To see this, write

$$E[(W - c)^2] = E(W^2 - 2cW + c^2) = E(W^2) - 2cEW + c^2 \qquad (1.45)$$

Setting to 0 the derivative of the right-hand side with respect to c, we find that indeed, $c = EW$.

Now use the Law of Total Expectation (Section 1.19.5):

$$\text{MSPE} = E[(Y - f(X))^2] = E\left[E((Y - f(X))^2 | X)\right] \qquad (1.46)$$

In the inner expectation, X is a constant, and from the statement following (1.45) we know that the minimizing value of $f(X)$ is "EW," in this case $E(Y|X)$, i.e. $\mu(X)$. Since that minimizes the inner expectation for any $\mathbf{X}$, the overall expectation is minimized too.

1.19.4 $\mu(t)$ Minimizes the Misclassification Rate

We are concerned here with the classification context. It shows that if we know the population distribution — we don't, but are going through this exercise to guide our intuition — the conditional mean provides the optimal action in the classification context.

Remember, in this context, $\mu(t) = P(Y | X = t)$, i.e. the conditional mean reduces to the conditional probability. Now plug in X for t, and we have the following.

Claim: *Consider all rules based on X that produce a guess $\widehat{Y}$, taking on values 0 and 1. The one that minimizes the overall misclassification rate $P(\widehat{Y} \neq Y)$ is*

$$\widehat{Y} = \begin{cases} 1, & \text{if } \mu(X) > 0.5 \\ 0, & \text{if } \mu(X) \leq 0.5 \end{cases} \qquad (1.47)$$

The claim is completely intuitive, almost trivial: After observing X, how should we guess Y? If conditionally Y has a greater than 50% chance of being 1, then guess it to be 1!

(Note: In some settings, a "false positive" may be worse than a "false negative," or *vice versa*. The reader should ponder how to modify the material here for such a situation. We'll return to this issue in Chapter 5.)

Think of this simple situation: There is a biased coin, with *known* probability of heads p. The coin will be tossed once, and we are supposed to guess the outcome.

Let's name your guess g (a nonrandom constant), and let C denote the as-yet-unknown outcome of the toss (1 for heads, 0 for tails). Then the reader should check that, no matter whether we choose 0 or 1 for g, the probability that we guess correctly is

$$
\begin{aligned}
P(C = g) &= P(C = 1)g + P(C = 0)(1 - g) & (1.48) \\
&= pg + (1 - p)(1 - g) & (1.49) \\
&= [2p - 1]g + 1 - p & (1.50)
\end{aligned}
$$

Now remember, p is known. How should we choose g, 0 or 1, in order to maximize (1.50), the probability that our guess is correct? Inspecting (1.50) shows that maximizing that expression will depend on whether $2p-1$ is positive or negative, i.e., whether $p > 0.5$ or not. In the former case we should choose $g = 1$, while in the latter case g should be chosen to be 0.

The above reasoning gives us the very intuitive — actually trivial, when expressed in English — result:

> If the coin is biased toward heads, we should guess heads. If the coin is biased toward tails, we should guess tails.

Now to show the original claim, we use The Law of Total Expectation. This will be discussed in detail in Section 1.19.5, but for now, it says this:

$$
E(V) = E[E(V|U)] \qquad (1.51)
$$

i.e. the expected value of a conditional random variable is the unconditional expectation. In the case where V is an indicator random variable, the above reduces to

$$
P(A) = E[P(A \mid U)] \qquad (1.52)
$$

Returning to our original claim, write

$$P(\widehat{Y} = Y) = E\left[P(\widehat{Y} = Y \mid X)\right] \tag{1.53}$$

In that inner probability, "p" is

$$P(Y = 1 \mid X) = \mu(X) \tag{1.54}$$

which completes the proof.

1.19.5   Some Properties of Conditional Expectation

Since the regression function is defined as a conditional expected value, as in (1.3), for mathematical analysis we'll need some properties. First, a definition.

1.19.5.1   Conditional Expectation As a Random Variable

For any random variables U and V with defined expectation, either of which could be vector-valued, define a new random variable W, as follows. First note that the conditional expectation of V given $U = t$ is a function of t,

$$\mu(t) = E(V \mid U = t) \tag{1.55}$$

This is an ordinary function, just like, say, $\sqrt{t}$. But we can turn that ordinary function into a random variable by plugging in a random variable, say Q, for t: $R = \sqrt{Q}$ is a random variable. Thinking along these lines, we define the *random variable* version of conditional expectation accordingly. In the *function* $\mu(t)$ in (1.55), we plug in U for t:

$$W = E(V|U) = \mu(U) \tag{1.56}$$

This W is a random variable. As a simple example, say we choose a number U at random from the numbers 1 through 5. We then randomly choose a second number V, from the numbers 1 through U. Then

$$\mu(t) = E(V \mid U = t) = \frac{1+t}{2} \tag{1.57}$$

We now form a new random variable $W = (1 + U)/2$.

And, since W is a random variable, we can talk of *its* expected value, which turns out to be an elegant result:

1.19.5.2   The Law of Total Expectation

A property of conditional expected value, proven in many undergraduate probability texts, is

$$E(V) = EW = E[E(V \mid U)] \qquad (1.58)$$

The foreboding appearance of this equation belies the fact that it is actually quite intuitive, as follows. Say you want to compute the mean height of all people in the U.S., and you already have available the mean heights in each of the 50 states. You cannot simply take the straight average of those state mean heights, because you need to give more weight to the more populous states. In other words, the national mean height is a *weighted* average of the state means, with the weight for each state being its proportion of the national population.

In (1.58), this corresponds to having V as height and U as state. State coding is an integer-valued random variable, ranging from 1 to 50, so we have

$$
\begin{aligned}
EV &= E[E(V \mid U)] & (1.59) \\
   &= EW & (1.60) \\
   &= \sum_{i=1}^{50} P(U = i)\, E(V \mid U = i) & (1.61)
\end{aligned}
$$

The left-hand side, EV, is the overall mean height in the nation; $E(V \mid U = i)$ is the mean height in state i; and the weights in the weighted average are the proportions of the national population in each state, $P(U = i)$.

Not only can we look at the mean of W, but also its variance. By using the various familiar properties of mean and variance, one can derive a similar relation for variance:

1.19.5.3   Law of Total Variance

For scalar V,

$$Var(V) = E[Var(V|U)] + Var[E(V|U)] \qquad (1.62)$$

One might initially guess that we only need the first term. To obtain the national variance in height, we would take the weighted average of the state variances. But this would not take into account that the mean heights vary from state to state, thus also contributing to the national variance in height, hence the second term.

This is proven in Section 2.12.8.3.

1.19.5.4   Tower Property

Now consider conditioning on two variables, say U_1 and U_2. One can show that

$$E\left[E(V|U_1, U_2) \mid U_1\right] = E(V \mid U_1) \qquad (1.63)$$

Here is an intuitive interpretation of that in the height example above. Take V, U_1 and U_2 to be height, state and gender, respectively, so that $E(V|U_1, U_2)$ is the mean height of all people in a certain state and of a certain gender. If we then take the mean of all these values for a certain state — i.e. take the average of the two gender-specific means in the state — we get the mean height in the state without regard to gender.

Again, note that we take the straight average of the two gender-specific means, because the two genders have equal proportions. If, say, U_2 were race instead of gender, we would need to compute a *weighted* average of the race-specific means, with the weights being the proportions of the various races in the given state.

This is proven in Section 7.8.1.

1.19.5.5   Geometric View

There is an elegant way to view all of this in terms of abstract vector spaces — (1.58) becomes the Pythagorean Theorem! — which we will address later in Mathematical Complements Sections 2.12.8 and 7.8.1.

1.20 Computational Complements

1.20.1 CRAN Packages

There are thousands of useful contributed R packages available on CRAN, the Comprehensive R Archive Network, *https://cran.r-project.org.* The easiest way to install them is from R's interactive mode, e.g.

```
> install.packages('freqparcoord','~/R')
```

Here I have instructed R to download the **freqparcoord** package, installing it in ∼/**R**, the directory where I like to store my packages.

(If you are using RStudio or some other indirect interface to R, all this can be done from a menu, rather than using **installing.packages**.)

Official R parlance is *package*, not *library*, even though ironically one loads a package using the **library()** function! For instance,

```
> library(freqparcoord)
```

One can learn about the package in various ways. After loading it, for instance, you can list its objects, such as

```
> ls('package:freqparcoord')
[1] "freqparcoord" "knndens"      "knnreg"
"posjitter"    "regdiag"
[6] "regdiagbas"   "rmixmvnorm"   "smoothz"
"smoothzpred"
```

where we see objects (functions here) **knndens()** and so on. There is the **help()** function, e.g.

```
> help(package=freqparcoord)
```

```
Information on package freqparcoord

Description:
```

Package:	freqparcoord
Version:	1.1.0
Author:	Norm Matloff <normmatloff@gmail.com> and Yingkang Xie <yingkang.xie@gmail.com>
Maintainer:	Norm Matloff <normmatloff@gmail.com>

. . .

Some packages have *vignettes*, extended tutorials. Type

```
> vignette ()
```

to see what's available.

1.20.2   The Function tapply() and Its Cousins

In Section 1.6.2 we had occasion to use R's `tapply()`, a highly useful feature
of the language. To explain it, let's start with useful function, **split()**.

Consider this tiny data frame:

```
> x
   gender  height
1       m      66
2       f      67
3       m      72
4       f      63
```

Now let's split by gender:

```
> xs <- split (x, x$gender)
> xs
$f
   gender  height
2       f      67
4       f      63
5       f      63

$m
   gender  height
1       m      66
3       m      72
```

Note the types of the objects:

- **xs** is an R list

- **xs$f** and **xs$m** are data frames, the male and female subsets of **x**

We *could* then find the mean heights for each gender this way:

```
> mean(xs$f$height)
[1]  64.33333
> mean(xs$m$height)
[1]  69
```

But with **tapply()**, we can combine the two operations:

```
> tapply(x$height, x$gender, mean)
       f          m
64.33333  69.00000
```

The first argument of `tapply()` must be a vector, but the function that is applied can be vector-valued. Say we want to find not only the mean but also the standard deviation. We can do this:

```
> tapply(x$height, x$gender, function(w)  c(mean(w), sd(w)))
$f
[1]  64.333333   2.309401

$m
[1]  69.000000   4.242641
```

Here our function, which we defined "on the spot," within our call to **tapply()**, produces a vector of two components. We asked **tapply()** to call that function on our vector of heights, doing so separately for each gender.

As noted in the title of this section, **tapply()** has "cousins." Here is a brief overview of some of them:

```
# form a matrix by binding the rows (1,2) and (3,4)
> m <- rbind(1:2, 3:4)
> m
     [,1] [,2]
[1,]    1    2
[2,]    3    4
# apply the sum() function to each row
> apply(m, 1, sum)
[1]  3 7
# apply the sum() function to each column
> apply(m, 2, sum)
[1]  4 6

> l <- list(a = c(3,8), b = 12)
> l
$a
```

```
[1]  3  8
$b
[1]  12
# apply sum() to each element of the list ,
# forming a new list
> lapply(l ,sum)
$a
[1]  11
$b
[1]  12
# do the same, but try to reduce the result
# to a vector
> sapply(l ,sum)
  a   b
 11  12
```

1.20.3   The Innards of the k-NN Code

Here are simplified versions of the code:

```
preprocessx <- function(x,kmax,xval=FALSE) {
    result$x <- x
    tmp <- FNN::get.knnx(data=x,  query=x,  k=kmax+xval)
    nni <- tmp$nn.index
    result$idxs <- nni[,(1+xval):ncol(nni)]
    result$xval <- xval
    result$kmax <- kmax
    class(result) <- 'preknn'
    result
}
```

The code is essentially just a wrapper for calls to the **FNN** package on CRAN, which does nearest-neighbor computation.

```
knnest <- function(y,xdata ,k, nearf=meany)
{
    idxs <- xdata$idxs
    idx <- idxs [,1:k]
    # set idxrows[[i]] to row i of idx, the indices of
    # the neighbors of the i-th observation
    idxrows <- matrixtolist (1,idx)
    # now do the kNN smoothing
```

```
# first , form the neighborhoods
x <- xdata$x
xy <- cbind(x,y)
nycol <- ncol(y)  # how many cols in xy are y?
# ftn to form one neighborhood (x and y vals)
form1nbhd <- function(idxrow) xy[idxrow,]
# now form all the neighborhoods
nearxy <-
    lapply(idxrows,function(idxrow) xy[idxrow,])
# now nearxy[[i]] is the rows of x corresponding to
# neighbors of x[i,], together with the associated
# Y values

# now find the estimated regression function values
# at each point in the training set
regest <- sapply(1:nrow(x),
    function(i) nearf(x[i,],nearxy[[i]]))
regest <-
    if (nycol > 1) t(regest) else as.matrix(regest)
xdata$regest <- regest
xdata$nycol <- nycol
xdata$y <- y
xdata$k <- k
class(xdata) <- 'knn'
xdata
}
```

1.20.4 Function Dispatch

The return value from a call to **lm()** is an object of R's S3 class structure; the class, not surprisingly, is named '**lm**'. It turns out that the functions **coef()** and **vcov()** mentioned in this chapter are actually related to this class, as follows.

Recall our usage, on the baseball player data:

```
> lmout <- lm(mlb$Weight ~ mlb$Height)
> coef(lmout) %*% c(1,72)
          [,1]
[1,]  193.2666
```

The call to **coef** extracted the vector of estimated regression coefficents (which we also could have obtained as **lmout$coefficents**). But here is what happened behind the scenes:

The R function **coef()** is a *generic function*, which means it's just a placeholder, not a "real" function. When we call it, the R interpreter says,

> This is a generic function, so I need to relay this call to the one associated with this class, **'lm'**. That means I need to check whether we have a function **coef.lm()**. Oh, yes we do, so let's call that.

That relaying action is referred to in R terminology as the original call being *dispatched* to **coef.lm()**.

This is a nice convenience. Consider another generic R function, **plot()**. No matter what object we are working with, the odds are that some kind of plotting function has been written for it. We can just call **plot()** on the given object, and leave it to R to find the proper call. (This includes the **'lm'** class; try it on our **lmout** above!)

Similarly, there are a number of R classes on which **coef()** is defined, and the same is true for **vcov()**.

One generic function we will use quite often, and indeed have already used in this chapter, is **summary()**. As its name implies, it summarizes (what the function's author believes) are the most important characteristics of the object. So, when this generic function is called on an **'lm'** object, the call is dispatched to **summary.lm()**, yielding estimated coefficients, standard errors and so on.

Another generic function to be used often here is **predict()**, from Section 1.10.3. In the example there, **lmout** was of class **'lm'**, so the call to **predict()** was dispatched to **predict.lm()**.

1.21   Centering and Scaling

It is common in many statistical methods to *center and scale* the data. Here we subtract from each variable the sample mean of that variable. This process is called *centering*. Typically one also *scales* each predictor, i.e. divides each predictor by its sample standard deviation. Now all variables will have mean 0 and standard deviation 1.

It is clear that this is very useful for k-NN regression. Consider the example later in this chapter involving Census data. Without at least scaling, variables that are very large, such as income, would dominate the nearest-neighbor computations, and small but important variables such as age would essentially be ignored. The **knnest()** function that we will be using does do centering and scaling as preprocessing for the predictor variables.

In a parametric setting such as linear models, centering and scaling has the goal of reducing numerical roundoff error.

In R, the centering/scaling operation is done with the **scale()** function. In order to be able to reverse the process later, the means and standard deviations are recorded as R *attributes*:

```
> m <- rbind(1:2,3:4)
> m
     [,1] [,2]
[1,]    1    2
[2,]    3    4
> m1 <- scale(m)
> m1
          [,1]       [,2]
[1,] -0.7071068 -0.7071068
[2,]  0.7071068  0.7071068
attr(,"scaled:center")
[1] 2 3
attr(,"scaled:scale")
[1] 1.414214 1.414214
> attr(m1,'scaled:center')
[1] 2 3
```

1.22 Exercises: Data, Code and Math Problems

Data problems:

1. In Section 1.12.1.2, the reader was reminded that the results of a cross-validation are random, due to the random partitioning into training and test sets. Try doing several runs of the linear and k-NN code in that section, comparing results.

2. Extend (1.28) to include interaction terms for age and gender, and age^2

and gender. Run the new model, and find the estimated effect of being female, for a 32-year-old person with a Master's degree.

3. Consider the **bodyfat** data mentioned in Section 1.2. Use **lm()** to form a prediction equation for **density** from the other variables (skipping the first three), and comment on whether use of indirect methods in this way seems feasible.

4. In Section 1.19.5.2, we gave this intuitive explanation:

> In other words, the national mean height is a *weighted* average of the state means, with the weight for each state being its proportion of the national population. Replace state by gender in the following.

  (a) Write English prose that relates the overall mean height of people and the gender-specific mean heights.

  (b) Write English prose that relates the overall proportion of people taller than 70 inches to the gender-specific proportions.

Mini-CRAN and other computational problems:

5. In Section 1.12, we used R's negative-index capability to form the training/test set partitioning. Show how we could use the R function **setdiff()** to do this as an alternate approach.

6. We saw in this chapter, e.g., in Figure 1.3, how R's **abline()** function can be used to add a straight line to a plot. What about adding a quadratic function?

  (a) Write an R function with call form

    abccurve (**coef** , xint)

  where **coef** is a vector of the coeficients a, b and c in the polynomial

$$a + bt + ct^2 \tag{1.64}$$

  and **xint** is a 2-element vector that gives the range of the horizontal axis for t. The function superimposes the quadratic curve onto the existing graph. Hint: Use R's **curve()** function.

  (b) Fit a quadratic model to the click-through data, and use your **abccurve()** function on the scatter plot for that data.

Math problems:

7. Suppose the joint density of (X, Y) is $3s^2 e^{-st}$, $1 < s < 2, 0 < t < \infty$. Find the regression function $\mu(s) = E(Y|X = s)$.

8. For (X, Y) in the notation of Section 1.19.3, show that the predicted value $\mu(X)$ and the predicton error $Y - \mu(X)$ are uncorrelated.

9. Suppose X is a scalar random variable with density g. We are interested in the nearest neighbors to a point t, based on a random sample $X_1, ..., X_n$ from g. Find L_k, the cumulative distribution function of the distance of the kth-nearest neighbor to t.

Chapter 2

Linear Regression Models

In this chapter we go into the details of linear models. Let's first set some notation, to be used here and in the succeeding chapters.

2.1 Notation

Some notation here will be used throughout the book, so it is important to get a firm understanding. (Note the List of Symbols and Abbreviations at the front of this book, for easy reference.)

Let Y be our response variable, and let $X = (X^{(1)}, X^{(2)}, ..., X^{(p))})'$ denote the vector of our p predictor variables. Using our weight/height/age baseball player example from Chapter 1 as our running example here, we would have $p = 2$, and Y, $X^{(1)}$ and $X^{(2)}$ would be weight, height and age, respectively.

The quantities $Y, X^{(1)}, ..., X^p$ denote the values of these random variables in the population.[1] For instance, in the baseball example, Y is weight, so it represents the weight of a player chosen randomly from the population. But there is also notation for the values of these variables in our data, thought of as a sample from the population:

Our sample consists of n data points, $X_1, X_2, ..., X_n$, each a p-element predictor vector, and $Y_1, Y_2, ..., Y_n$, associated scalars. In the baseball example, n was 1015. Also, the third player had height 72, was of age 30.78, and

---

[1]Sometimes the population is rather conceptual, as discussed in Section 1.5.

weighed 210. So,

$$X_3 = \begin{pmatrix} 72 \\ 30.78 \end{pmatrix} \tag{2.1}$$

and

$$Y_3 = 210 \tag{2.2}$$

Write the X_i in terms of their components:

$$X_i = (X_i^{(1)}, ..., X_i^{(p)})' \tag{2.3}$$

So, again using the baseball player example, the height, age and weight of the third player would be $X_3^{(1)}$, $X_3^{(2)}$ and Y_3, respectively.

And just one more piece of notation: We sometimes will need to augment a vector with a 1 element at the top, such as we did in (1.9). Our notation for this will consist of a tilde above the symbol, For instance, (2.1) becomes

$$\widetilde{X_3} = \begin{pmatrix} 1 \\ 72 \\ 30.78 \end{pmatrix} \tag{2.4}$$

So, our linear model is, for a p-element vector $t = (t_1, ..., t_p)'$,

$$\mu(t) = \beta_0 + \beta_1 \ t_1 + ... + \beta_p \ t_p = \widetilde{t}' \ \beta \tag{2.5}$$

In the baseball example, with both height and weight as predictors:

$$
\begin{aligned}
\mu(\text{height,age}) \quad &= \quad \beta_0 + \beta_1 \ \text{height} + \beta_2 \ \text{age} \tag{2.6} \\
&= \quad (1, \text{height}, \text{age})' \begin{pmatrix} \beta_0 \\ \beta_1 \\ \beta_2 \end{pmatrix} \tag{2.7}
\end{aligned}
$$

2.2 The "Error Term"

Define

$$\epsilon = Y - \mu(X) \tag{2.8}$$

This is the error we would make if we were to predict Y from X, if we somehow knew the population regression function $\mu(t)$. It is also interpretable as **the collective effect of all predictors of Y that are not in our model** (and for which we usually do not have data).

We could then write (2.5) as

$$Y = \beta_0 + \beta_1 \ t_1 + ... + \beta_p \ t_p + \epsilon \tag{2.9}$$

This is the more common way to define the linear regression model, and ϵ, a term in that model, is called the *error term*. We will sometimes use this formulation, but the primary one is (2.5).

2.3 Random- vs. Fixed-X Cases

We will usually consider the X_i and Y_i to be random samples from some population. If we have data on people and are predicting weight from height and age, for instance, that means weight, height and age are random variables, since we are sampling people at random. Thus $(X_1, Y_1), ..., (X_n, Y_n)$ are independent and identically distributed (i.i.d.), with their distribution being that of the population. If, for instance, 23% of people in our population are taller than 72 inches, then

$$P(X_i^{(1)} > 75) = 0.23 \tag{2.10}$$

according to the population. This is a *random-X* setting, meaning that both the X_i and Y_i are random.

But there are some situations in which the X values are fixed by design, known as a *fixed-X* setting. This might be the case in chemistry research for instance, in which we decide in advance to perform experiments at specific levels of concentration of some chemicals.

Recall that the regression function is

$$\mu(t) = E(Y \mid X = t) \tag{2.11}$$

i.e., we are dealing with the *conditional* distribution of Y given X. So, in many cases, it doesn't matter whether our data arose in an random-X vs. fixed-X setting; in the random-X case, once we condition on X, we are in the same situation as in the fixed-X case. It does matter in some situations, though, as will be seen in Chapter 5.

By the way, in fixed-X settings, the X values are often chosen to form an *orthogonal design*. In this context, it is assumed that each predictor variable has mean 0 (by subtracting their means, if necessary), and that the data vectors for different predictor variables are orthogonal in the linear algebra sense: The inner product of the vector of values of predictor j and the one for predictor k is 0:

$$\sum_{i=1}^{n} X_i^{(j)} X_i^{(k)} = 0, \quad \text{for all } j \neq k \tag{2.12}$$

This simplified computation in the pre-computer days, and may result in better interpretability of the β_i.

The mean-0 property also implies (Section 2.12.4) that the regression model does not have a constant term, i.e., for $t = (t_1, ..., t_p)'$,

$$\mu(t) = \beta_1 \, t_1 + ... + \beta_p \, t_p \tag{2.13}$$

We will take this to be part of our definition of the term *orthogonal design*.

2.4   Least-Squares Estimation

Linear regression analysis is sometimes called *least-squares estimation*. Let's first look at how this evolved.

2.4.1   Motivation

This discussion will begin at the population level. As noted in Section 1.19, setting $f(X) = \mu(X)$ minimizes the mean squared prediction error, i.e., minimizes

$$E[(Y - f(X))^2] \tag{2.14}$$

over all functions f. And since our assumption is that $\mu(t) = \widetilde{t}\,'\beta$, we can set

$$f(t) = \widetilde{t}\,'b \tag{2.15}$$

above, and thus also say that $b = \beta$ minimizes

$$E[(Y - \widetilde{X}'b))^2] \tag{2.16}$$

over all vectors b.

Now let us consider sample analogs of the above. If $W_1, ..., W_n$ is a sample from a population having mean EW, the sample analog of EW is $\overline{W} = (\sum_{i=1}^{n} W_i)/n$; one is the average value of W in the population, and the other is the average value of W in the sample. Let's write this correspondence as

$$EW \longleftrightarrow \overline{W} \tag{2.17}$$

It will be crucial to always keep in mind the distinction between population values and their sample estimates, especially when we discuss overfitting in detail.

Similarly, for any fixed b, (2.16) is a population quantity, the average squared error using b for prediction in the population (recall Section 1.5). The population/sample correspondence here is

$$E[(Y - \widetilde{X}'b))^2] \longleftrightarrow \frac{1}{n} \sum_{i=1}^{n} (Y_i - \widetilde{X}_i\,'b)^2 \tag{2.18}$$

where the right-hand side is the average squared error using b for prediction in the sample.

So, since β is the value of b minimizing (2.16), it is intuitive to take our estimate, $\widehat{\beta}$, to be the value of b that minimizes (2.18). Hence the term *least squares*.

To find the minimizing b, we could apply calculus, taking the partial derivatives of (2.18) with respect to b_i, $i = 0, 1, ..., p$, set them to 0 and solve.
Fortunately, R's **lm()** does all that for us, but it's good to know what is
happening inside. Also, this will give the reader more practice with matrix
expressions, which will be important in some parts of the book.

2.4.2   Matrix Formulations

Use of matrix notation in linear regression analysis greatly compactifies
and clarifies the presentation. You may find that this requires a period of
adjustment at first, but it will be well worth the effort.

As usual, use p for the number of predictor variables, and let A denote the
$n \times (p + 1)$ matrix of X values in our sample,

$$A = \begin{pmatrix} \widetilde{X}_1' \\ \widetilde{X}_2' \\ ... \\ \widetilde{X}_n' \end{pmatrix} \tag{2.19}$$

and let D be the $n \times 1$ vector of Y values,[2]

$$D = \begin{pmatrix} Y_1 \\ Y_2 \\ ... \\ Y_n \end{pmatrix} \tag{2.20}$$

In the baseball example, row 3 of A is

$$(1, 30.78, 72) \tag{2.21}$$

and the third element of D is 210.

---

[2]The matrix is typically called X in regression literature, but there are so many
symbols here using "X" that it is clearer to call the matrix something else. The same
comment applies to the vector D.

2.4.3 (2.18) in Matrix Terms

Our first order of business will be to recast the right-hand side of (2.18) as a matrix expression. To start, look at the quantities $\widetilde{X_i}' b$, $i = 1, ..., n$ there in (2.18). Stringing them together in matrix form, we get

$$\begin{pmatrix} \widetilde{X_1}' \\ \widetilde{X_2}' \\ ... \\ \widetilde{X_n}' \end{pmatrix} b = Ab \tag{2.22}$$

Now consider the n summands in (2.18), before squaring. Stringing them into a vector as in Section A.9, we get

$$D - Ab \tag{2.23}$$

We need just one more step: Recall (see (A.15)) that for a vector $a = (a_1, ..., a_k)'$,

$$\sum_{i=1}^{k} a_k^2 = a'a \tag{2.24}$$

In other words, (2.18) (except for the $1/n$ factor) is actually

$$(D - Ab)'(D - Ab) \tag{2.25}$$

Now that we have this in matrix form, we can go about finding the optimal b.

2.4.4 Using Matrix Operations to Minimize (2.18)

Remember, we will set $\widehat{\beta}$ to whatever value of b minimizes (2.25). Thus we need to take the derivative of that expression with respect to b, and set the result to 0. Here we can draw upon the matrix derivative material in Section A.10.

Specifically, in the context of (A.40), set $u = D - Ab$, $M = -A$, $w = D$ and $v = b$. This tells us that the derivative of (2.25) with respect to b is

$$-2A'(D - Ab) \tag{2.26}$$

Setting this to 0, we have

$$A'D = A'Ab \tag{2.27}$$

Solving for b we have our answer:

$$\widehat{\beta} = (A'A)^{-1}A'D \tag{2.28}$$

This is what **lm()** calculates! In essence, it does

solve (t (a) %*% a , t (a) %*% d)

(See further comment on the calculation in Section 2.13.3.)

2.4.5   Models without an Intercept Term

In some cases, it is appropriate to omit the β_0 term from (2.5). The derivation in this setting is the same as before, except that (2.19) becomes

$$A = \begin{pmatrix} X_1' \\ X_2' \\ ... \\ X_n' \end{pmatrix} \tag{2.29}$$

which differs from (2.19) only in that this new matrix does not have a column of 1s.

This may arise when modeling some physical or chemical process, for instance, in which theoretical considerations imply that $\mu(0) = 0$. A much more common use of such a model occurs as follows.

In computation for linear models, the data are typically first centered and scaled (Section 1.21). It turns out that centering forces the intercept term, $\widehat{\beta}_0$, to 0. This is shown in Section 2.12.4. but our point now is that that action does not change the other $\widehat{\beta}_i$, $i > 0$. So, centering does no harm —

if our goal is Description, we usually are not interested in the uncentered value of β_0 — and it may be computationally beneficial as noted above.

In R, we can request that **lm()** omit the intercept term in the model by including the symbol -1 in the predictor list. For example,

```
> x <- rnorm(1000,25,8)
> y <- x + rnorm(1000)
> lm(y ~ x-1)

Call:
lm(formula = y ~ x - 1)

Coefficients:
(Intercept)              x
    0.04575        0.99927

> lm(y ~ x-1)

Call:
lm(formula = y ~ x - 1)

Coefficients:
      x
1.001
```

2.5   A Closer Look at lm() Output

Since the last section was rather abstract, let's get our bearings by taking a closer look at the output in the baseball example:[3]

```
> lmout <- lm(mlb$Weight ~ mlb$Height + mlb$Age)
> summary(lmout)
...
Coefficients:
              Estimate  Std. Error  t value  Pr(>|t|)
(Intercept)  -187.6382    17.9447    -10.46   < 2e-16
mlb$Height      4.9236     0.2344     21.00   < 2e-16
mlb$Age         0.9115     0.1257      7.25   8.25e-13
```

---

[3]Note the use of the ellipsis ..., indicating that portions of the output have been omitted, for clarity.

```
(Intercept)  ***
mlb$Height   ***
mlb$Age      ***
```

Signif. **codes**:
    0 ******* 0.001 ****** 0.01
    * 0.05 . 0.1   1
. . .

Multiple **R**-squared: 0.318,
Adjusted **R**-squared: 0.3166
. . .

There is a lot here! Let's get an overview, so that the material in the coming
sections will be better motivated.

2.5.1   Statistical Inference

The **lm()** output is heavily focused on *statistical inference* — forming confidence intervals and performing significance tests — and the first thing you
may notice is all those asterisks: The estimates of the intercept, the height
coefficient and the age coefficient all are marked with three stars, indicating
a p-value of less than 0.001.

Those p-values correspond to tests of the hypothesis

$$H_0 : \beta_i = 0 \tag{2.30}$$

under assumptions to be discussed shortly. But first, what are the practical
implications of these p-values?

Look at the coefficient for height, for example. The test of the hypothesis
that $\beta_1 = 0$, i.e., no height effect on weight, has a p-value of 2×10^{-10},
extremely small. Thus the hypothesis would be resoundingly rejected, and
one could say, "Height has a significant effect on weight." Not surprising at
all, though the finding for age might be more interesting, in that we expect
athletes to keep fit, even as they age.

We could form a confidence interval for β_2, for instance, by adding and
subtracting 1.96 times the associated standard error,[4] which is 0.1257 in
this case. Our resulting CI would be about (0.66,1.16), indicating that

----
[4]The standard error of an estimator was defined in Section 1.6.3.

the average player gains between 0.66 and 1.16 pounds per year. So even baseball players gain weight over time!

We will return to this vital topic of misuse of p-values in Section 2.10.

2.6 Assumptions

But where did this come from? Surely there must be some assumptions underlying these statistical inference procedures. What are they?

2.6.1 Classical

The classical assumptions, to which the reader may have some prior exposure, are:

- **Linearity:** There is some vector β for which Equation (2.5) holds for all t.

- **Normality:** The assumption is that, conditional on the vector of predictor variables X, the response variable Y has a normal distribution.

  In the weight/height/age example, this would mean, for instance, that within the subpopulation of all baseball players of height 72 and age 25, weight is normally distributed.

- **Homoscedasticity:** Just as we define the regression function in terms of the conditional mean,

$$\mu(t) = E(Y \mid X = t) \qquad (2.31)$$

we can define the conditional variance function

$$\sigma^2(t) = Var(Y \mid X = t) \qquad (2.32)$$

The homoscedasticity assumption is that $\sigma^2(t)$ does not depend on t.

In the weight/height/age example, this would say that the variance in weight among, say, 70-inches-tall 22-year-olds is the same as that among the subpopulation of those of height 75 inches and age 32.

- **Independence:** The Y_i are (conditionally) independent.

By the way, note that the above assumptions all concern the structure of the *population*. In addition, we assume that in the *sample* drawn from that population, the observations are independent.

2.6.2   Motivation: the Multivariate Normal Distribution Family

We'll discuss the real-world propriety of the above three assumptions

- Linearity

- Normality (conditional, of Y given X)

- Homoscedasticity

in later chapters, but it is worthwhile to take a look now at the motivation for those assumptions. This will help us understand where the assumptions came from originally, and help us to visualize how departures from the assumptions can arise.

For simplicity, we will focus here on the case $p = 1$, i.e., just one predictor variable.

All of the assumptions are intuitively reasonable as choices for a simple model (which, by the way, is extremely widely used). But they also all follow if the vector $(Y, X)'$ has a *bivariate normal* distribution. What is this?

Consider the example in which Y and X are human weight and height, respectively. To say Y is normally distributed means its density follows the familiar "bell-shaped curve," with equation[5]

$$f(t) = \frac{1}{\sqrt{2\pi}d} e^{-\frac{1}{2}((t-c)/d)^2} \tag{2.33}$$

where c and d are the population mean and standard deviation. The same would be true for a normal distribution for X. But what does it mean for Y and X to be *jointly* normal, i.e., have a *bivariate normal* distribution?

The bivariate normal density takes the shape of a three-dimensional bell, as in Figure 2.1. (The figure is adapted from Romaine Francois' old R

---

[5]There are other families of bell-shaped curves, such as the *Cauchy*, so the normal density form is not "the" bell-shaped one.

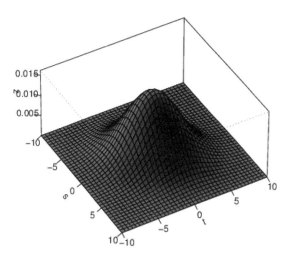

Figure 2.1: Bivariate normal density, $\rho = 0.2$ (see color insert)

Graphics Gallery.) A density function is a population-level entity, but if we were to draw a sample from such a population and form a two-variable histogram from it, the result would look like this 3-D bell shape.

The mathematical density form is pretty complex,

$$f(s,t) = \frac{1}{2\pi\sigma_1\sigma_2\sqrt{1-\rho^2}} e^{-\frac{1}{2(1-\rho^2)}\left[\frac{(s-\mu_1)^2}{\sigma_1^2} + \frac{(t-\mu_2)^2}{\sigma_2^2} - \frac{2\rho(s-\mu_1)(t-\mu_2)}{\sigma_1\sigma_2}\right]}, \quad (2.34)$$

for $-\infty < s, t < \infty$, where the μ_i and σ_i are the means and standard deviations of X and Y, and ρ is the correlation between X and Y (Section 2.12.1).

Now, let's see how this relates to our linear regression assumptions above, Linearity, Normality and Homoscedasticity. Since the regression function by definition is the conditional mean of Y given X, we need the conditional density. That means holding X constant, so we treat s in (2.34) as a constant. (This will give us the conditional density, except for a multiplicative constant.) We'll omit the messy algebraic details, but the point is that in the end, (2.34) reduces to a function that

- has the form (2.33), thus giving us the (conditional) Normality property;

- has mean as a linear function of s, yielding the Linearity property; and

- has variance independent of s, thus giving us Homoscedasticity.

The specific linear function in the second bullet above can be shown to be

$$E(Y \mid X = s) = \rho \frac{\sigma_2}{\sigma_1}(s - \mu_1) + \mu_2 \qquad (2.35)$$

and the constant-variance property in the third bullet is specifically

$$Var(Y \mid X = s) = (1 - \rho)^2 \sigma_2^2 \qquad (2.36)$$

which is indeed independent of s.

In other words, if X and Y have a bivariate normal distribution, all of our linear regression assumptions follow. This is the original motivation for the assumptions, though as mentioned they are plausible in many applications.

One can gain further insight by examining the effect of ρ. Toward this end, take another look at Figure 2.1, in particular the *level sets* of the graphed function, meaning the points of constant height. In other words, what happens if we slice that mound horizontally, parallel to the s-t plane?

That's equivalent to setting the exponent in (2.34) to some constant. Some readers may recognize this as the equation of an ellipse. Then the major, i.e., longer, axis of that ellipse turns out to be the regression line!

Moreover, consider what will happen as we increase ρ toward 1.0 (in the picture, $\rho = 0.2$). The minor axis will shrink, and the 3-D bell shape will become flatter and flatter, closing in on the regression line. See the picture for $\rho = 0.8$, Figure 2.2. This is reflected in the conditional variance in (2.36) going to 0 as ρ goes to 1.

These issues, and properties of the general multivariate normal distribution, are discussed in Section 2.12.2.

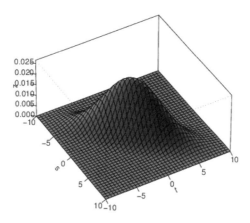

Figure 2.2: Bivariate normal density, $\rho = 0.8$ (see color insert)

2.7 Unbiasedness and Consistency

Here we discuss two statistical properties of the least-squares estimator $\widehat{\beta}$.

2.7.1 $\widehat{\beta}$ Is Unbiased

One of the central concepts in the early development of statistics was *unbiasedness*. As you'll see, to some degree it is only historical baggage, but on the other hand it does become quite relevant in some contexts here.

To explain the concept, say we are estimating some population value θ, using an estimator $\widehat{\theta}$ based on our sample. Remember, $\widehat{\theta}$ is a random variable — if we take a new sample, we get a new value of $\widehat{\theta}$. So, some samples will yield a $\widehat{\theta}$ that overestimates θ, while in other samples $\widehat{\theta}$ will come out too low.

The pioneers of statistics believed that a nice property for $\widehat{\theta}$ to have would be that *on average*, i.e., averaged over all possible samples, $\widehat{\theta}$ comes out "just right":

$$E\widehat{\theta} = \theta \tag{2.37}$$

This seems like a nice property for an estimator to have (though far from mandatory, as we'll see below), and sure enough, our least-squares estimator has that property:

$$E\widehat{\beta} = \beta \tag{2.38}$$

Note that since this is a vector equation, the unbiasedness is meant for the individual components. In other words, (2.38) is a compact way of saying

$$E\widehat{\beta}_j = \beta_j, \; j = 0, 1, ..., p \tag{2.39}$$

This is derived in the Mathematical Complements portion of this chapter, Section 2.12.5. Note that we do not need the Normality or Homoscedasticity assumption for this result.

2.7.2   Bias As an Issue/Nonissue

Arguably the pioneers of statistics shouldn't have placed so much emphasis on unbiasedness. Most statistical estimators have some degree of bias (though the amount of bias is usually small and goes to 0 as the sample size n grows). Even the much-heralded unbiased nature of the classical definition of sample variance, Equation (2.41) below, is somewhat misleading, as the sample standard deviation, arguably just as important a quantity, is biased (Section 2.12.7). And other than least-squares, none of the regression function estimators in common use, such as k-NN, is unbiased.

Nevertheless, bias can be an issue in some contexts, as will be seen later in this chapter.

2.7.3 $\widehat{\beta}$ Is Statistically Consistent

In contrast to unbiasedness, which as argued above may not be a generally necessary goodness criterion for an estimator, there is a more basic property that we would insist that almost any estimator have, *consistency*: As the sample size n goes to infinity, then the sample estimate $\widehat{\theta}$ goes to θ. This is not a very strong property, but it is a minimal one. It is shown in Section 2.12.6 that the least-squares estimator $\widehat{\beta}$ is indeed a consistent estimator of β. Again, we do not need the Normality or Homoscedasticity assumption for this result.

2.8 Inference under Homoscedasticity

Let's see what the homoscedasticity assumption gives us.

2.8.1 Review: Classical Inference on a Single Mean

You may have noticed the familiar Student-t distribution mentioned in the output of **lm()** above. Before proceeding, it will be helpful to review this situation from elementary statistics.

We have a random sample $W_1, ..., W_n$ from a population having mean $\nu = EW$ and variance η^2. Suppose W is normally distributed in the population. Form

$$\overline{W} = \frac{1}{n} \sum_{i=1}^{n} W_i \qquad (2.40)$$

and

$$S^2 = \frac{1}{n-1} \sum_{i=1}^{n} (W_i - \overline{W})^2 \qquad (2.41)$$

Then

$$T = \frac{\overline{W} - \nu}{S/\sqrt{n}} \qquad (2.42)$$

has a Student-t distribution with $n - 1$ degrees of freedom (df).

This is then used for statistical inference on ν. We can form a 95% confidence interval by adding and subtracting $c \times S/\sqrt{n}$ to $\overline{W}$, where c is the point of the upper-0.025 area for the Student-t distribution with $n - 1$ df.

Under the normality assumption, such inference is exact; a 95% confidence interval, say, has exactly 0.95 probability of containing ν.

2.8.2   Back to Reality

The normal distribution model is just that, a model, not expected to be exact. It rarely happens, if ever at all, that a population distribution is exactly normal. Human weight, for instance, cannot be negative and cannot be a million pounds; it is bounded, unlike normal distributions, whose support is $(-\infty, \infty)$. So "exact" inference using the Student-t distribution as above is not exact after all.

If n is large, the assumption of a normal population becomes irrelevant: The Central Limit Theorem (CLT, Section 2.12.3) tells us that

$$\frac{\overline{W} - \nu}{\eta/\sqrt{n}} \tag{2.43}$$

has an approximate N(0,1) distribution *even though the distribution of W is not normal*. We then must show that if we replace η by S in (2.43), the result will still be approximately normal. This follows from something called *Slutsky's Theorem* and the fact that S goes to η as $n \to \infty$.[6] Thus we can perform (approximate) statistical inference on ν using (2.42) and N(0,1), again *without assuming that W has a normal distribution*.

For instance, since the upper 2.5% tail of the N(0,1) distribution starts at 1.96, an approximate 95% confidence interval for ν would be

$$\overline{W} \pm 1.96 \frac{S}{\sqrt{n}} \tag{2.44}$$

What if n is small? We could use the Student-t distribution anyway, but we would have no idea how accurate it would be. We could not even use the data to assess the normality assumption on which the t-distribution is based, as we would have too little data to do so.

The normality assumption for the W_i, then, is of rather little value, and as explained in the next section, is of even less value in the regression context.

One possible virtue, though, of using Student-t would be that it gives a wider interval than does N(0,1). For example, for $n = 28$, our confidence interval would be

$$\overline{W} \pm 2.04 \frac{s}{\sqrt{n}} \tag{2.45}$$

---

[6]In its simpler form, the theorem says that if U_n converges to a normal distribution and $V_n \to v$ as $n \to \infty$, then U_n/V_n also is asymptotically normal.

instead of (2.44). The importance of this is that using S instead of η adds further variability to (2.42), which goes away as $n \to \infty$ but makes (2.43) overly narrow. Using a Student-t value might compensate for that, though it may also overcompensate.

2.8.3 The Concept of a Standard Error

The following concept will be used repeatedly throughout the book.

> If $\widehat{\theta}$ is an approximately normally distributed estimator of a population value θ, then an approximate 95% confidence interval for θ is
>
> $$\widehat{\theta} \pm 1.96 \; s.e.(\widehat{\theta}) \qquad (2.46)$$
>
> whereas in Section 1.6.3, the notation s.e.() denotes "standard error of."

In (2.44), the standard error of $\overline{W}$ is $S/\sqrt{n}$.

2.8.4 Extension to the Regression Case

The discussion in the last section concerned inference for a mean. What about inference for regression functions (which are conditional means)?

The first point to note is this:

- Under the classical assumption that the conditional distribution of Y given X is normal, then $\widehat{\beta}$ has an exact multivariate normal distribution. (This follows from the properties in Section 2.12.2.)

- The distribution of the least-squares estimator $\widehat{\beta}$ is approximately $(p+1)$-variate normal, *without* assuming normality.[7]

That second bullet again follows from the CLT. (Since we are looking at fixed-X regression here, we need a non-identically distributed version of the

---

[7]The statement is true even without assuming homoscedasticity, but we won't drop that assumption until the next chapter.

CLT.) Consider for instance a typical component of $A'D$ in (2.28),

$$\sum_{i=1}^{n} X_i^{(j)} Y_i \tag{2.47}$$

This is a sum of independent terms, thus approximately normal. In (2.28), we are working with a random vector, D, and in such a context, the corresponding CLT result is multivariate normal. So, $\widehat{\beta}$ has an asymptotic $(p+1)$-variate normal distribution. (A more formal derivation is presented in Section 2.12.11.)

To perform statistical inference, we need the approximate covariance matrix (Section 2.12.1) of $\widehat{\beta}$, from which we can obtain standard errors of the $\widehat{\beta}_j$. *The standard way to do this is by assuming homoscedasticity.*

So, **lm()** assumes that in (2.32), the function $\sigma^2(t)$ is constant in t. For brevity, then, we will simply refer to it as σ^2. Note that this plus our independence assumption implies the following about the (conditional) covariance matrix of D:

$$Cov(D|A) = \sigma^2 I \tag{2.48}$$

where I is the identity matrix.

To avoid (much) clutter, define

$$B = (A'A)^{-1} A' \tag{2.49}$$

Then by the properties of covariance matrices (Equation (2.79)),

$$
\begin{aligned}
Cov(\widehat{\beta} \,|A) &= Cov(BD) & (2.50)\\
&= B \, Cov(D|A) \, B' & (2.51)\\
&= \sigma^2 BB' & (2.52)
\end{aligned}
$$

Fortunately, the various properties of matrix transpose (Section A.3) can be used to show that

$$BB' = (A'A)^{-1} \tag{2.53}$$

Thus

$$Cov(\widehat{\beta}) = \sigma^2 (A'A)^{-1} \tag{2.54}$$

That's a nice (surprisingly) compact expression, but the quantity σ^2 is an unknown population value. It thus must be estimated, as we estimated η^2 by S^2 in Section 2.8.1. And again, an unbiased estimator is available. So, we take as our estimator of σ^2

$$s^2 = \frac{1}{n - p - 1} \sum_{i=1}^n (Y_i - \widetilde{X}_i'\widehat{\beta})^2 \tag{2.55}$$

which can be shown to be unbiased.

If the normality assumption were to hold, then quantities like

$$\frac{\widehat{\beta}_i - \beta_i}{s\sqrt{a_{ii}}} \tag{2.56}$$

would have an exact Student-t distribution with $n-p-1$ degrees of freedom, where a_{ii} is the (i, i) element of $(A'A)^{-1}$.[8]

But as noted, this is usually an unrealistic assumption, and we instead rely on the CLT. Putting the above together, we have:

> The conditional distribution of the least-squares estimator $\widehat{\beta}$, given A, is approximately multivariate normal (Section 2.6.2) with mean β and approximate covariance matrix
>
> $$s^2 (A'A)^{-1} \tag{2.57}$$
>
> Thus the standard error of $\widehat{\beta}_j$ is the square root of element (j, j) of this matrix (counting the top-left element as being in row 0, column 0).
>
> Similarly, suppose we are interested in some linear combination $\lambda'\beta$ of the elements of β, estimating it by $\lambda'\widehat{\beta}$ Section (A.4). By (2.80), the standard error is then the square root of

---

[8]A common interpretation of the number of degrees of freedom here is, "We have n data points, but must subtract one degree of freedom for each of the $p + 1$ estimated parameters."

$$s^2 \lambda'(A'A)^{-1}\lambda \qquad\qquad (2.58)$$

And as before, we might as well calculate s^2 with a denominator of n, as opposed to the $n - p - 1$ expression above.

Recall from Chapter 1, by the way, that R's **vcov()** function gives us the matrix (2.57), both for **lm()** and also for some other regression modeling functions that we will encounter later.

Before going to some examples, note that the conditional nature of the statements above is not an issue, even in random-X settings. Say for instance we form a 95% confidence interval for some quantity, conditional on A. Let V be an indicator variable for the event that the interval contains the quantity of interest. Then

$$P(V = 1) = E[P(V = 1 \mid A)] = E(0.95) = 0.95 \qquad (2.59)$$

Thus the unconditional coverage probability is still 0.95.

2.8.5   Example: Bike-Sharing Data

Let's form some confidence intervals from the bike-sharing data.

```
> lmout <- lm(reg ~ temp+temp2+workingday+clearday ,
      data=shar)
> summary(lmout)
. . . .
Coefficients:
             Estimate Std. Error t value Pr(>|t|)
(Intercept)  -1362.56     232.82  -5.852 1.09e-08
temp         11059.20     988.08  11.193  < 2e-16
temp2        -7636.40    1013.90  -7.532 4.08e-13
workingday     685.99      71.00   9.661  < 2e-16
clearday       518.95      69.52   7.465 6.34e-13
. . .
Multiple R-squared:  0.6548, Adjusted R-squared:   0.651
```

We estimate that a working day adds about 686 riders to the day's ridership. An approximate 95% confidence interval for the population value for this effect is

$$685.99 \pm 1.96 \times 71.00 = (546.83, 825.15) \qquad (2.60)$$

This is a disappointingly wide interval, but it shouldn't surprise us. After all, it is based on only 365 data points.

Given the nonlinear effect of temperature in our model, finding a relevant confidence interval here is a little more involved. Let's compare the mean ridership for our example in the last chapter — 62 degree weather, a Sunday and sunny — with the same setting but with 75 degrees.

The difference in (population!) mean ridership levels between these two settings is[9]

$$(\beta_0 + \beta_1 0.679 + \beta_2 0.679^2 + \beta_3 0 + \beta_1 1) - (\beta_0 + \beta_1 0.525 + \beta_2 0.525^2 + \beta_3 0 + \beta_1 1)$$

$$= \beta_1 0.154 + \beta_2 0.186$$

Our sample estimate for that difference in mean ridership between the two types of days is then obtained as follows:

```
> lamb <- c(0,0.154,0.186,0,0)
> t(lamb) %*% coef(lmout)
        [,1]
[1,]  282.7453
```

or about 283 more riders on the warmer day. For a confidence interval, we need a standard error. So, in (2.58), take $\lambda = (0, 0.154, 0.186, 0, 0)'$. Our standard error is then obtained via

```
> sqrt(t(lamb) %*% vcov(lmout) %*% lamb)
        [,1]
[1,]  47.16063
```

Our confidence interval for the difference between 75-degree and 62-degree days is

$$282.75 \pm 1.96 \cdot 47.16 = (190.32, 375.18) \qquad (2.61)$$

---

[9]Recall that the dataset here uses a scaled version of temperature; see page 37.

Again, a very wide interval, but it does appear that a lot more riders show up on the warmer days.

The value of s is itself probably not of major interest, as its use is usually indirect, in (2.57). However, we can determine it if need be, as **lmout$residuals** contains the *residuals*, i.e., the sample prediction errors

$$Y_i - \widetilde{X_i}'\widehat{\beta}, \ i = 1, 2, ..., n \tag{2.62}$$

Using (2.55), we can find s:

```
> s <- sqrt(sum(lmout$residuals^2) / (365-4-1))
> s
> s
[1]  626.303
```

2.9   Collective Predictive Strength of the $X^{(j)}$

The R^2 quantity in the output of **lm()** is a measure of how well our model predicts Y. Yet, just as $\widehat{\beta}$, a sample quantity, estimates the population quantity β, one would reason that the R^2 value printed out by **lm()** must estimate a population quantity too. In this section, we'll make that concept precise, and deal with a troubling bias problem.

We will also introduce an alternative form of the cross-validation notion discussed in Section 1.12.

2.9.1   Basic Properties

Note carefully that we are working with population quantities here, generally unknown, but existent nonetheless. Note too that, for now, we are NOT assuming normality or homoscedasticity. In fact, even the assumption of having a linear regression function will be dropped for the moment. The context, by the way, is random-X regression (Section 2.3).

Suppose we somehow knew the exact population regression function $\mu(t)$. Whenever we would encounter a person/item/day/etc. with a known X but unknown Y, we would predict the latter by $\mu(X)$. Define ϵ to be the prediction error

$$\epsilon = Y - \mu(X) \tag{2.63}$$

It can be shown (Section 2.12.9) that $\mu(X)$ and ϵ are uncorrelated, i.e., have zero covariance. We can thus write

$$Var(Y) = Var[\mu(X)] + Var(\epsilon) \tag{2.64}$$

With this partitioning, it makes sense to say:

The quantity

$$\omega = \frac{Var[\mu(X)]}{Var(Y)} \tag{2.65}$$

is the proportion of variation of Y explainable by X.

Section 2.12.9 goes further:

Define

$$\rho = \sqrt{\omega} \tag{2.66}$$

Then ρ is the correlation between our predicted value $\mu(X)$ and the actual Y.

Again, the normality and homoscedasticity assumptions are NOT needed for these results. In fact, *they hold for any regression function, not just one satisfying the linear model.* This includes the "fancy" nonparametric techniques such as CART (Chapter 10) and SVM and neural networks (Chapter 11).

2.9.2   Definition of R^2

The quantity R^2 output by **lm()** is the sample analog of ρ^2:

> R^2 is the squared sample correlation between the actual re-
> sponse values Y_i and the predicted values $\tilde{X}'_i \,\widehat{\beta}$. R^2 is a consis-
> tent estimator of ρ^2.

Exactly how is R^2 defined? From (2.64) and (2.65), we see that

$$\rho^2 = 1 - \frac{Var[\epsilon]}{Var(Y)} \tag{2.67}$$

Since $E\epsilon = 0$, we have

$$Var(\epsilon) = E(\epsilon^2) \tag{2.68}$$

The latter is the average squared prediction error in the population, whose
sample analog is the average squared error in our sample. In other words,
using our "correspondence" notation from before,

$$E(\epsilon^2) \longleftrightarrow \frac{1}{n}\sum_{i=1}^{n}(Y_i - \tilde{X}'_i\widehat{\beta})^2 \tag{2.69}$$

Now considering the denominator in (2.67), the sample analog is

$$Var(Y) \longleftrightarrow \frac{1}{n}\sum_{i=1}^{n}(Y_i - \overline{Y})^2 \tag{2.70}$$

where of course $\overline{Y} = (\sum_{i=1}^{n} Y_i)/n$.

And that is R^2:

$$R^2 = 1 - \frac{\frac{1}{n}\sum_{i=1}^{n}(Y_i - \tilde{X}'_i\widehat{\beta})^2}{\frac{1}{n}\sum_{i=1}^{n}(Y_i - \overline{Y})^2} \tag{2.71}$$

(Yes, the $1/n$ factors do cancel, but it will be useful to leave them there.)

As a sample estimate of the population ρ^2, the quantity R^2 would appear
to be a very useful measure of the collective predictive ability of the $X^{(j)}$.

However, the story is not so simple, and curiously, the problem is actually bias.

2.9.3   Bias Issues

R^2 can be shown to be biased upward, not surprising in light of the fact that we are predicting on the same data that we had used to calculate $\widehat{\beta}$. In the extreme, we could fit an $n - 1$ degree polynomial in a single predictor, with the curve passing through each data point, producing $R^2 = 1$, even though our ability to predict future data would be very weak.

The bias can be severe if p is a substantial portion of n. (In the above polynomial example, we would have $p = n - 1$, even though we started with $p = 1$.) This is the *overfitting* problem mentioned in the last chapter, to be treated in depth in later chapters. But for now, let's see how bad the bias can be, using the following simulation code:

```
simr2 <- function(n,p,nreps) {
    r2s <- vector(length=nreps)
    for (i in 1:nreps) {
        x <- matrix(rnorm(n*p),ncol=p)
        y <- x %*% rep(1,p) + rnorm(n,sd=sqrt(p))
        r2s[i] <- getr2(x,y)
    }
    hist(r2s)
}

getr2 <- function(x,y) {
    smm <- summary(lm(y ~ x))
    smm$r.squared
}
```

Here we are simulating a population in which

$$Y = X^{(1)} + ... + X^{(p)} + \epsilon \tag{2.72}$$

so that β consists of a 0 followed by p 1s. We set the $X^{(j)}$ to have variance 1, and ϵ has variance p. This gives $\rho^2 = 0.50$. Hopefully R^2 will usually be near this value. To assess this, I ran **simr2(25,8,1000)**, i.e., $n = 25$ and $p = 8$, with 1000 repetitions of the experiment. The result is shown in Figure 2.3.

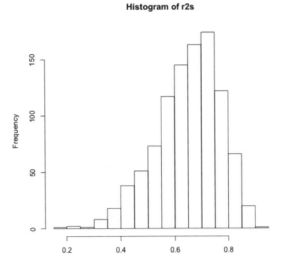

Figure 2.3: Plotted R^2 values, n = 25

These results are not encouraging at all! The R^2 values are typically around 0.7, rather than 0.5 as they should be. In other words, R^2 is typically giving us much too rosy a picture as to the predictive strength of our $X^{(j)}$.

Of course, it should be kept in mind that I deliberately chose a setting that produced substantial overfitting — 8 predictors for only 25 data points, which is probably too many predictors.

Running the simulation with $n = 250$ should show much better behavior. The results are shown in Figure 2.4. This is indeed much better. Note, though, that the upward bias is still evident, with values more typically above 0.5 than below it.

Note too that R^2 seems to have large variance, even in the case of $n = 250$. Thus in samples in which p/n is large, we should not take our sample's value of R^2 overly seriously.

2.9.4   Adjusted-R^2

The adjusted-R^2 statistic is aimed at serving as a less biased version of ordinary R^2. Its derivation is actually quite simple, though note that we

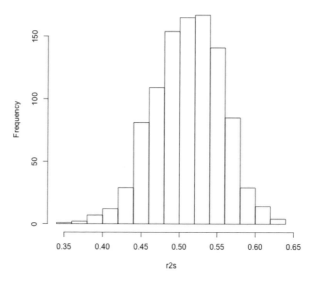

Figure 2.4: Plotted R^2 values, n $=$ 250

do need to assume homoscedasticity.

Under the latter assumption, $Var(\epsilon) = \sigma^2$ in (2.67). Then the numerator in (2.71) is biased, which we know from (2.55) can be fixed by using the factor $1/(n-p-1)$ instead of $1/n$. Similarly, we know that the denominator will be unbiased if we divide by $1/(n-1)$ instead of $1/n$. Those changes do NOT make (2.71) unbiased; the ratio of two unbiased estimators is generally biased. However, the hope is that this new version of R^2, called *adjusted* R^2, will have less bias than the original.

The formula is

$$R_{adj}^2 = 1 - \frac{\frac{1}{n-p-1}\sum_{i=1}^n (Y_i - \widetilde{X}_i'\widehat{\beta})^2}{\frac{1}{n-1}\sum_{i=1}^n (Y_i - \overline{Y})^2} \tag{2.73}$$

We can explore this using the same simulation code as above. We simply change the line

smm\$r.squared

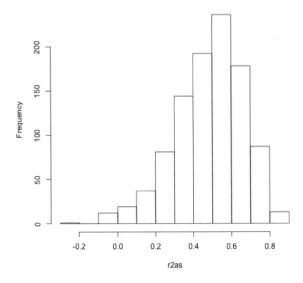

Figure 2.5: Plotted adjusted R^2 values, n = 25

to

smm$adj.r.squared

Rerunning **simr2(25,8,1000)**, we obtain the result shown in Figure 2.5. This is a good sign! The values are more or less centered around 0.5, as they should be (though there is still a considerable amount of variation).

See the Computational Complements section at the end of this chapter for a note on how R^2 and the adjusted version are computed from the output of **lm()**.

2.9.5   The "Leaving-One-Out Method"

Our theme here in Section 2.9 has been assessing the predictive ability of our model, with the approach described so far being the R^2 measure. But recall that we have another measure: Section 1.12 introduced the concept of *cross-validation* for assessing predictive ability. We will now look at a variant of that method.

First, a quick review of cross-validation: Say we have n observations in our data set. With cross-validation, we randomly partition the data into a training set and a validation set, of k and $n - k$ observations, respectively. We fit our model to the training set, and use the result to predict the response variable in the validation set, and then see how well those predictions turned out.

Clearly there is an issue of the choice of k. If k is large, our validation set will be too small to obtain an accurate estimate of predictive ability. That is not a problem if k is small, but then we have a subtler problem: We are getting an estimate of strength of our model when constructed on k observations, but in the end we wish to use all n observations.

One solution is the Leaving One Out Method (LOOM). Here we set $k = n - 1$, but apply the training/validation process to *all* possible $(n - 1, 1)$ partitions. The name alludes to the fact that LOOM repeatedly omits one observation, predicting it from fitting the model to the remaining observation. It is hoped that this gives us "the best of both worlds": We have n validation points, the best possible, and the training sets are of size $n - 1$, i.e., nearly full-sized.

2.9.6 Extensions of LOOM

Instead of the Leaving One Out Method, we might leave out k observations instead of just 1, known as *k-fold cross-validation*. In other words, for each possible subsct of k observations, we predict those k by the remaining $n - k$. This gives us many more test sets, at a cost of more computation.

There is theoretical evidence [128] that as the sample size n goes to infinity, cross-validation will only provide statistical consistency if k-fold cross-validation is used with $k/n \to 1$.

2.9.7 LOOM for k-NN

LOOM is directly invocable in the **regtools** package. If in calling the **preprocesx()** function, one specifies **xval = TRUE**, the computation of nearest neighbors to a data point won't include the point itself.

2.9.8   Other Measures

A number of other measures of predictive ability are in common use, notably *Mallows' C_p* and the *Akaike Information Criterion*. These will be treated in Chapter 9.

2.10   The Practical Value of p-Values — Small OR Large

"Sir Ronald [Fisher] has befuddled us, mesmerized us, and led us down the primrose path" — Paul Meehl, professor of psychology and the philosophy of science

When the concept of significance testing, especially the 5% value for α, was developed in the 1920s by Sir Ronald Fisher, many prominent statisticians opposed the idea — for good reason, as we'll see below. But Fisher was so influential that he prevailed, and thus significance testing became the core operation of statistics.

So, today significance testing is deeply entrenched in the field, and even though it is widely recognized as faulty. many continue to engage in the practice.[10] Most modern statisticians understand this, It was eloquently stated in a guide to statistics prepared for the U.S. Supreme Court by two prominent scholars, one a statistician and the other a law professor [80]:

> Statistical significance depends on the p-value, and p-values depend on sample size. Therefore, a 'significant' effect could be small. Conversely, an effect that is 'not significant' could be large. By inquiring into the magnitude of an effect, courts can avoid being misled by p-values. To focus attention where it belongs — on the actual size of an effect and the reliability of the statistical analysis — interval estimates may be valuable. Seeing a plausible range of values for the quantity of interest helps describe the statistical uncertainty in the estimate.

The basic problem is that a significance test is answering the wrong question. Say in a regression analysis we are interested in the relation between

---

[10]Many are forced to do so, e.g., to comply with government standards in pharmaceutical testing. My own approach in such situations is to quote the test results but then point out the problems, and present confidence intervals as well.

$X^{(1)}$ and Y. Our test might have as null hypothesis

$$H_0 : \beta_1 = 0 \tag{2.74}$$

But we probably know *a priori* that there is at least *some* relation between the two variables; β_1 cannot be 0.000000000... to infinitely many decimal places. So we already know that H_0 is false.[11] The better approach is to form a confidence interval for β_1, so that we can gauge the *size* of β_1, i.e., the strength of the relation.

Note carefully that this does not mean avoiding making a decision. The point is to make an *informed* decision, rather than letting the machine make a decision for you that may not be useful.

2.10.1 Misleadingly Small p-Values

Many researchers are ecstatic when they find a tiny p-value. But actually, that p-value may be rather meaningless.

2.10.1.1 Example: Forest Cover Data

For instance, consider another UCI data set, Forest Cover, which involves a remote sensing project. The goal was to predict which one of seven types of ground cover exists in a certain inaccessible location, using variables that can be measured by satellite. One of the variables is Hillside Shade at Noon (HS12).

For this example, I restricted the data to Cover Types 1 and 2, and took a random subset of 1000 observations to keep the example manageable. I named the resulting data frame **f2512**. The logistic model here is

$$P(\text{Cover Type 2}) = \frac{1}{1 + e^{-(\beta_0 + \beta_1 \ \text{HS12})}} \tag{2.75}$$

Here is the **glm()** output, with column 8 being HS12 and column 56 being a dummy variable indicating Cover Type 2:

```
> glmout <-
    glm( f2512 [ ,56] ~ f2512 [ ,8] , family=binomial )
```

---

[11]A similar point holds for the F-test in **lm()** output, which tests that *all* the β_i are 0, i.e., $H_0 : \beta_1 = \beta_2 = \ldots \beta_p = 0$.

```
> summary(glmout)
...
Coefficients:
              Estimate Std. Error  z value  Pr(>|z|)
(Intercept)  -2.147856   0.634077   -3.387  0.000706 ***
f2512[, 8]    0.014102   0.002817    5.007  5.53e-07 ***
...
```

The triple-star result for β_1 would indicate that HS12 is a "very highly significant" predictor of cover type. Yet we see that $\widehat{\beta_1}$, 0.014102, is tiny. HS12 is in the 200+ range, with sample means 227.1 and 223.4 for the two cover types, differing only by 3.7. Multiplying the latter by 0.014102 gives a value of about 0.052, which is swamped in (2.75) by the $\widehat{\beta_0}$ term, -2.147856. In plain English: HS12 has almost no predictive power for Cover Type, yet the test declares it "very highly significant."

The confidence interval for β_1 here is

$$0.014102 \pm 1.96 \cdot 0.002817 = (0.00858, 0.01962) \qquad (2.76)$$

The fact that the interval excludes 0 is irrelevant. The real value of the interval here is that it shows that β_1 is quite small; even the right-hand end point is tiny.

2.10.1.2   Example: Click Through Data

In the informal analysis of the Click-Through Rate data in Section 1.8, we stated that it appeared that educational level and CTR have no substantial relation. Now that we have formal tools available, let's revisit that.

```
> summary(lm(ctr$CTR ~ ctr$College_Grad))
...
Coefficients:
                    Estimate Std. Error  t value
(Intercept)        0.0141233  0.0005969   23.659
ctr$College_Grad  -0.0137300  0.0024334   -5.642
                    Pr(>|t|)
(Intercept)        < 2e-16 ***
ctr$College_Grad  8.28e-07 ***
...
Mult. R-squared: 0.3938,  Adj. R-squared: 0.3815
...
```

Again putting aside the issue of whether this data on U.S. states can be considered a random sample from some population, the computer output here declares the education variable to be "very highly significant."

And indeed, the R-squared values are rather good. In fact, recalling that R-squared is the squared sample correlation between Y and X (in the one-predictor case), we see that that correlation is about -0.63. That value was cited in the online article from which I obtained this data [53], which stated "This data presents a stunning -0.63 rate between [educational level and CTR]."

Yet all of those numbers are leading us astray. As we saw in Section 1.8, the effect of education on CTR is quite negligible. So, again, one should treat significance tests with great skepticism.

Note too our "sample size" n here seems small, only 51. On the one hand, that would suggest a substantial upward bias in R-squared, as discussed previously.

On the other hand, each CTR value is based on hundreds of thousands of clicks, more than 31 million in all. In this sense we have a similar large-data problem as in the forest cover data in the last section.

2.10.2 Misleadingly LARGE p-Values

Just as it is wrong to treat very small p-values as automatically showing "significance," it is equally dangerous to treat large p-values as evidence of no important effect in the given quantity of interest, say a coeficient β_i. We might simply not have enough data to say anything at all about that quantity.

This can occur even in very large samples. A notable case is dummy variables. Say $X^{(5)}$ is a dummy variable and we are interested in estimating β_5, to gauge the impact this variable has on Y. But we might have only a few observations in which $X^{(5)} = 1$. Intuitively this will prevent us from obtaining a good estimate of β_5, even if n is very large. This makes sense analytically, as follows.

The sample variance of $X^{(5)}$ will be $q(1 - q)$, where $q = P(X^{(6)} = 1)$. The latter will be small, resulting in a small corresponding element in $A'A$ in Section 2.8, if we center the data. That produces a large value in $(A'A)^{-1}$, thus a large standard error for $\widehat{\beta_5}$.

So, a large p-value doesn't necessarily indicate that the effect of interest is small.

2.10.3   The Verdict

So, a small p-value does not necessarily imply an important effect, and a large p-value likewise should not be treated as showing lack of an important effect. Very misleading results can occur by relying on p-values.

This book recommends using confidence intervals instead of p-values. The two key points are:

- The *width* of the interval tells us the accuracy of our estimator.

- The *location* of the interval gives us an estimate of the size of the effect.

Say our interval for β_i excludes 0 but the location is very near 0. Then the effect is probably small, and we should probably not call the effect "signifcant." On the other hand, if the interval is near 0 or even contains it, but the interval is very wide, the latter property tells us that we just don't have much to say about the effect.

Some would object to our claim above that we almost always know *a priori* that H_0 is false. They might point to tests of the form

$$H_0 : \theta \le c, \quad H_1 : \theta > c \tag{2.77}$$

One major problem with that argument is that there is almost always measurement error, due say to either finite-precision machine measurements or sampling bias (sampling a narrower or skewed population than we had intended). In addition, there is still the problem in which, say, $\theta > c$ but with $\theta - c$ being so tiny that the difference is negligible.

2.11   Missing Values

The "dirty little secret" about data analysis is that most data is dirty. Some data is erroneous, and/or it is often missing altogether.

On the one hand, R is very good about missing values, which are coded as NA. It checks data for NAs (which comes at a cost of somewhat slower execution), and tries to accommodate them, typically by omitting observations that have at least one NA value. R actually has a function, **complete.cases()**, to flag such observations.

Though seemingly reasonable, there can be real problems with the complete-cases approach. The problem is that it can create biases. If we are studying income and the richest people tend not to fill out the income blank in a survey, the bias effect is obvious. But it can be more subtle than that.

For instance, one of the earliest methods for dealing with missing data is to simply replace an NA by the mean of that variable. But this attentuates the relations between variables. And again, in the case of the richer people tending to omit the income portion of a survey, clearly replacing NA by the mean income produces biases as well.

This problem has been the subject of intense study by top researchers [91], yet there is no really good solution. The problem is that one must take into account the mechanism that led to the NA values. Various levels of randomness have been proposed, with the easiest one to solve being Missing Completely at Random (MCAR). As its name implies, the fact that the value Q_{ij} of variable j in observation i is NA is assumed to be statistically independent of both Q_{ij} and the other Q_{ik} — a very stringent assumption. Less stringent assumptions have been formulated as well, but are difficult to verify.

The **regtools** package includes several functions that assume MCAR, for the linear model, principal components and the log-linear model. Several packages that attack the problem in a more sophisticated way are available on CRAN, such as Amelia II. Further discussion is beyond the scope of this book.

2.12 Mathematical Complements

2.12.1 Covariance Matrices

The *covariance* between random variables X and Y is defined as

$$Cov(X, Y) = E[(X - EX)(Y - EY)] \tag{2.78}$$

Suppose that typically when X is larger than its mean, Y is also larger than its mean, and vice versa for below-mean values. Then (2.78) will likely be positive. In other words, if X and Y are positively correlated (a term we will define formally below but keep intuitive for now), then their covariance is positive. Similarly, if X is often smaller than its mean whenever Y is larger than its mean and *vice versa*, the covariance and correlation between them will be negative. All of this is roughly speaking, of course, since it depends on *how much* and *how often* X is larger or smaller than its mean, etc.

For a random vector $U = (U_1, ..., U_k)'$, its *covariance matrix* $Cov(U)$ is the $k \times k$ matrix whose (i, j) element is $Cov(U_i, U_j)$. Also, for a constant matrix A with k columns, AU is a new random vector, and one can show that

$$Cov(AU) = A \ Cov(U) \ A' \qquad (2.79)$$

In the special case in which A is a row vector a', this reduces to

$$Var(a'U) = a' \ Cov(U) \ a \qquad (2.80)$$

Covariance does measure how much or little X and Y vary together, but it is hard to decide whether a given value of covariance is "large" or not. For instance, if we are measuring lengths in feet and change to inches, then (2.78) shows that the covariance will increase by 12 if the unit change is just in X, and by $12^2 = 144$ if the change is in Y as well. Thus it makes sense to scale covariance according to the variables' standard deviations. Accordingly, the *correlation* between two random variables X and Y is defined by

$$\rho(X, Y) = \frac{Cov(X, Y)}{\sqrt{Var(X)}\sqrt{Var(Y)}} \qquad (2.81)$$

So, correlation is unitless, i.e. does not involve units like feet, pounds, etc. And it can be shown (Section 2.12.8.1) that'

$$-1 \leq \rho(X, Y) \leq 1 \qquad (2.82)$$

These are all population values. The sample analog of (2.78) is

$$\frac{1}{n}\sum_{i=1}^{n}(X_i - \overline{X})(Y_i - \overline{Y}) \qquad (2.83)$$

This is calculated by R's **cov()** function, though as explained in Section 2.7.2, the function uses $n-1$ instead of n as the divisor, following tradition. An example of **cov()** is given in Section 2.13.2.1.

2.12.2   The Multivariate Normal Distribution Family

In Section 2.6.2, we discussed the bivariate normal distribution family, and its role in motivating the classical assumptions of linear regression analysis.

Consider the case of general p. If $(Y, X')'$ has a $p+1$-variate normal distribution, it can be shown that the Linearity, Normality and Homoscedasticity assumptions hold for the regression of Y on any subset of the predictors in X. In fact, by the symmetry of the situation, the same is true for regressing any $X^{(i)}$ on any of the other $X^{(j)}$ and/or Y. This fact will be useful at some points in coming chapters.

Matrix notation allows a compact representation of the multivariate normal density:

$$c\ e^{-0.5(t-\mu)'\Sigma^{-1}(t-\mu)} \tag{2.84}$$

where μ and Σ are the mean vector and covariance matrix of the given random vector, and c is a constant needed to make the density integrate to 1.0.

The multivariate normal distribution family has many interesting properties:

- **Property A:**

  Suppose the random vector V has a multivariate normal distribution with mean ν and covariance matrix Γ. Let A be a constant matrix with the same number of columns as the length of V. Then the random vector $W = AV$ is also multivariate normal distributed, with mean $A\nu$ and covariance matrix

  $$Cov(W) = A\Gamma A' \tag{2.85}$$

  Note carefully that the remarkable part of that last statement is that W, the new random vector, also has a multivariate normal distribution, "inheriting" it from V. The statement about W's mean and covariance matrix are true even if V does not have a multivariate normal distribution, as we saw in Section 2.12.1.

- **Property B:**

  In general, if two random variables T and U have 0 correlation, this does not imply they are independent. However, if they have a bivariate normal distribution, the independence does hold.

- **Property C:**

  Suppose B is a $k \times k$ *idempotent* matrix, i.e. $B^2 = B$, and suppose U is a k-variate normally distributed random vector with $B \; EU = 0$ and with covariance matrix $\sigma^2 I$. Then the quantity

$$U'BU/\sigma^2 \tag{2.86}$$

  has a chi-squared distribution with degrees of freedom equal to rank(B).

R functions for multivariate normal distributions are discussed in Section 2.13.2.

2.12.3   The Central Limit Theorem

Roughly speaking, the *Central Limit Theorem* states that sums of random variables have an approximately normal distribution. More formally, if U_i, $i = 1, 2, 3, \ldots$ are i.i.d., each with mean μ and variance σ^2, then the cumulative distribution function of

$$\frac{U_1 + \ldots + U_n - n\mu}{\sigma\sqrt{n}} \tag{2.87}$$

goes to that of N(0,1) as $n \to \infty$.

The multivariate version is true as well. A sum of i.i.d. random vectors will have an approximately multivariate normal distribution.

There are versions of the CLT for independent, non-identically distributed random variables as well, under assumptions such as the *Lindeberg Condition*.

2.12.4   Details on Models Without a Constant Term

Here we will fill in the details of some claims made in Section 2.4.5. First, let's see why $\widehat{\beta}_0$ is forced to 0 if we center the data. Expand (2.18) (without the $1/n$ factor) to

$$\sum_{i=1}^{n}(Y_i - b_0 - b_1X_i^{(1)} - ... - b_pX_i^{(p)})^2 \tag{2.88}$$

Set the partial derivative with respect to b_0 to 0:

$$0 = \sum_{i=1}^{n}(Y_i - b_0 - b_1X_i^{(1)} - ... - b_pX_i^{(p)})(-1) \tag{2.89}$$

But any centered variable will sum to 0, in this case Y and each $X^{(j)}$, so (2.89) becomes

$$0 = nb_0 \tag{2.90}$$

Since we take $\widehat{\beta}$ to be the value of b that minimizes (2.88), we see that

$$\widehat{\beta}_0 = 0 \tag{2.91}$$

demonstrating the claimed result.

2.12.5 Unbiasedness of the Least-Squares Estimator

We will show that $\widehat{\beta}$ is conditionally unbiased,

$$E(\widehat{\beta} \mid X_1, ..., X_n) = \beta \tag{2.92}$$

under the linearity assumption

$$E(Y \mid X) = \mu(X) = \widetilde{X}'\beta \tag{2.93}$$

This approach has the advantage of including the fixed-X case, and it also implies the unconditional case for random-X, since

$$E\widehat{\beta} = E[E(\widehat{\beta} \mid X_1, ..., X_n)] = E\beta = \beta \tag{2.94}$$

by the Law of Total Expectation, (1.58).

So let's derive (2.92). First note that Equation (2.93) tells us that

$$E(D \mid A) = A\beta \qquad\qquad (2.95)$$

where A and D are as in Section 2.4.2.

Now using (2.28) we have

$$
\begin{aligned}
E(\widehat{\beta} \mid X_1, ..., X_n) &= E[\widehat{\beta} \mid A] & (2.96)\\
&= E[(A'A)^{-1}A'D \vert A] & (2.97)\\
&= (A'A)^{-1}A'E(D \mid A) & (2.98)\\
&= (A'A)^{-1}A'A\beta) & (2.99)\\
&= \beta & (2.100)
\end{aligned}
$$

thus showing that $\widehat{\beta}$ is unbiased.

2.12.6   Consistency of the Least-Squares Estimator

We'll make use of a famous theorem:

Strong Law of Large Numbers (SLLN): Say $W_1, W_2, ...$ are i.i.d. (scalar or vector) with common mean EW. Then

$$\lim_{n \to \infty} \frac{1}{n} \sum_{i=1}^{n} W_i = EW, \quad \text{with probability 1} \qquad (2.101)$$

Though independent, non-identically distributed versions of the SLLN do exist, they have rather technical conditions, so we will assume the random-X setting here.

Below, the role of W_i will sometimes be played by the vectors

$$(Y_i, X_i^{(1)}, ..., X_i^{(p)})' \qquad\qquad (2.102)$$

and sometimes by individual scalars. We will assume that the various expectations exist, e.g., $E(XX')$ below.

Armed with that fundamental theorem in probability theory, rewrite (2.28) as

$$\hat{\beta} = \left(\frac{1}{n}A'A\right)^{-1}(\frac{1}{n}A'D) \tag{2.103}$$

To avoid clutter, we will not use the $\widetilde{X}$ notation here for augmenting with a 1 element at the top of a vector. Assume instead that the 1 is $X^{(1)}$.

By the SLLN, the (i,j) element of $\frac{1}{n}A'A$ converges as $n \to \infty$:

$$\frac{1}{n}(A'A)_{ij} = \frac{1}{n}\sum_{k=1}^{n}X_k^{(i)}X_k^{(j)} \to E[X^{(i)}X^{(j)}] = [E(XX')]_{ij} \tag{2.104}$$

i.e.,

$$\frac{1}{n}A'A \to E(XX') \tag{2.105}$$

The vector $A'D$ is a linear combination of the columns of A' (Section A.4), with the coefficients of that linear combination being the elements of the vector D. Since the columns of A' are X_k, $k = 1, ..., n$, we then have

$$A'D = \sum_{k=1}^{n}Y_kX_k \tag{2.106}$$

and thus

$$\frac{1}{n}A'D \to E(YX) \tag{2.107}$$

The latter quantity is

$$
\begin{aligned}
E[E(YX \mid X)] &= E[XE(Y \mid X)] &\quad (2.108)\\
&= E[X(X'\beta)] &\quad (2.109)\\
&= E[X(X'I\beta)] &\quad (2.110)\\
&= E[(XX')I\beta)] &\quad (2.111)\\
&= E(XX')\,\beta &\quad (2.112)\\
&&\quad (2.113)
\end{aligned}
$$

So, we see that $\widehat{\beta}$ converges to

$$[E(XX')]^{-1}E(XY) = [E(XX')]^{-1}E(XX'\beta) = [E(XX')]^{-1}E(XX')\beta = \beta \tag{2.114}$$

2.12.7   Biased Nature of S

It was stated in Section 2.7.2 that S, even with the $n-1$ divisor, is a *biased* estimator of η, the population standard deviation. We'll derive that here.

$$\begin{aligned} 0 \quad &< \quad Var(S) &\tag{2.115} \\ &= \quad E(S^2) - (ES)^2 &\tag{2.116} \\ &= \quad \eta^2 - (ES)^2 &\tag{2.117} \end{aligned}$$

since S^2 is an unbiased estimator of η^2. So,

$$ES < \eta \tag{2.118}$$

2.12.8   The Geometry of Conditional Expectation

Readers with a good grounding in vector spaces may find the material in this section helpful to their insight. It is recommended that the reader review Section 1.19.5 before continuing.[12]

2.12.8.1   Random Variables As Inner Product Spaces

Consider the set of all scalar random variables U defined in some probability space that have finite second moment, i.e. $E(U^2) < \infty$. This forms a linear space: The sum of two such random variables is another random variable with finite second moment, as is a scalar times such a random variable.

---

[12]It should be noted that the treatment here will not be fully mathematically rigorous. For instance, we bring in projections below, without addressing the question of the conditions for their existence.

We can define an inner product on this space. For random variables S and T in this space, define

$$(S, T) = E(ST) \tag{2.119}$$

This defines the norm

$$||S|| = (S, S)^{1/2} = \sqrt{E(S^2)} \tag{2.120}$$

So, if $ES = 0$, then

$$||S|| = \sqrt{Var(S)} \tag{2.121}$$

Many properties for regression analysis can be derived quickly from this vector space formulation. Let's start with (2.82).

The famous *Cauchy-Schwartz Inequality* for inner product spaces states that for any vectors x and y, we have

$$|(x, y)| \le ||x|| \, ||y|| \tag{2.122}$$

It is left as an exercise to the reader to show that this implies (2.82).

2.12.8.2 Projections

Inner product spaces also have the notion of a *projection*. Suppose we have an inner product space $\mathcal{V}$, and subspace $\mathcal{W}$. Then for any vector x, the projection z of x onto $\mathcal{W}$ is defined to be the closest vector to x in $\mathcal{W}$. An important property is that we have a "right triangle," i.e.

$$(z, x - z) = 0 \tag{2.123}$$

We say that z and $x - z$ are *orthogonal*. And the Pythagorean Theorem holds:

$$||x||^2 = ||z||^2 + ||x - z||^2 \tag{2.124}$$

2.12.8.3 Conditional Expectations As Projections

In regression terms, the discussion in Section 1.19.3 shows that the regression function, $E(Y \mid X) = \mu(X)$ has the property that

$$\mu(X) = \operatorname{argmin}_g E[(Y - g(X))^2] = \operatorname{argmin}_g \|Y - g(X)\|^2 \qquad (2.125)$$

as **g** ranges over all functions of X. Therefore, by definition, $\mu(X)$ is the projection of Y onto the subspace consisting of all random variables with finite variance that are functions of X. This view can be very useful.

We can also use (2.124) to derive the Law of Total Variance, (1.62). For convenience in present notation, rewrite that equation as

$$Var(Y) = E[Var(Y|X)] + Var[E(Y|X)] \qquad (2.126)$$

The derivation will be less cluttered if we restrict attention to the case $EY = 0$. (For the general case, define a new random variable $W = Y - EY$, and apply the mean-0 result, left as an exercise for the reader.) Note that by the Law of Total Expectation (Section 1.19.5.2), this implies the $\mu(X)$ also has mean 0.

Then (2.124) and (2.121) say that

$$Var(Y) = E[\mu(X)^2] + E\left[(Y - \mu(X))^2\right] \qquad (2.127)$$

Recalling that $E\mu(X) = 0$, the first term in (2.127) is

$$Var[\mu(X)] = Var[E(Y|X)] \qquad (2.128)$$

which is exactly the second term in (1.62).

Now rewrite the second term in (2.127) using (1.58):

$$
\begin{aligned}
E\left[(Y - \mu(X))^2\right] &= E\{E\left[(Y - \mu(X))^2 \mid X\right]\} & (2.129) \\
&= E[Var(Y|X)] & (2.130)
\end{aligned}
$$

And, that last expression is exactly the first term in (1.62)! So, we are done with the derivation.

2.12.9 Predicted Values and Error Terms Are Uncorrelated

Assume a random-X context, and take x in (2.123) to be Y, so in that equation

$$z = \mu(X) \tag{2.131}$$

and thus

$$E[\mu(X)(Y - \mu(X))] = 0 \tag{2.132}$$

In other words, our prediction $\mu(X)$ is uncorrelated with our prediction error, $Y - \mu(X)$.

The above concerns the population level, but a similar argument can be made at the sample level for linear models. Here we will assume a fixed-X model (conditioning on X in the random-X case), and once again use the notation of Section 2.4.2.

Define

$$\widehat{\epsilon}_i = Y_i - \widetilde{X}_i \widehat{\beta} \tag{2.133}$$

Also define $\widehat{\epsilon}$ to be the vector of the $\widehat{\epsilon}_i$.

The claim is then that the correlation between $\widehat{\epsilon}_i$ and $\widehat{\beta}_j$ is 0 for any i and j. Again, a vector space argument can be made. In this case, take the full vector space to be $\mathcal{R}^n$, the space in which D roams, and the subspace will be that spanned by the columns of A.

The vector $A\widehat{\beta}$ is in that subspace, and because $b = \widehat{\beta}$ minimizes (2.25), $A\widehat{\beta}$ is then the projection of D onto that subspace. Again, that makes $D - A\widehat{\beta}$ and $A\widehat{\beta}$ orthogonal, i.e.

$$\widehat{\epsilon}' A\widehat{\beta} = (D - A\widehat{\beta})' A\widehat{\beta} = 0 \tag{2.134}$$

Since this must hold for all A, we see that each $\widehat{\epsilon}_i$ is uncorrelated with any component of $\widehat{\beta}$.

2.12.10   Classical "Exact" Inference

Again, assume a fixed-X setting. Classically we assume not only the linear model, independence of the observations and homoscedasticity, but also normality: Conditional on X_i, the response Y_i has a normal distribution. Since A is constant, (2.28) shows that $\widehat{\beta}$ has an exact multivariate normal distribution.

All this gives rise to the classical hypothesis testing structure, such as the use of the Student-t test for $H_0 : \beta_i = 0$. How does this work?

Our test statistic for $H_0 : \beta_i = 0$ is (2.56), i.e.

$$\frac{\widehat{\beta}_i - \beta_i}{s\sqrt{a_{ii}}} \tag{2.135}$$

Does this actually have a Student-t distribution under H_0? That distribution is defined as the ratio $Z/\sqrt{Q/r}$, where Z is a N(0,1) random variable, Q is a chi-squared random variable with r degrees of freedom, and Z and Q are independent.

Here, the roles of Z and Q will be played by

$$Z = \frac{\widehat{\beta}_i - \beta_i}{\sigma\sqrt{a_{ii}}} \tag{2.136}$$

and

$$Q = (n - p + 1)s^2/\sigma^2 \tag{2.137}$$

from (2.55). Z is clearly N(0,1)-distributed, so it remains to show that Z and Q are independent, and that Q has the claimed chi-squared distribution.

Are Z and Q independent? Yes: First note that the numerator in (2.55) is $\widehat{\epsilon}'\widehat{\epsilon}$. Then recall that we found in Section 2.12.9 that $\widehat{\epsilon}$ and $\widehat{\beta}$ are uncorrelated. Since we are now assuming normality, that uncorrelatedness implies that $\widehat{\epsilon}$ is independent of $\widehat{\beta}$.

We then must show that Q from (2.137) is chi-squared distributed. This can be seen as follows.

The numerator in (2.55) is

$$(D - A\widehat{\beta})'(D - A\widehat{\beta}) \tag{2.138}$$

But

$$D - A\widehat{\beta} = (I - H)D \tag{2.139}$$

where H is the famous *hat matrix*,

$$H = A(A'A)^{-1}A' \tag{2.140}$$

(Exercise 8). Furthermore,

$$(I - H)\,ED = E(D - A\widehat{\beta}) = A\beta - A\beta = 0 \tag{2.141}$$

by the linear model and the unbiasedness of $\widehat{\beta}$.

Finally, since A has full rank $p + 1$, the same will hold for H. Meanwhile, I has rank n. Thus $I - H$ will have rank $n - p + 1$ (Exercise (9)).

We then apply Property C in Section (2.12.2), establishing that Q has a chi-squared distribution.

2.12.11 Asymptotic $(p + 1)$-Variate Normality of $\widehat{\beta}$

Here we show that asymptotically $\widehat{\beta}$ has a $(p + 1)$-vartiate normal distribution, and importantly, derive the corresponding asymptotic covariance matrix, even without the normality assumption for Y given X, and for that matter, *without the homoscedasticity assumption*. We assume the random-X setting,[13] and as in Section 2.12.6, avoid clutter by incorporating the 1 element of $\widetilde{X}$ into X.

First, define the actual prediction errors we would have if we knew the true population value of β and were to predict the Y_i from the X_i,

$$\epsilon_i = Y_i - X_i'\beta \tag{2.142}$$

---

[13] The derivation could be done for the fixed-X case, but we would need to use a CLT for non-identically distributed random variables, and it would get messy.

Let G denote the vector of the ϵ_i:

$$G = (\epsilon_1, ..., \epsilon_n)' \tag{2.143}$$

Then

$$D = A\beta + G \tag{2.144}$$

We first show that the distribution of $\sqrt{n}(\widehat{\beta} - \beta)$ converges to $(p+1)$-variate normal with mean 0.

Multiplying both sides of (2.144) by $(A'A)^{-1}A'$, we have

$$\widehat{\beta} = \beta + (A'A)^{-1}A'G \tag{2.145}$$

Thus

$$\sqrt{n}(\widehat{\beta} - \beta) = (A'A)^{-1}\sqrt{n} \; A'G \tag{2.146}$$

Using Slutsky's Theorem and (2.105), the right-hand side has the same asymptotic distribution as

$$[E(XX')]^{-1}\sqrt{n} \; (\frac{1}{n}A'G) \tag{2.147}$$

We also have that

$$A'G = \sum_{i=1}^{n} \epsilon_i X_i \tag{2.148}$$

This is a sum of i.i.d. terms with mean 0, the latter fact coming from

$$E(\epsilon X) = E[E(\epsilon X | X)] = 0 \tag{2.149}$$

since $E(\epsilon | X) = 0$. So the CLT says that $\sqrt{n} \; \cdot (A'G/n)$ is asymptotically normal with mean 0 and covariance matrix equal to that of ϵX.

Putting this information together with (2.146), we have:

$\widehat{\beta}$ is asymptotically $(p + 1)$-variate normal with mean β and covariance matrix

$$\frac{1}{n} \, [E(XX')]^{-1} Cov(\epsilon X)[E(XX')]^{-1} \tag{2.150}$$

But we can go still further: By the properties of covariance matrices (Exercise 12),

$$Cov(\epsilon X) = E(\epsilon^2 XX') - E(\epsilon X) \, E(\epsilon X)' = E(\epsilon^2 XX') \tag{2.151}$$

So, the general, heteroscedastic asymptotic covariance matrix of $\widehat{\beta}$ is

$$\frac{1}{n} \, [E(XX')]^{-1} E[\epsilon^2 XX'][E(XX')]^{-1} \tag{2.152}$$

This version will turn out to be useful in Chapter 3.

2.13 Computational Complements

2.13.1 Details of the Computation of (2.28)

For the purpose of reducing roundoff error, linear regression software typically uses the *QR decomposition* in place of the actual matrix inversion seen in (2.28). See Section A.5.

There is also the issue of whether the matrix inverse in (2.28) exists. Consider again the example of female wages in Section 1.16.1. Suppose we construct dummy variables for *both* male and female, and say, also use age as a predictor:

```
> data(prgeng)
> prgeng$female <- as.integer(prgeng$sex == 1)
> prgeng$male <- as.integer(prgeng$sex == 2)
> lm(wageinc ~ age + male + female, data=prgeng)
...
Coefficients:
(Intercept)          age         male       female
    44178.1        489.6     -13098.2           NA
```

How did that NA value arise? Think of the matrix A, denoting its column j as c_j. We have

- c_1 consists of all 1s

- c_2 contains the age values

- c_3 has 1s for the men, 0s for the women

- c_4 has 0s for the men, 1s for the women

Here is the key point:

$$c_1 = c_3 + c_4 \tag{2.153}$$

In other words,

$$1 \cdot c_1 + (-1) \cdot c_3 + (-1) \cdot c_4 = 0 \tag{2.154}$$

Thus (the reader may wish to review Section A.7) A is of less than full rank, as are A' and $A'A$. Therefore $A'A$ is not invertible.

There is the notion of *generalized matrix inverse* to deal with this issue, useful in *Analysis of Variance Models*, but for our purposes, proper choice of dummy variables solves the problem.

2.13.2 R Functions for the Multivariate Normal Distribution Family

In R the density, cdf and quantiles of the multivariate normal distribution are given by the functions **dmvnorm()**, **pmvnorm()** and **qmvnorm()** in the library **mvtnorm**. You can simulate a multivariate normal distribution by using **mvrnorm()** in the library **MASS**.

2.13.2.1 Example: Simulation Computation of a Bivariate Normal Quantity

Consider a vector $X = (X_1, X_2)'$ having a bivariate normal distribution with mean vecor $(1, 1)'$, with standard deviations 1 for each component, and correlation 0.5 between the components. Say we are interested in the quantity $E(|X_1 - X_2|)$. This cannot be determined analytically, but it is easy via simulation:

```
> library (MASS)
> mu <- c(1,1)
> sig <- rbind(c(1,0.5),c(0.5,1))
> sig
     [,1] [,2]
[1,]  1.0  0.5
[2,]  0.5  1.0
> x <- mvrnorm(n=100, mu=mu, Sigma=sig)
> head(x)
           [,1]        [,2]
[1,]  2.5384881   2.8800789
[2,]  1.4566831   2.1817883
[3,] -0.3286932  -0.2016951
[4,]  1.6158710   1.2448996
[5,]  2.0325496   0.1370805
[6,]  0.9100862   0.9779601
> mean(abs(x[,1] - x[,2]))
[1]  0.8767933
```

So $E(|X_1 - X_2|)$ is about 0.88. Note the word **about**, though, as this is only our sample estimate, not the population. To see this more concretely, let's get the estimate of the covariance matrix:

```
> cov(x)
          [,1]        [,2]
[1,]  1.134815  0.3996990
[2,]  0.399699  0.9384828
```

This is substantially off the correct values:

$$\begin{pmatrix} 1 & 0.5 \\ 0.5 & 1 \end{pmatrix} \tag{2.155}$$

Since this is a simulation and thus we have the luxury of knowing the exact covariance matrix and setting n, we see that we need to set a much larger value of n.

2.13.3   More Details of 'lm' Objects

Since a call to **summary()** on an **"lm"** object yields, among other things, R^2 and adjusted R^2, one might think that these quantities are components of the object. There actually are no such components, and these quantities are computed by **summary.lm()**.

This will give us an opportunity to learn more about R objects, especially those of **"lm"** class. Let's take a look.

```
> lmout <- lm(Weight ~ Height + Age, data=mlb)
> names(lmout)
 [1] "coefficients"  "residuals"
 [3] "effects"       "rank"
 [5] "fitted.values" "assign"
 [7] "qr"            "df.residual"
 [9] "xlevels"       "call"
[11] "terms"         "model"
```

Some of these are themselves names of S3 objects, yes, objects-within-objects!

```
> lmout$model
    Weight Height    Age
1      180     74  22.99
2      215     74  34.69
3      210     72  30.78
...
> str(lmo$model)
'data.frame':   3 obs. of   2 variables:
 $ y: num   1 2 4
 $ x: int   1 2 3
 - attr(*, "terms")=Classes 'terms',
       'formula' length 3 y ~ x
   .. ..- attr(*, "variables")= language list(y, x)
...
```

Another way to inspect an object is **str()**

```
> str(lmout)
List of 12
 $ coefficients : Named num [1:3]  -187.638 4.924 0.912
   ..- attr(*, "names")= chr [1:3] "(Intercept)"
   "Height" "Age"
 $ residuals    : Named num [1:1015]  -17.66 6.67 15.08
```

```
   10.84  −16.34  . . .
 ..−  attr(∗, "names")=  chr  [1:1015]  "1"  "2"  "3"  "4"  . . .
$ effects        : Named num  [1:1015]  −6414.8  −352.5
   −124.8  10.8  −16.5  . . .
 ..−  attr(∗, "names")=  chr  [1:1015]  "(Intercept)"
 "Height"  "Age"  ""  . . .
$ rank           : int  3
$ fitted.values: Named num  [1:1015]  198  208  195
   199  204  . . .
 ..−  attr(∗, "names")=  chr  [1:1015]  "1"  "2"  "3"
 "4"  . . .
$ assign         : int  [1:3]  0  1  2
$ qr            : List of 5
 ..$ qr    : num  [1:1015, 1:3]  −31.8591  0.0314  0.0314
   0.0314  0.0314  . . .
 .. ..−  attr(∗, "dimnames")=List  of  2
. . .
```

The **fitted.values** component will be used occasionally in this book. The ith one is

$$\widehat{\mu}(X_i) \tag{2.156}$$

which can be viewed in two ways:

- It is the value of the estimated regression function at the ith observation in our data set.

- It is the value that we would predict for Y_i if we did not know Y_i. (We do know it, of course.)

In the context of the second bullet, our prediction error for observation i is

$$Y_i - \widehat{\mu}(X_i) \tag{2.157}$$

Recall that this is known as the ith residual. These values are available to us in **lmout$residuals**.

With these various pieces of information in **lmout**, we can easily calculate R^2 in (2.71). The numerator there, for instance, involves the sum of the squared residuals. The reader can browse through these and other computations by typing

```
> edit (summary.lm)
```

2.14   Exercises: Data, Code and Math Problems

Data problems:

1. Consider the census data in Section 1.16.1.

   (a) Form an approximate 95% confidence interval for β_6 in the model (1.28).

   (b) Form an approximate 95% confidence interval for the gender effect for Master's degree holders, $\beta_6 + \beta_7$, in the model (1.28).

2. The full bikeshare dataset spans 3 years' time. Our analyses here have only used the first year. Extend the analysis in Section 2.8.5 to the full data set, adding dummy variables indicating the second and third year. Form an approximate 95% confidence interval for the difference between the coefficients of these two dummies.

3. Suppose we are studying growth patterns in children, at k particular ages. Denote the height of the ith child in our sample data at age j by H_{ij}, with $H_i = (H_{i1}, ..., H_{ik})'$ denoting the data for child i. Suppose the population distribution of each H_i is k-variate normal with mean vector μ and covariance matrix Σ. Say we are interested in successive differences in heights, $D_{ij} = H_{i,j+1} - H_{ij}$, $j = 1, 2, ..., k - 1$. Define $D_i = (D_{i1}, ..., D_{ik})'$. Explain why each D_i is $(k-1)$-variate normal, and derive matrix expressions for the mean vector and covariance matrices.

4. In the simulation in Section 2.9.3, it is claimed that $\rho^2 = 0.50$. Confirm this through derivation.

5. In the census example in Section 1.16.2, find an appropriate 95% confidence interval for the difference in mean incomes of 50-year-old men and 50-year-old women. Note that the data in the two subgroups will be independent.

Mini-CRAN and other problems:

6. Write a function with call form

mape (lmout)

where **lmout** is an 'lm' object returned from a call to **lm()**. The function will compute the mean absolute prediction error,

$$\frac{1}{n} \sum_{i=1}^{n} |Y_i - \widehat{\mu}(X_i)| \tag{2.158}$$

You'll need the material in Section 2.13.3.

Math problems:

7. Show that (2.122) implies (2.82).

8. Derive (2.138).

9. Suppose B is a $k \times k$ symmetric, idempotent matrix. Show that

$$\text{rank}(I - B) = k - \text{rank}(B) \tag{2.159}$$

(Suggestion: First show that the eigenvalues of an idempotent matrix must be either 0 or 1.)

10. In the derivation of (1.62) in Section (2.12.8.3), we assumed for convenience that $EY = 0$. Extend this to the general case, by defining $W = Y - EY$ and then applying (1.62) to W and X. Make sure to justify your steps.

11. Suppose we have random variables U and V, with equal expected values and each with variance 1. Let ρ denote the correlation between them. Show that

$$\lim_{\rho \to 1} P(|U - V| < \epsilon) = 1 \tag{2.160}$$

for any $\epsilon > 0$. (Hint: Use *Markov's Inequality*, $P(T > c) \le ET/c$ for any nonnegative random variable T and any positive constant c.)

12. For a scalar random variable U, a famous formula is

$$Var(U) = E(U^2) - (EU)^2 \tag{2.161}$$

Derive the analog for the covariance matrix of a random vector U,

$$Cov(U) = E(UU') - (EU)(EU)' \qquad (2.162)$$

Chapter 3

Homoscedasticity and Other Assumptions in Practice

This chapter will take a practical look at the classical assumptions of linear regression models. While most of the assumptions are not very important for the Prediction goal, assumption (d) below both matters for Description *and* has a remedy. Again, **this is crucial for the Description goal, because otherwise our statistical inference may be quite inaccurate.**

To review, the assumptions are:

(a) Linearity: The conditional mean is linear in parameters β.

$$E(Y \mid \widetilde{X} = \widetilde{t}) = \widetilde{t}'\beta \qquad (3.1)$$

(b) Normality: The conditional distribution of Y given X is normal.

(c) Independence: The data (X_i, Y_i) are independent across i.

(d) Homoscedasticity:

$$Var(Y \mid X = t) \qquad (3.2)$$

is constant in t.

Verifying assumption (a), and dealing with substantial departures from it, form the focus of an entire chapter, Chapter 6. So, this chapter will focus on assumptions (b)-(d).

The bulk of the material will concern (d).

3.1   Normality Assumption

We already discussed (b) in Section 2.8.4, but the topic deserves further comment. First, let's review what was found before. (We continue to use the notation from that chapter, as in Section 2.4.2.)

Neither normality nor homoscedasticity is needed for $\widehat{\beta}$ to be unbiased, consistent and asymptotically normal. Standard statistical inference procedures do assume homoscedasticity, but we'll return to the latter issue in Section 3.3. For now, let's concentrate on the normality assumption. Retaining the homoscedasticity assumption for the moment, we found in the last chapter that:

> The conditional distribution of the least-squares estimator $\widehat{\beta}$, given A, is approximately multivariate normal distributed with mean β and approximate covariance matrix
>
> $$s^2(A'A)^{-1} \tag{3.3}$$
>
> Thus the standard error of $\widehat{\beta}_j$ is the square root of element j of this matrix (counting the top-left element as being in row 0, column 0).
>
> Similarly, suppose we are interested in some linear combination $\lambda'\beta$ of the elements of β (Section A.4), estimating it by $\lambda'\widehat{\beta}$. The standard error is the square root of
>
> $$s^2\lambda'(A'A)^{-1}\lambda \tag{3.4}$$

The reader should not overlook the word *asymptotic* in the above. Without assumption (b) above, our inference procedures (confidence intervals, significance tests) are indeed valid, but only approximately. On the other hand, the reader should be cautioned (as in Section 2.8.1) that so-called "exact" inference methods assuming normal population distributions, such as the Student-t distribution and the F distribution, are themselves only approximate, since true normal distributions rarely if ever exist in real life.

In other words:

> We must live with approximations one way or the other, and the end result is that the normality assumption is not very important.

3.2 Independence Assumption — Don't Overlook It

Statistics books tend to blithely say things like "Assume the data are independent and identically distributed (i.i.d.)," without giving any comment to (i) how they might be nonindependent and (ii) what the consequences are of using standard statstical methods on nonindependent data. Let's take a closer look at this.

3.2.1 Estimation of a Single Mean

Note the denominator $S/\sqrt{n}$ in (2.42). This is the standard error of $\overline{W}$, i.e., the estimated standard deviation of that quantity. That in turn comes from a derivation you may recall from statistics courses,

$$Var(\overline{W}) \quad = \quad \frac{1}{n^2} Var(\sum_{i=1}^{n} W_i) \tag{3.5}$$

$$= \quad \frac{1}{n^2} \sum_{i=1}^{n} Var(W_i)) \tag{3.6}$$

$$= \quad \frac{1}{n}\sigma^2 \tag{3.7}$$

and so on. In going from the first equation to the second, we are making use of the usual assumption that the W_i are independent.

But suppose the W_i are correlated. Then the correct equation is

$$Var(\sum_{i=1}^{n} W_i) = \sum_{i=1}^{n} Var(W_i) + 2 \sum_{1 \le i < j \le n} Cov(W_i, W_j) \tag{3.8}$$

It is often the case that our data are positively correlated. Many data sets, for instance, consist of multiple measurements on the same person, say 10 blood pressure readings for each of 100 people. In such cases, the covariance terms in (3.8) will be positive, and (3.7) will yield too low a value. Thus the denominator in (2.42) will be smaller than it should be. That means that our confidence interval (2.44) will be too small (as will be p-values), a serious problem in terms of our ability to do valid inference.

Here is the intuition behind this: Although we have 1000 blood pressure readings, the positive intra-person correlation means that there is some degree of repetition in our data. Thus we don't have "1000 observations worth" of data, i.e., our effective n is less than 1000. Hence our confidence interval, computed using $n = 1000$, is overly optimistic.

Note that $\overline{W}$ will still be an unbiased and consistent estimate of ν.[1] In other words, $\overline{W}$ is still useful, even if inference procedures computed from it may be suspect.

3.2.2   Inference on Linear Regression Coefficients

All of this applies to inference on regression coefficients as well. If our data is correlated, i.e., rows within (A, D) are not independent, then (2.54) will be incorrect, because the off-diagonal elements in $Cov(Y)$ won't be 0s. And if they are positive, (2.54) will be "too small," and the same will be true for (3.3). Again, the result will be that our confidence intervals and p-values will be too small, i.e., overly optimistic. In such a situation, then our $\widehat{\beta}$ will still be useful, but our inference procedures will be suspect.

3.2.3   What Can Be Done?

This is a difficult problem. Some possibilities are:

- Simply note the dependency problem, e.g., in our report to a client, and state that though our estimates are valid (in the sense of statistical consistency), we don't have reliable standard errors.

- Somehow model the dependency, i.e., model the off-diagonal elements of $Cov(Y)$. For instance, in the blood pressure case, we might try *mixed effects models* [77] [52]. Or if the dependency is due to a time variable, one might use time series models.

---

[1]Mathematically, this claim about consistency would have to be posed more rigorously, with conditions on the quantities $Cov(W_i, W_j)$, say as $]i - j| \to \infty$.

- Collapse the data in some way to achieve independence.

An example of this last point is presented in the next section.

3.2.4   Example: MovieLens Data

The MovieLens data (*http://grouplens.org/*) consists of ratings of various movies by various users. The main file of the 100K version, which we'll analyze here, consists of columns User ID, Movie ID, Rating and Timestamp. There is one row per rating. If a user has rated, say eight movies, then he/she will have eight rows scattered about in the data matrix. Of course, most users have not rated most movies. There is also a file for user data, with age, gender and so on,[2] and another for movie data, showing title, genres etc.

Let Y_{ij} denote the ratings of user i, $j = 1, 2, ..., N_i$, where N_i is the number of movies rated by this user. In the example here, we are not taking into account which movies the user rates, just analyzing general user behavior. We are treating the users as a random sample from a conceptual population of all potential users, and similarly for the movies.

As with the blood pressure example above, for fixed i, the Y_{ij} are not independent, since they come from the same user. Some users tend to give harsher ratings, others tend to give favorable ones. But we can form

$$T_i = \frac{1}{N_i} \sum_{j=1}^{N_i} Y_{ij} \qquad (3.9)$$

the average rating given by user i, and now we have *independent* random variables. And, if we treat the N_i as random too, and i.i.d., then the T_i are i.i.d., enabling standard statistical analyses.

For instance, we can run the model, say,

$$\text{mean rating} = \beta_0 + \beta_1 \text{ age} + \beta_2 \text{ gender} \qquad (3.10)$$

and then pose questions such as "Do older people tend to give lower ratings?" Let's see what this gives us.

---

[2]Unfortunately, in recent editions of the data, this is no longer included.

The data are in two separate files: **u.data** contains the ratings, and **u.user** contains the demographic information. Let's read in the first file:

```
> ud <- read.table('u.data',header=FALSE,sep='\t')
> head(ud)
    V1   V2 V3          V4
1 196 242  3 881250949
2 186 302  3 891717742
3  22 377  1 878887116
4 244  51  2 880606923
5 166 346  1 886397596
6 298 474  4 884182806
```

The first record in the file has user 196 rating movie 242, giving it a 3, and so on. (The fourth column is a timestamp.) There was no header on this file (nor on the next one we'll look at below), and the field separator was a TAB.

Now, let's find the T_i. For each unique user ID, we'll find the average rating given by that user, making use of the **tapply()** function (Section 1.20.2):

```
> z <- tapply(ud$V3,ud$V1,mean)
> head(z)
       1        2        3        4        5        6
3.610294 3.709677 2.796296 4.333333 2.874286 3.635071
```

We are telling R, "Group the ratings by user, and find the mean of each group." So user 1 (not to be confused with the user in the first line of **u.data**, user 196) gave ratings that averaged 3.610294 and so on.

Now we'll read in the demographics file:

```
> uu <- read.table('u.user',header=F,sep='|')
# no names in the orig data, so add some
> names(uu) <- c('userid','age','gender',
      'occup','zip')
> head(uu)
  userid age gender      occup   zip
1      1  24      M technician 85711
2      2  53      F      other 94043
3      3  23      M     writer 32067
4      4  24      M technician 43537
5      5  33      F      other 15213
6      6  42      M  executive 98101
```

Now append our T_i to do this latter data frame, and run the regression:

```
> uu$gender <- as.integer(uu$gender == 'M')
> uu$avg_rat <- z
> head(uu)
  userid age gender      occup   zip  avg_rat
1      1  24      0 technician 85711 3.610294
2      2  53      0      other 94043 3.709677
3      3  23      0     writer 32067 2.796296
4      4  24      0 technician 43537 4.333333
5      5  33      0      other 15213 2.874286
6      6  42      0  executive 98101 3.635071
> q <- lm(avg_rat ~ age + gender, data=uu)
> summary(q)
. . .
Coefficients:
              Estimate Std. Error t value Pr(>|t|)
(Intercept) 3.4725821  0.0482655  71.947  < 2e-16 ***
age         0.0033891  0.0011860   2.858  0.00436 **
gender      0.0002862  0.0318670   0.009  0.99284
. . .
Multiple R-squared: 0.008615,
    Adjusted R-squared: 0.006505
```

This is again an example of how misleading signficance tests can be. The age factor here is "double star," so the standard response would be "Age is a highly signficant predictor of movie rating." But that is not true at all. A 10-year difference in age only has an impact of about 0.03 on ratings, which are on the scale 1 to 5. And the R-squared values are tiny. So, while the older users tend to give somewhat higher ratings, the effect is negligible.

On the other hand, let's look at what factors may affect which kinds of users post more. Consider the model

$$\text{mean number of ratings} = \beta_0 + \beta_1 \text{ age } + \beta_2 \text{ gender} \qquad (3.11)$$

Run the analysis:

```
# get ratings count for each user
> uu$ni <- tapply(ud$V3, ud$V1, length)
> summary(lm(ni ~ age + gender, data=uu))
. . .
Coefficients:
              Estimate Std. Error t value Pr(>|t|)
```

```
(Intercept)  120.6782    10.9094   11.062  < 2e-16 ***
age           -0.7805     0.2681   -2.912   0.00368 **
gender        16.8124     7.2029    2.334   0.01980 *
...
```

There is a somewhat more substantial age effect here, with older people posting fewer ratings. A 10-year age increase brings about 8 fewer postings. Is that a lot?

```
> mean(uu$ni)
[1]  106.0445
```

So there is a modest decline in posting activity.

But look at gender. Women rate an average of 16.8 fewer movies than men, rather substantial.

3.3   Dropping the Homoscedasticity Assumption

For an example of problems with the homoscedasticity assumption, again consider weight and height. It is intuitive that tall people have more variation in weight than do short people, for instance. We can confirm that in our baseball player data. Let's find the sample standard deviations for each height group (restricting to the groups with over 50 observations), seen in Section 1.6.2:

```
> library(freqparcoord)
> data(mlb)
> m70 <- mlb[mlb$Height >= 70 & mlb$Height <= 77,]
> sds <- tapply(m70$Weight, m70$Height, sd)
> plot(70:77, sds)
```

The result is shown in Figure 3.1. The upward trend is clearly visible, and thus the homoscedasticity assumption is not reasonable.

The vector $\widehat{\beta}$ in (2.28) is called the *ordinary least-squares* (OLS) estimator of β, in contrast to *weighted least-sqaures* (WLS), to be discussed shortly. Statistical inference on β using OLS is based on (2.54),

$$Cov(\widehat{\beta}) = \sigma^2 (A'A)^{-1} \qquad (3.12)$$

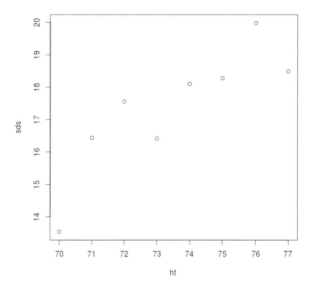

Figure 3.1: Standard deviations of weight, by height group

which is in turn based on the homoscedasticity assumption — that $Var(Y \mid X = t)$ is constant in t. Yet that assumption is rarely if ever valid.

Given the inevitable nonconstancy of (2.32), there are questions that must be raised:

- Do departures from constancy of (2.32) in t substantially impact the validity of statistical inference procedures that are based on (2.54)?

- Can we somehow estimate the function $\sigma^2(t)$, and then use that information to perform a WLS analysis?

- Can we somehow modify (2.54) for the heteroscedastic case?

These points will be addressed in this section.

3.3.1 Robustness of the Homoscedasticity Assumption

In statistics parlance, we ask,

Is classical inference on the vector of regression coefficients β *robust* to the homoscedasticity assumption, meaning that there is not much effect on the validity of our inference procedures (confidence intervals, significance tests) unless the setting is quite profoundly heteroscedastic?

We can explore this idea via simulation in known settings. For instance, let's investigate settings in which

$$\sigma(t) = |\mu(t)|^q \tag{3.13}$$

where q is a parameter to vary in the investigation. We've discussed previously that typically in practice, $\sigma^2(t)$ should increase with $\mu(t)$ if the latter is positive, so for $q > 0$ Equation (3.13) should serve as a good test bed in which to explore the effects of heteroscedasticity

This includes several important cases in which the value of q is implied by famous models:

- $q = 0$: Homoscedasticity.

- $q = 0.5$: Conditional distribution of Y given X is Poisson.

- $q = 1$: Conditional distribution of Y given X is exponential.

Below is the code. It simulates a setting in which

$$E(Y \mid X = t) = \beta_1 t_1 + \ldots + \beta_p t_p \tag{3.14}$$

and with conditional variance as in (3.13). We construct a nominal 95% confidence interval for β_1 in each simulated sample of size **n**, and then calculate the proportion of the **nreps** intervals that contain 1.0, the true value of β_1.

```
simhet <- function(n,p,nreps,sdpow) {
    bhls <- vector(length=nreps)
    ses <- vector(length=nreps)
    for (i in 1:nreps) {
        x <- matrix(rnorm(n*p),ncol=p)
        meany <- x %*% rep(1,p)
        sds <- abs(meany)^sdpow
        y <- meany + rnorm(n,sd=sds)
```

n	p	q	conf. lvl.
100	5	0.0	0.94683
100	5	0.5	0.92359
100	5	1.0	0.90203
100	5	1.5	0.87889
100	5	2.0	0.86129

Table 3.1: Heteroscedasticity effect simulation

```
      lmout <- lm(y ~ x)
      bh1s[i] <- coef(lmout)[2]
      ses[i] <- sqrt(vcov(lmout)[2,2])
   }
   mean(abs(bh1s - 1.0) < 1.96*ses)
}
```

The simulation finds the true confidence level (providing **nreps** is set to a large value) corresponding to a nominal 95% confidence interval. Table 3.1 shows the results of a few runs, all with **nreps** set to 100,000. We see that there is indeed an effect on the true confidence level.

3.3.2   Weighted Least Squares

If one knows the function $\sigma^2(t)$ (at least up to a constant multiple), one can perform a weighted least-squares (WLS) analysis. Here, instead of minimizing (2.18), one minimizes

$$\frac{1}{n} \sum_{i=1}^{n} \frac{1}{w_i} (Y_i - \widetilde{X}_i' b)^2 \tag{3.15}$$

(without the $1/n$ factor, of course), where

$$w_i = \sigma^2(X_i) \tag{3.16}$$

Just as one can show that in the homoscedastic case, OLS gives the optimal (minimum-variance unbiased) estimator, the same is true for WLS in heteroscedastic settings, provided we know the function $\sigma^2(t)$.

R's **lm()** function has an optional **weights** argument for specifying the w_i. But needless to say, this situation is not common. To illustrate this point, consider the classical inference procedure for a single mean, reviewed in Section 2.8.1. If we don't know the population mean ν, we are even less likely to know the population variance η^2. The same holds in the regression context, concerning conditional means and conditional variances.

One option would be to estimate the function $\sigma(t)$ using nonparametric regression techniques. This was first proposed in [122] and later studied extensively in work by Raymond Carroll, including [32].

For instance, we can use our k-NN function **knnest()** in the **regtools** package [97], with **nearf = vary**. The latter specifies to calculate the sample variances of Y in the neighborhoods of our X values. This gives us estimates of $\sigma^2(X_i)$, exactly what we need.

With the baseball data, though, we can just estimate the variances as we did for Figure 3.1. Let's run the analysis first with, and then without, weights, and check how much difference weighting makes:

```
> mlb <- mlb[,c(4,6,5)]
> m70 <- mlb[mlb$Height >= 70 & mlb$Height <= 77,]
> vars <- tapply(m70$Weight,m70$Height,var)
> wts <- 1 / vars[m70$Height-69]
> summary(lm(m70$Weight ~ m70$Height, weights=wts))
...
Coefficients:
              Estimate  Std. Error  t value  Pr(>|t|)
(Intercept)  -160.5523    21.7870   -7.369  3.78e-13
m70$Height      4.9119     0.2972   16.528  < 2e-16

(Intercept) ***
m70$Height  ***
...
> summary(lm(m70$Weight ~ m70$Height))
...
Coefficients:
              Estimate  Std. Error  t value  Pr(>|t|)
(Intercept)  -162.8544    22.7182   -7.168  1.54e-12
m70$Height      4.9438     0.3087   16.013  < 2e-16

(Intercept) ***
m70$Height  ***
```

. . .

The weighted analysis, the "true" one (albeit with the weights being only approximate), did give slightly different results than those of OLS, including the standard errors.

It should be noted that the estimated conditional variances seem to flatten somewhat toward the right end, and are based on smaller sample sizes at both ends:

```
> vars
        70        71        72        73        74
183.3702  270.0965  308.4762  269.3699  327.7612
        75        76        77
333.9579  399.2606  341.7576
> tapply (m70$Weight , m70$Height , length )
 70   71   72   73   74   75   76   77
 51   89  150  162  173  155  101   55
```

3.3.3 A Procedure for Valid Inference

Fortunately, there exists a rather simple procedure, originally developed by Eicker and later refined by White [43] [137]. This section will present the methodology, and test it on data.

3.3.4 The Methodology

It was found in Section 2.12.11 that the general, heteroscedastic asymptotic covariance matrix of $\widehat{\beta}$ is

$$\frac{1}{n} \, [E(XX')]^{-1} E[\epsilon^2 XX'][E(XX')]^{-1} \tag{3.17}$$

Each of the factors is easily estimated by sample analogs:

$$\widehat{E}(XX') = \frac{1}{n} A'A \tag{3.18}$$

$$\widehat{E}(\epsilon^2 XX') = \frac{1}{n} \sum_{i=1}^{n} \widehat{\epsilon}_i^2 X_i X_i' \tag{3.19}$$

where as before

$$\widehat{\epsilon}_i = Y_i - \tilde{X}_i \widehat{\beta} \tag{3.20}$$

The expression in (3.19) is rather unwieldy, but it is easy to program, and most important, we're in business! We can now conduct valid statistical inference even in the heteroscedastic case.

Code implementing (3.19) (or refinements of it) is available in R's **car** [49] and **sandwich** packages [95], as the functions **hccm()** and **vcovHC()**, respectively. (These functions also offer various refinements of the method.) They are drop-in replacements for the standard **vcov()**, so that for instance we could use **vcovHC()** in place of **vcov()** on page 87.

3.3.5   Example: Female Wages

Let's try this on the census data, Section 1.16.1. We'll find the standard error of $\widehat{\beta}_6$, the coefficient of the **fem** variable.

Keep in mind that in β and $\widehat{\beta}$, the first element is number 0, so $\widehat{\beta}_6$ is the seventh. From Section 2.8.4, then, the estimated variance of $\widehat{\beta}_6$ is given in the (7,7) element of the estimated covariance matrix of $\widehat{\beta}$, a matrix obtainable through R's **vcov()** function under the homoscedasticity assumption. The standard error of $\widehat{\beta}$ is then the square root of that (7,7) element. As mentioned, **vcovHC()** is a drop-in replacement for **vcov()** in using the Eicker method, without assuming homoscedasticity.

Continuing with the data frame **prgeng** in that example, we have

```
> lmout <- lm(wageinc ~ age+age2+wkswrkd+ms+phd+fem,
        data=prgeng)
> sqrt(vcov(lmout)[7,7])
[1]   705.2994
> library(sandwich)
> sqrt(vcovHC(lmout)[7,7])
[1]   593.85
```

That is quite a difference. Apparently, using OLS was conservative in this case, i.e., was causing us to have wider confidence intervals than necessary and was inflating p-values.

n	p	q	conf. lvl.
100	5	0.0	0.95176
100	5	0.5	0.94928
100	5	1.0	0.94910
100	5	1.5	0.95001
100	5	2.0	0.95283

Table 3.2: Heteroscedasticity simulation

3.3.6   Simulation Test

Let's see if it works, at least in the small simulation experiment in Section 3.3.1. We use the same code as before, simply replacing the call to **vcov()** by one to **vcovHC()**. The results, shown in Table 3.2, are excellent.

3.3.7   Variance-Stabilizing Transformations

Most classical treatments of regression analysis devote a substantial amount of space to *transformations* of the data. For instance, one might replace Y by $\ln Y$, and possibly apply the log to the predictors as well. There are several reasons why this might be done:

(a) The distribution of Y given X may be skewed, and applying the log may make it more symmetric, thus more normal-like.

(b) Log models may have some meaning relevant to the area of application, such as *elasticity* models in economics.

(c) Applying the log may convert a heteroscedastic setting to one that is close to homoscedastic.

One of the themes of this chapter has been that the normality assumption is not of much practical importance, which would indicate that Reason (a) above may not be so useful. Reason (b) is domain-specific, and thus outside the scope of this book. But Reason (c) relates directly to our current discussion on heteroscedasticity. Here is how transformations come into play.

The Delta Method (Section 3.6.1) says, roughly, that if the random variable W is approximately normal with a small *coefficient of variation* (ratio of standard deviation to mean), and g is a smooth function, then the new random variable $g(W)$ is also approximately normal, with mean $g(EW)$ and variance

$$[g'(EW)]^2 Var(W) \tag{3.21}$$

Let's consider that in the context of (3.13). Assuming that the regression function is always positive, (3.13) reduces to

$$\sigma(t) = \mu^q(t) \tag{3.22}$$

Now, suppose (3.22) holds with $q = 1$. Take $g(t) = \ln(t)$. Then since

$$\frac{d}{dt}\ln t = \frac{1}{t} \tag{3.23}$$

we see that (3.21) becomes

$$\frac{1}{\mu^2(t)} \cdot \mu^2(t) = 1 \tag{3.24}$$

In other words $Var(\ln Y \mid X = t)$ is approximately 1 for all t, and we are back to the homoscedastic case. Similarly, if $q = 0.5$, then setting $g(t) = \sqrt{t}$ would give us approximate homoscedasticity.

However, this method has real drawbacks: Distortion of the model, difficulty interpreting the coefficients and so on.

Let's look at a very simple model that illustrates the distortion issue. (It is further explored in Section 3.6.2.) Suppose X takes on the values 1 and 2. Given $X = 1$, Y is either 2 or $1/2$, with probability $1/2$ each. If $X = 2$, then Y is either 4 or $1/4$, with probability $1/2$ each. Let $U = \log_2 Y$.

Let μ_Y and μ_U denote the regression functions of Y and U on X. Then

$$\mu_U(1) = 0.5 \cdot 1 + 0.5 \cdot (-1) = 0 \tag{3.25}$$

and similarly $\mu_U(2) = 0$ as well. So, there is no relation between U and X at all! Yet the relation between Y and X is quite substantial. The transformation has destroyed the latter relation.

Of course, this example is highly contrived, and one can construct examples with the opposite effect. Nevertheless, it shows that a log transformation can indeed bring about considerable distortion. This is to be expected in a sense, since the log function flattens out as we move to the right. Indeed, the U.S. Food and Drug Administration once recommended against using transformations.[3]

3.3.8 The Verdict

While the examples here do not constitute a research study (the reader is encouraged to try the code in other settings, simulated and real), an overall theme is suggested.

In principle, WLS provides more efficient estimates and correct statistical inference. What are the implications?

If our goal is Prediction, then forming correct standard errors is typically of secondary interest, if at all. And unless there is really strong variation in the proper weights, having efficient estimates is not so important. In other words, for Prediction, OLS may be fine.

The picture changes if the goal is Description, in which case correct standard errors may be important. Variance-stabilizing transformations may cause problems, and while one might estimate WLS weights via nonparametric regression methods as mentioned above, these may be too sensitive to sampling variation. But the method of Section 3.3.3 is now commonly available in statistical software packages, and is likely to be the best way to cope with heteroscedasticity.

3.4 Further Reading

The MovieLens dataset (Section 3.2.4) is a very popular example for work in the field of *recommender systems*, in which user ratings of items are predicted. A comprehensive treatment is available in [1].

---

[3]Quoted in The Log Transformation Is Special, *Statistics in Medicine*, Oliver Keene, 1995, 811-819. That author takes the opposite point of view.

3.5   Computational Complements

3.5.1   The R merge() Function

The MovieLens data set in Section 3.2.4 is interesting in that it consists of three main files, not one. We've already discussed the first two files; the third, **u.item**, contains data on the movies (name, release date, genre).

In Section 3.2.4, we had not collapsed the **u.data** file by taking averages of movie ratings across each user. Had we wanted to retain the ability to distinguish between movies, paying attention to genre and so on, we would have needed to combine the **u.data** and **u.item** files. How can that be done?

The key point is that the two files have a column in common, the user ID. R's **merge()** function can then be used to exploit this.

The **merge()** function does as its name implies, merging two data frames. In the simple case, there must be a column having the same name in each of the two input data frames. Let's call that column k. Then whenever that column has the same value in row i of one of the input data frames and row j of the other, we form an output row that concatenates those two rows. (Of course, the common column is not duplicated.) In computer science, this is called a *join* operation.

The code for the movie join is seen here:

```
> ud <- read.table('u.data',header=F,sep='\t')
> uu <- read.table('u.user',header=F,sep='|')
> udu <- merge(ud,uu,by.x=1,by.y=1)
> head(udu)
  V1 V2.x V3.x     V4.x V2.y V3.y        V4.y    V5
1  1    1    1  5 874965758   24   M technician 85711
2  1   23    4  875072895   24   M technician 85711
3  1  223    5  876892918   24   M technician 85711
4  1  171    5  889751711   24   M technician 85711
5  1   16    5  878543541   24   M technician 85711
6  1   73    3  876892774   24   M technician 85711
```

Here **by.x** and **by.y** specify the position of the common column within the two input data frames.

The first four columns of the output data frame are those of **ud**, while the last four come from **uu**. The latter actually has five columns, but the first of them is the user ID, the column of commonality between **uu** and **ud**.

Using this merged data frame, we could now do various analyses involving user-movie interactions, with demographic variables as covariates. Note, though, that the people who assembled the MovieLens dataset were wise to make three separate files, because one unified file would have lots of redundant information, and thus would take up much more room. This may not be an issue with this 100K version of the data, but MovieLens also has a version of size 20 million.

3.6 Mathematical Complements

3.6.1 The Delta Method

Say we have an estimator $\widehat{\theta}$ of a scalar population quantity θ, such that the estimator is asymptotically normally distributed. Denote the asymptotic variance of $\widehat{\theta}$ by $AVar(\widehat{\theta})$, meaning that the cumulative distribution function of $\sqrt{n}(\widehat{\theta} - \theta)$ converges to the cdf of $N(0, AVar(\widehat{\theta}))$ as $n \to \infty$.

For some suitable function f, we form a new estimator,

$$W = f(\widehat{\theta}) \tag{3.26}$$

for the quantity $f(\theta)$. We will assume at first for convenience that θ is scalar-valued. Then roughly speaking, the Delta Method gives us a way to show that W is also asymptotically normal, and most importantly, provides us with the asymptotic variance of W:

$$AVar(W) = [g(\theta)]^2 AVar(\widehat{\theta}) \tag{3.27}$$

where g is the derivative of f.

A standard error for W is then computed by substituting $\widehat{\theta}$ for θ in (3.27), and then taking the square root.

The method extends to the case of multivariate θ. Here (3.27) becomes

$$G' ACov(\widehat{\theta}) G \tag{3.28}$$

where G is the gradient (column vector of partial derivatives) of f, evaluated at θ.

An intuitive derivation of (3.27) is simple. We expand $\mathbf{f}$ into a Taylor series, taking the first terms:

$$f(t) \approx f(a) + g(a)(t-a) \tag{3.29}$$

for t near a. Setting $t = \widehat{\theta}$ and $a = \theta$, we have

$$W \approx f(\theta) + g(\theta)(\widehat{\theta} - \theta) \tag{3.30}$$

Taking variances of both sides, and recalling that θ is a constant, we obtain (3.27). Equation (3.28) then follows the same reasoning, accompanied by (2.85). We will stop with the intuition here and not go through the formal details.

3.6.2   Distortion Due to Transformation

Consider this famous inequality:

> *Jensen's Inequality:* Suppose h is a convex function,[4] and V is a random variable for which the expected values in (3.31) exist. Then

$$E[h(V)] \geq h(EV) \tag{3.31}$$

In our context, h is our transformation in Section 3.3.7, and the $E()$ are conditional means, i.e., regression functions. In the case of the log transform (and the square-root transform), h is concave-down, so the sense of the inequality is reversed:

$$E[\ln Y | X = t] \leq \ln(E(Y | X = t) \tag{3.32}$$

Since equality will hold only in trivial cases, we see that the regression function of $\ln Y$ will be smaller than the log of the regression function of Y.

---

[4]This is "concave up," in the calculus sense.

Say we assume that

$$E(Y|X = t) = e^{\beta_0 + \beta_1 t} \qquad (3.33)$$

and reason that this implies that a linear model would be reasonable for $\ln Y$:

$$E(\ln Y|X = t) = \beta_0 + \beta_1 t \qquad (3.34)$$

Jensen's Inequality tells us that such reasoning may be risky. In fact, if we are in a substantially heteroscedastic setting (for Y, not $\ln Y$), the discrepancy between the two sides of (3.32) could vary a lot with t, potentially producing quite a bit of distortion to the shape of the regression curve. This follows from a result of Robert Becker [12]. who expresses the difference between the left- and right-hand sides of (3.31) in terms of $Var(V)$.

3.7 Exercises: Data, Code and Math Problems

Data problems:

1. This problem concerns the **InstEval** dataset in the **lme4** package, which is on CRAN [9]. Here users are students, and "items" are instructors, who are rated by students. Perform an analysis similar to that in Section 3.2.4. Note that you may need to do some data wrangling first, e.g., change R factors to numeric type.

Mini-CRAN and other computational problems:

2. In Section 3.3.2, the standard errors for the OLS and WLS estimates differed by about 8%. Using R's **pnorm()** function, explore how much impact this would have on true confidence level in a nominal 95% confidence interval.

3. This problem concerns the bike-sharing data (Section 1.1).

(a) Check heteroscedasticity here, using the nonparametric regression approach of Section 3.3.2: Run **knnest()** twice, once to estimate $\mu(t)$ and then to estimate $\sigma^2(t)$, evaluated for $t = X_1, ..., X_n$. Plot the latter values against former ones.

(b) Write general code for this, in the form of a function with call form

   plotmusig2 (y , xdata , k)

where the arguments are as in **knnest()**.

Math problems:

4. Consider a fixed-X regression setting (Section 2.3) with *replication*, meaning that more than one Y is observed for each X value. (Here we have just one predictor, $p = 1$.) So our data are

$$(X_i, Y_{i1}, ..., Y_{in_i}), \quad i = 1, ..., r \tag{3.35}$$

Assume that $X > 0$ and

$$Var(Y \mid X = t) = X^q \sigma^2 \tag{3.36}$$

for known q.

Find an unbiased estimator for σ^2.

5. (Presumes background in Maximum Likelihood estimation.) Assume the setting of Problem 4, but with q unknown and to be estimated from the data. The conditional distribution of Y given X is assumed normal with mean $\beta_0 + \beta_1 X$. Also, for simplicity, assume $n_1 = ..., n_r = m$. Write an R function with call form

lmq (y , x)

with **y** being the $r \times m$ matrix of Y values and x being the vector of X values. The code will estimate β, σ^2 and q by the method of Maximum Likelihood, by calling the R function **mle()**.

6. Suppose we have one predictor variable, but our response variable Y is actually a vector, of length 2,

$$Y = (Y^{(1)}, Y^{(2)})' \tag{3.37}$$

This gives us two regression functions,

$$\mu_i(t) = E(Y^{(i)} \mid X = t), \quad i = 1, 2 \tag{3.38}$$

If we use a linear model, our β coefficients now are doubly subscripted,

$$\mu_i(t) = \beta_{0i} + \beta_{1i}t, \quad i = 1, 2 \tag{3.39}$$

Now suppose we are interested in the ratios $\mu_2(t)/\mu_1(t)$, which we estimate by

$$\frac{\widehat{\beta}_{02} + \widehat{\beta}_{12}t}{\widehat{\beta}_{01} + \widehat{\beta}_{11}t} \tag{3.40}$$

Use the Delta Method to derive the asympotic variance of this estimator.

Chapter 4

Generalized Linear and Nonlinear Models

Consider our bike-sharing data (e.g., Section 1.1), which spans a time period of several years. On the assumption that ridership trends are seasonal, and that there is no other time trend (e.g., no long-term growth in the program), then there would be a periodic relation between ridership R and G, the day in our data; here G would take the values 1, 2, 3, ..., with the top value being, say, $3 \times 365 = 1095$ for three consecutive years of data.[1] Assuming that we have no other predictors, we might try fitting the model with a sine term:

$$\text{mean } R = \beta_0 + \beta_1 \sin(2\pi \cdot G/365) \tag{4.1}$$

Just as adding a quadratic term didn't change the linearity of our model in Section 1.16.1 with respect to β, the model (4.1) is linear in β too. In the notation of Section 2.4.2, as long as we can write our model as

$$\text{mean } D = A \, \beta \tag{4.2}$$

then by definition the model is linear: Multiplying β by a constant changes

---

[1]We'll ignore the issue of leap years here, to keep things simple.

mean D by that constant. In the bike data model above, A would be

$$A = \begin{pmatrix} 1 & \sin(2\pi \cdot 1/365) \\ 1 & \sin(2\pi \cdot 2/365) \\ \cdots & \\ 1 & \sin(2\pi \cdot 1095/365) \end{pmatrix} \tag{4.3}$$

But in this example, we have a known period, 365. In some other periodic setting, the period might be unknown, and would need to be estimated from our data. Our model might be, say,

$$\text{mean } Y = \beta_0 + \beta_1 \sin(2\pi \cdot X/\beta_2) \tag{4.4}$$

where β_2 is the unknown period. This does not correspond to (4.2). The model is still parametric, but is nonlinear.

Nonlinear parametric modeling, then, is the topic of this chapter. We'll develop procedures for computing least squares estimates, and forming confidence intervals and p-values, again without assuming homoscedasticity. The bulk of the chapter will be devoted to the *Generalized Linear Model* (GLM), which is a widely-used broad class of nonlinear regression models. Two important special cases of the GLM will be the *logistic* model introduced briefly in Section 1.1, and *Poisson regression*.

4.1   Example: Enzyme Kinetics Model

Data for the famous Michaelis-Menten enzyme kinetics model is available in the **nlstools** package on CRAN [11]. For the data set **vmkm**, we predict the reaction rate V from substrate concentration S. The model used was suggested by theoretical considerations to be

$$E(V \mid S = t) = \frac{\beta_1 t}{\beta_2 + t} \tag{4.5}$$

In the second data set, **vmkmki**,[2] an addiitonal predictor I, inhibitor concentration, was added:

> **data** (vmkmki)

---

[2]There were 72 observation in this data, but the last 12 appear to be anomalous (gradient 0 in all elements), and thus were excluded.

```
> head (vmkmki)
     S      I     v
1  200   0.00  18.1
2  200   0.00  18.8
3  200   6.25  17.7
4  200   6.25  18.1
5  200  12.50  16.4
6  200  12.50  17.6
```

Their new model was

$$E(V \mid S = t, I = u) = \frac{\beta_1 t}{t + \beta_2 \left(1 + u/\beta_3\right)} \quad (4.6)$$

We'll fit the model using R's **nls()** function:

```
> library (nlstools)
> data (vmkmki)
> regftn <- function (t, u, b1, b2, b3)
      b1 * t / (t + b2 * (1 + u/b3))
```

All nonlinear least-squares algorithms are iterative: We make an initial guess at the least-squares estimate, and from that, use the data to update the guess. Then we update the update, and so on, iterating until the guesses converge. In **nls()**, we specify the initial guess for the parameters, using the **start** argument, an R list.[3] Let's set that up, and then run the analysis:

```
> bstart <- list (b1=1,b2=1, b3=1)
```

The values 1 here were arbitrary, not informed guesses at all. Domain expertise can be helpful.

```
> z <- nls (v ~ regftn (S, I, b1, b2, b3), data=vmkmki,
    start=list (b1=1,b2=1, b3=1))
> z
Nonlinear regression model
  model: v ~ regftn (S, I, b1, b2, b3)
   data: vmkmki
    b1     b2     b3
18.06  15.21  22.28
 residual sum-of-squares: 177.3
```

---

[3] This also gives the code a chance to learn the names of the parameters, needed for computation of derivatives.

```
Number of iterations to convergence: 11
Achieved convergence tolerance: 4.951e−06
```

So, $\widehat{\beta_1} = 18.06$ etc.

We can apply **summary()**, **coef()** and **vcov()** to the output of **nls()**, just as we did earlier with **lm()**. For example, here is the approximate covariance matrix of the coefficient vector:

```
> vcov(z)
          b1         b2          b3
b1 0.4786776   1.374961   0.8930431
b2 1.3749612   7.568837  11.1332821
b3 0.8930431  11.133282  29.1363366
```

This assumes homoscedasticity. Under that assumption, an approximate 95% confidence interval for β_1 would be

$$18.06 \pm 1.96 \sqrt{0.4786776} \tag{4.7}$$

One can use the approach in Section 3.3.3 to adapt **nls()** to the heteroscedastic case, and we will do so in Section 4.5.2.

4.2   The Generalized Linear Model (GLM)

GLMs are *generalized* versions of linear models, in the sense that, although the regression function $\mu(t)$ is of the form $t'\beta$, it is some function of a linear function of $t'\beta$.

4.2.1   Definition

To motivate GLMs, first recall again the logistic model, introduced in (1.36). We are dealing with a classification problem, so the Y takes on the values 0 and 1. Let $X = (X^{(1)}, X^{(2)}, ..., X^{(p)})'$ denote the vector of our predictor variables.

Our model is

$$\mu(t) = E(Y \mid X^{(1)} = t_1, ..., X^{(p)} = t_p) = q(\beta_0 + \beta_1 t_1 + ... + \beta_p t_p) \tag{4.8}$$

where

$$q(s) = \frac{1}{1 + e^{-s}} \tag{4.9}$$

and $t = (t_1, ..., t_p)'$.[4]

The key point is that even though the right-hand side of (4.8) is not linear in t, it is *a function of* a linear expression in β, hence the term *generalized linear model* (GLM).

So, GLMs actually form a broad class of models. We can use many different functions $q()$ instead of the one in (4.9); for each such function, we have a different GLM. The particular GLM we wish to use is specified by the inverse of the function $q(t)$, called the *link function*.

In addition, GLM assumes a specified parametric class for the conditional distribution of Y given X, which we will denote $F_{Y|X}$. In the logistic case, the assumption is trivially that the distribution is *Bernoulli*, i.e., binomial with number of trials equal to 1. Having such assumptions enables Maximum Likelihood estimation (Section 4.8.1).

4.2.2 Poisson Regression

As another example, the *Poisson regression* model assumes that $F_{Y|X}$ is a Poisson distribution and sets $q() = \exp()$, i.e.,

$$\mu(t) = E(Y \mid X^{(1)} = t_1, ..., X^{(p)} = t_p) = e^{\beta_0 + \beta_1 t_1 + ... + \beta_p t_p} \tag{4.10}$$

Why the use of exp()? The model for the most part makes sense without the exponentiation, i.e.,

$$\mu(t) = E(Y \mid X^{(1)} = t_1, ..., X^{(p)} = t_p) = \beta_0 + \beta_1 t_1 + ... + \beta_p t_p \tag{4.11}$$

But many analysts hesitate to use (4.11) as the model, as it may generate negative values. They view that as a problem, since $Y \geq 0$ (and $P(Y > 0) > 0$), so we have $\mu(t) > 0$. The use of exp() in (4.10) meets that objection.

---

[4]Recall from Section 1.17.1 that the classification problem is a special case of regression.

Many who work with Poisson distributions find the Poisson relation

$$\sigma^2(t) = \mu(t) \tag{4.12}$$

to be overly restrictive. They would like to use a model that is "Poisson-like" but with $>$ instead of $=$ in (4.12),

$$\sigma^2(t) > \mu(t) \tag{4.13}$$

a condition called *overdispersion*. This arises if we have a *mixture* of Poisson distributions, i.e., the variable in question has a Poisson distribution in various subpopulations of our population (Exercise 4).

One such model is that of the *negative binomial*. Though this distribution is typically presented as the number of Bernoulli trials to attain r successes, in this context we simply use it as a general distribution model. It can be shown that a gamma-distributed mixture of Poisson random variables has a negative binomail distribution.

The function **glm.nb** in the **MASS** package performs a negative binomial regression analysis.

4.2.3   Exponential Families

In general, the core of GLM assumes that $F_{Y|X}$ belongs to an *exponential family* [136]. This is formally defined as a parametric family whose probability density/mass function has the form

$$\exp[\eta(\theta)T(x) - A(\theta) + B(x)] \tag{4.14}$$

where θ is the parameter vector and x is a value taken on by the random variable. Though this may seem imposing, it suffices to say that the above formulation includes many familiar distribution families such as Bernoulli, binomial, Poisson, exponential and normal. In the Poisson case, for instance, setting $\eta(\theta) = \log \lambda$, $T(x) = x$, $A(\theta) = -\lambda$ and $B(x) = -\log(x!)$ yields the expression

$$\frac{e^{-\lambda}\lambda^x}{x!} \tag{4.15}$$

the famous form of the Poisson probability mass function.

As previously noted, GLM terminology centers around the *link function*, which is the functional inverse of our function $q()$ above. For Poisson regression, the link function is the inverse of exp(), i.e., log(). For logit, set $u = q(s) = (1 + \exp(-s))^{-1}$, and solve for s, giving us the link function,

$$\text{link}(u) = \ln \frac{u}{1-u} \tag{4.16}$$

4.2.4 R's glm() Function

Of course, the **glm()** function does all this for us. For ordinary usage, the call is the same as for **lm()**, except for one extra argument, **family**. In the Poisson regression case, for example, the call looks like

glm(y ~ x, **family** = **poisson**)

The **family** argument actually has its own online help entry:

```
> ?family
family                          package:stats
R Documentation

Family Objects for Models

Description:
...

Usage:

    family(object, ...)

    binomial(link = "logit")
    gaussian(link = "identity")
    Gamma(link = "inverse")
    inverse.gaussian(link = "1/mu^2")
    poisson(link = "log")
    quasi(link = "identity", variance = "constant")
    quasibinomial(link = "logit")
    quasipoisson(link = "log")

...
```

Ah, so the **family** argument is a function! There are built-in ones we can use, such as the **poisson** one we used above, or a user could define her own custom function.

Well, then, what are the arguments to such a function? A key argument is **link**, which is obviously the link function $q^{-1}()$ discussed above, which we found to be **log()** in the Poisson case.

For a logistic model, as noted earlier, $F_{Y|X}$ is binomial with number of trials m equal to 1. Recall that the variance of a binomial random variable with m trials is $mr(1-r)$, where r is the "success" probability on each trial, Recall too that the mean of a 0-1-valued random variable is the probability of a 1. Putting all this together, we have

$$\sigma^2(t) = \mu(t)[1 - \mu(t)] \tag{4.17}$$

Sure enough, this appears in the code of the built-in function **binomial()**:

```
> binomial
function (link = "logit")
{
...
      variance <- function(mu) mu * (1 - mu)
```

Let's now turn to details of two of the most widely-used models, the logistic and Poisson.

4.3   GLM: the Logistic Model

The logistic regression model, introduced in Section 1.1, is by far the most popular nonlinear regression method. Here we are predicting a response variable Y that takes on the values 1 and 0, indicating which of two classes our unit belongs to. As we saw in Section 1.17.1, this indeed is a regression situation, as $E(Y \mid X = t)$ reduces to $P(Y = 1 \mid X = t)$.

The model, again, is

$$P(Y = 1 \mid X = (t_1, ..., t_p)) = \frac{1}{1 + e^{-(\beta_0 + \beta_1 t_1 + ... + \beta_p t_p)}} \tag{4.18}$$

4.3.1 Motivation

We noted in Section 1.1 that the logistic model is appealing for two reasons: (a) It takes values in [0,1], as a model for probabilities should, and (b) it is monotone in the predictor variables, as in the case of a linear model, a common situation in practice.

But there's even more reason to choose the logistic model. It turns out that the logistic model is implied by many familiar distribution families. In other words, there is often good theoretical justification for using the logit.

To illustrate that, consider a very simple example of text classification, involving Twitter tweets. Suppose we wish to automatically classify tweets into those involving financial issues and all others. We'll do that by having our code check whether a tweet contains words from a list of financial terms we've set up for this purpose, say *bank*, *rate* and so on.

Here Y is 1 or 0, for the financial and nonfinancial classes, and X is the number of occurrences of terms from the list. Suppose that from past data we know that *among financial tweets*, the number of occurrences of words from this list has a Poisson distribution with mean 1.8, while for nonfinancial tweets the mean is 0.2. Mathematically, that says that $F_{X|Y=1}$ is Poisson with mean 1.8, and $F_{X|Y=0}$ is Poisson with mean 0.2. (Be sure to distinguish the situation here, in which $F_{X|Y}$ is a Poisson distribution, from Poisson regression, in which it is assumed that $F_{Y|X}$ is Poisson.) Finally, suppose 5% of all tweets are financial.

Recall once again (Section 1.17.1) that in the classification case, our regression function takes the form

$$\mu(t) = P(Y = 1 \mid X = t) \tag{4.19}$$

Let's calculate this function:

$$
\begin{aligned}
P(Y = 1 \mid X = t) &= \frac{P(Y = 1 \text{ and } X = t)}{PX = t)} \tag{4.20} \\
&= \frac{P(Y = 1 \text{ and } X = t)}{P(Y = 1 \text{ and } X = t \text{ or } Y = 0 \text{ and } X = t)} \\
&= \frac{\pi \ P(X = t \mid Y = 1)}{\pi \ P(X = t \mid Y = 1) + (1 - \pi) \ P(X = t \mid Y = 0)}
\end{aligned}
$$

where $\pi = P(Y = 1)$ is the population proportion of individuals in class 1.

The numerator in (4.20) is

$$0.05 \cdot \frac{e^{-1.8} \, 1.8^t}{t!} \tag{4.21}$$

and similarly the denominator is

$$0.05 \cdot \frac{e^{-1.8} \, 1.8^t}{t!} + 0.95 \cdot \frac{e^{-0.2} \, 0.2^t}{t!} \tag{4.22}$$

Putting this into (4.20) and simplifying, we get

$$P(Y = 1 \mid X = t) \;=\; \frac{1}{1 + 19 e^{1.6 (\frac{1}{9})^t}} \tag{4.23}$$

$$=\; \frac{1}{1 + \exp(\log 19 + 1.6 - t \log 9)} \tag{4.24}$$

That last expression is of the form

$$\frac{1}{1 + \exp[-(\beta_0 + \beta_1 t)]} \tag{4.25}$$

with

$$\beta_0 = -\log 19 - 1.6 \tag{4.26}$$

and

$$\beta_1 = \log 9 \tag{4.27}$$

In other words the setting in which $F_{X|Y}$ is Poisson implies the logistic model!

This is true too if $F_{X|Y}$ is an exponential distribution. Since this is a continuous distribution family rather than a discrete one, the quantities $P(X = t | Y = i)$ in (4.23) must be replaced by density values:

$$P(Y = 1 \mid X = t) =$$

$$\frac{\pi \, f_1(X = t \mid Y = 1)}{\pi \, f_1(X = t \mid Y = 1) + (1 - \pi) \, f_0(X = t \mid Y = 0)} \quad (4.28)$$

where the within-class densities of X are

$$f_i(w) = \lambda_i e^{-\lambda_i w}, \quad i = 0, 1 \quad (4.29)$$

After simplifying, we again obtain a logistic form.

Most important, consider the multivariate normal case (Section 2.6.2): Say for groups $i = 0, 1$, $F_{X|Y=i}$ is a multivariate normal distribution with mean vector μ_i and covariance matrix Σ, where the latter does *not* have a subscript i. This is a generalization of the classical two-sample t-test setting, in which two (scalar) populations are assumed to have possibly different means but the same variance.[5] Again using (4.28), and going through a lot of algebra, we find that again $P(Y = 1 \mid X = t)$ turns out to have a logistic form,

$$P(Y = 1 \mid X = t) = \frac{1}{1 + e^{-(\beta_0 + \overline{\beta}' t)}} \quad (4.30)$$

with

$$\beta_0 = \log(1 - \pi) - \log \pi + \frac{1}{2}(\mu_1' \mu_1 - \mu_0' \mu_0) \quad (4.31)$$

and

$$\overline{\beta} = (\mu_0 - \mu_1)' \Sigma^{-1} \quad (4.32)$$

where t is the vector of predictor variables, the β vector is broken down into $(\beta_0, \overline{\beta})$, and π is $P(Y = 1)$. The messy form of the coefficients here is not important; instead, the point is that we find that the multivariate normal model implies the logistic model, giving the latter even more justification.

In summary:

> Not only is the logistic model intuitively appealing because it is a monotonic function with values in (0,1), but also because it

---

[5]It is also the setting for *Fisher's Linear Discriminant Analysis*, to be discussed in Section 5.6.

is implied by various familiar parametric models for the within-class distribution of X.

No wonder the logit model is so popular!

4.3.2   Example: Pima Diabetes Data

Another famous UCI data set is from a study of the Pima tribe of Native Americans, involving factors associated with diabetes. There is data on 768 women.[6] Let's predict diabetes from the other variables:

```
> pima <- read.csv('pima-indians-diabetes.data')
> head(pima)
  NPreg Gluc BP Thick Insul  BMI Genet Age Diab
1     6  148 72    35     0 33.6 0.627  50    1
2     1   85 66    29     0 26.6 0.351  31    0
3     8  183 64     0     0 23.3 0.672  32    1
4     1   89 66    23    94 28.1 0.167  21    0
5     0  137 40    35   168 43.1 2.288  33    1
6     5  116 74     0     0 25.6 0.201  30    0
# Diab = 1 means has diabetes
> logitout <- glm(Diab ~ ., data=pima, family=binomial)
> summary(logitout)
...
Coefficients:
                 Estimate  Std. Error  z value
(Intercept)   -8.4046964   0.7166359  -11.728
NPreg          0.1231823   0.0320776    3.840
Gluc           0.0351637   0.0037087    9.481
BP            -0.0132955   0.0052336   -2.540
Thick          0.0006190   0.0068994    0.090
Insul         -0.0011917   0.0009012   -1.322
BMI            0.0897010   0.0150876    5.945
Genet          0.9451797   0.2991475    3.160
Age            0.0148690   0.0093348    1.593
                 Pr(>|z|)
(Intercept)     < 2e-16 ***
NPreg          0.000123 ***
Gluc            < 2e-16 ***
BP             0.011072 *
```

---

[6]The data set is at *https://archive.ics.uci.edu/ml/datasets/Pima+Indians+Diabetes.* I have added a header record to the file.

Thick	0.928515
Insul	0.186065
BMI	2.76e−09 ***
Genet	0.001580 **
Age	0.111192
...	

4.3.3 Interpretation of Coefficients

In nonlinear regression models, the parameters β_i do not have the simple marginal interpretation they enjoy in the linear case. Statements like we made in Section 1.9.1.2, "We estimate that, on average, each extra year of age corresponds to almost a pound in extra weight," are not possible here.

However, in the nonlinear case, the regression function is still defined as the conditional mean, which in the logit case reduces to the conditional probability of a 1. Practical interpretation is definitely still possible, if slightly less convenient.

Consider for example the estimated Glucose coefficient in our diabetes data above, 0.035. Let's apply that to the people similar to the first person in the data set:

```
> pima[1,]
  NPreg Gluc BP Thick Insul  BMI Genet Age Diab
1     6  148 72    35     0 33.6 0.627  50    1
```

Ignore the fact that this woman has diabetes. Let's consider the subpopulation of all women with the same characteristics as this one, i.e., all who have had 6 pregnancies, a glucose level of 148 and so on, through an age of 50. The estimated proportion of women with diabetes in this subpopulation is

$$\frac{1}{1 + e^{-(8.4047 + 0.1232 \cdot 6 + ... + 0.0149 \cdot 50)}} \tag{4.33}$$

We don't have to plug these numbers in by hand, of course:

```
> l <- function(t) 1/(1+exp(-t))
> pima1 <- unlist(pima[1,-9])
> l(coef(logitout) %*% c(1,pima1))
          [,1]
[1,] 0.7217266
```

Note that **pima[1,-9]** is actually a data frame (having been derived from a data frame), so in order to multiply it, we needed to make a vector out of it, using **unlist()**.

So, we estimate that about 72% of women in this subpopulation have diabetes. But what about the subpopulation of the same characteristics, but of age 40 instead of 50?

```
> w <- pima1
> w['Age'] <- 40
> l(coef(logitout) %*% c(1,w))
        [,1]
[1,]  0.6909047
```

Only about 69% of the younger women have diabetes.

So, there is an effect of age on developing diabetes, but only a mild one; a 10-year increase in age only increased the chance of diabetes by about 3.1%. However, note carefully that this was for women having a given set of the other factors, e.g., 6 pregnancies. Let's look at a different subpopulation, those with 2 pregnancies and a glucose level of 120, comparing 40- and 50-year-olds:

```
> u <- pima1
> u[1] <- 2
> u[2] <- 100
> v <- u
> v[8] <- 40
> l(coef(logitout) %*% c(1,u))
        [,1]
[1,]  0.2266113
> l(coef(logitout) %*% c(1,v))
        [,1]
[1,]  0.2016143
```

So here, the 10-year age effect was somewhat less, about 2.5%. A more careful analysis would involve calculating standard errors for these numbers, but the chief point here is that the effect of a factor in nonlinear situations depends on the values of the other factors.

$$P(Y = 0 \mid X^{(1)} = t_1, ..., X^{(p)} = t_p) = t_p) = 1 - \ell(\beta_0 + \beta_1 t_1 + ... + \beta_p t_p) \quad (4.34)$$

Some analysts like to look at the *log-odds ratio*,

$$\log \frac{P(Y = 1 \mid X^{(1)} = t_1, ..., X^{(p))} = t_p)}{P(Y = 0 \mid X^{(1)} = t_1, ..., X^{(p))} = t_p)} \tag{4.35}$$

in this case the logaritihm of the ratio of the probability of having and not having the disease. By Equation (4.8), this simplifies to

$$\beta_0 + \beta_1 t_1 + ... + \beta_p t_p \tag{4.36}$$

a linear function. Thus, in interpreting the coefficients output from a logistic analysis, it is convenient to look at this *log-odds ratio*, as it gives us a single marginal-effect number for each factor. This may be sufficient for the application at hand, but a more thorough analysis should consider the effects of the factors on the probabilities themselves.

4.3.4 The predict() Function Again

In the previous section, we evaluated the estimated regression function (and thus predicted values as well) the straightforward but messy way, e.g.,

```
> l(coef(logitout) %*% c(1,v))
```

The easy way is to use R's **predict()** function:

```
> predict(object=logitout, newdata=pima[1,-9],
    type='response')
        1
0.7217266
```

We saw that in Section 1.10.3 for objects of **'lm'** class. But in our case here, we invoked it on **logitout**. What is the class of that object?

```
> class(logitout)
[1] "glm" "lm"
```

So, it is an object of class **'glm'**, and, we see, the latter is a subclass of the class **'lm'**. For that subclass, the **predict()** function, i.e., **predict.glm()**, there is an extra argument (actually several), **type**. The value of that argument that we want here is **type** = **'response'**, alluding to the fact that we want a prediction on the scale of the response variable, Y.

4.3.5   Overall Prediction Accuracy

How well can we predict in the Pima example above? For the best measure, we should use cross validation or something similar, but we can obtain a quick measure as follows.

The value returned by **glm()** has class 'glm', which is actually a subclass of 'lm'. The latter, and thus the former, includes a component **fitted.values**, the ith of which is

$$\widehat{\mu}(X_i) \tag{4.37}$$

i.e., the estimated value of the regression function at observation i. If we did not know Y_i, we would predict it to be 1 or 0, depending on whether $\widehat{\mu}(X_i)$ is greater than or less than 0.5. In R terms, that predicted value is simply

round (logitout **$ fitted** . values [i])

Using the fact that the proportion of 1s in a vector of 1s and 0s is simply the mean value in that vector, we have that the overall probability of correct classification is

> **mean**(pima**$**Diab == **round** (logitout **$ fitted** . values))
0.7825521

That seems pretty good (though again, it is biased upward and cross validation would give us a better estimate), but we must compare it against how well we would do without the covariates. We reason as follows. First,

> **mean**(pima**$**Diab)
[1] 0.3489583

Most of the women do not have diabetes, so our strategy, lacking covariate information, would be to always guess that $Y = 0$. We will be correct a proportion

> 1 − 0.3489583
[1] 0.6510417

of the time. Thus our 78% accuracy using covariates does seem to be an improvement.

4.3.6 Example: Predicting Spam E-mail

One application of these methods is *text classification*. In our example here, the goal is machine prediction of whether an incoming e-mail message is *span*, i.e., unwanted mail, typically ads.

We'll use the **spam** dataset from the UCI Machine Learning Data Repository. It is also available from the CRAN package **ElemStatLearn** [64], which we will use here, but note that the UCI version includes a word list. It has data on 4601 e-mail messages, with 57 predictors. The first 48 of those predictors consist of frequencies of 48 words. Thus the first column, for instance, consists of the proportions of Word 1 in each of the 4601 messages, with the total number of words in a message as the base in each case. The remaining predictors involve measures such as the numbers of consecutive capital letters in words.

The last column is an R *factor* with levels *spam* and *e-mail*. This R type is explained in Section 4.7.2, and though **glm()** can handle such variables, for pedagogical reasons, let's use dummies for a while. (We will begin using factors directly in Section 5.6.3.)

Let's fit a logistic model.

```
> library(ElemStatLearn)
> data(spam)
> spam$spam <- as.integer(spam$spam == 'spam')
> glmout <- glm(spam ~ ., data=spam,
      family=binomial)
> summary(glmout)
...
```

```
Coefficients:
             Estimate  Std. Error  z value
(Intercept) −1.569e+00  1.420e−01  −11.044
A.1         −3.895e−01  2.315e−01   −1.683
A.2         −1.458e−01  6.928e−02   −2.104
A.3          1.141e−01  1.103e−01    1.035
A.4          2.252e+00  1.507e+00    1.494
A.5          5.624e−01  1.018e−01    5.524
A.6          8.830e−01  2.498e−01    3.534
A.7          2.279e+00  3.328e−01    6.846
A.8          5.696e−01  1.682e−01    3.387
...
             Pr(>|z|)
```

```
(Intercept)   < 2e−16  ***
A.1           0.092388  .
A.2           0.035362  *
A.3           0.300759
A.4           0.135168
A.5           3.31e−08  ***
A.6           0.000409  ***
A.7           7.57e−12  ***
A.8           0.000707  ***
...
```

Let's see how accurately we can predict with this model:

```
> mean(spam$spam == round(glmout$fitted.values))
[1]  0.9313193
```

Not bad at all. But much as we are annoyed by spam, we hope that a genuine message would not be likely to be culled out by our spam filter. Let's check:

```
> spamnot <- which(spam$spam == 0)
> mean(round(glmout$fitted.values[spamnot]) == 0)
[1]  0.956241
```

So if a message is real, it will have a 95% chance of getting past the spam filter.

4.3.7   Linear Boundary

In (4.18), which values of $t = (t_1, ..., t_p)'$ will cause us to gues $Y = 1$ and which will result in a guess of $Y = 0$? The boundary occurs when (4.18) has the value 0.5. In other words, the boundary consists of all t such that

$$\beta_0 + \beta_1 t_1 + ... + \beta_p t_p = 0 \tag{4.38}$$

So, the boundary has linear form, a *hyperplane* in p-dimensional space. This may seem somewhat abstract now, but it will have value later on.

4.4 GLM: the Poisson Regression Model

Since in the Pima data (Section 4.3.2) the number of pregnancies is a count, we might consider predicting it using Poisson regression.[7] Here's how we can do this with **glm()**:

```
> poisout <- glm(NPreg ~ ., data=pima, family=poisson)
> summary(poisout)
...
Coefficients:
```

	Estimate	Std. Error	z value
(Intercept)	0.2963661	0.1207149	2.455
Gluc	−0.0015080	0.0006704	−2.249
BP	0.0011986	0.0010512	1.140
Thick	0.0000732	0.0013281	0.055
Insul	−0.0003745	0.0001894	−1.977
BMI	−0.0002781	0.0027335	−0.102
Genet	−0.1664164	0.0606364	−2.744
Age	0.0319994	0.0014650	21.843
Diab	0.2931233	0.0429765	6.821

	$Pr(>|z|)$	
(Intercept)	0.01408	*
Gluc	0.02450	*
BP	0.25419	
Thick	0.95604	
Insul	0.04801	*
BMI	0.91896	
Genet	0.00606	**
Age	< 2e−16	***
Diab	9.07e−12	***

```
...
```

On the other hand, even if we believe that our count data follow a Poisson distribution, there is no law dictating that we use Poisson regression, i.e., the model (4.10). As mentioned following that equation, the main motivation for using exp() in that model is to ensure that our regression function is nonnegative, conforming to the nonnegative nature of Poisson random variables. This is not unreasonable, but as noted in a somewhat different context in Section 3.3.7, transformations — in this case, the use of exp() — can produce distortions. Let's try the "unorthodox" model, (4.11):

---

[7]It may seem unnatural to predict this, but as noted before, predicting any variable may be useful if data on that variable may be missing.

```
> quasiout <- glm(NPreg ~ .,data=pima,
    family=quasi(variance="mu^2"),start=rep(1,9))
```

This "quasi" family is a catch-all option, specifying a linear model but here allowing us to specify a Poisson variance function:

$$Var(Y \mid X = t) = [\mu(t)]^2 \tag{4.39}$$

with $\mu(t) = t'\beta$. This is (4.11), not the standard Poisson regression model, but worth trying anyway.

Well, then, which model performed better? As a rough, quick look, ignoring issues of overfitting and the like, let's consider R^2. This quantity is not calculated by **glm()**, but recall from Section 2.9.2 that R^2 is the squared correlation between the predicted and actual Y values. This quantity makes sense for any regression situation, so let's calculate it here:

```
> cor(poisout$fitted.values,poisout$y)^2
[1]  0.2314203
> cor(quasiout$fitted.values,quasiout$y)^2
[1]  0.3008466
```

The "unorthodox" model performed better than the "official" one! We cannot generalize from this, but it does show again that one must use transformations carefully.

4.5 Least-Squares Computation

A point made in Section 1.4 was that the regression function, i.e., the conditional mean, is the optimal predictor function, minimizing mean squared prediction error. This still holds in the nonlinear (and even nonparametric) case. The problem is that in the nonlinear setting, the least-squares estimator does not have a nice, closed-form solution like (2.28) for the linear case. Let's see how we can compute the solution through iterative approximation.

4.5.1 The Gauss-Newton Method

Denote the nonlinear model by

$$E(Y \mid X = t) = g(t, \beta) \tag{4.40}$$

where both t and β are possibly vector-valued. In (4.5), for instance, t is a scalar but β is a vector. The least-squares estimate $\widehat{\beta}$ is the value of b that minimizes

$$\sum_{i=1}^{n}[Y_i - g(X_i, b)]^2 \tag{4.41}$$

Many methods exist to minimize (4.41), most of which involve derivatives with respect to b. (The reason for the plural *derivatives* is that there is a partial derivative for each of the elements of b.)

The best intuitive explanation of derivative-based methods, which will also prove useful in a somewhat different context later in this chapter, is to set up a Taylor series approximation for $g(X_i, b)$ (as in Section 3.6.1):

$$g(X_i, b) \approx g(X_i, \widehat{\beta}) + h(X_i, \widehat{\beta})'(b - \widehat{\beta}) \tag{4.42}$$

where $h(X_i, b)$ is the derivative vector of $g(X_i, b)$ with respect to b, and the prime symbol, as usual, means matrix transpose (not a derivative). The value of $\widehat{\beta}$, is of course yet unknown, but let's put that matter aside for now. Then (4.41) is approximately

$$\sum_{i=1}^{n}[Y_i - g(X_i, \widehat{\beta}) + h(X_i, \widehat{\beta})'\widehat{\beta} - h(X_i, \widehat{\beta})' \, b]^2 \tag{4.43}$$

At iteration k we take our previous iteration b_{k-1} to be an approximation to $\widehat{\beta}$, and make that substitution in (4.43), yielding

$$\sum_{i=1}^{n}[Y_i - g(X_i, b_{k-1}) + h(X_i, b_{k-1})'b_{k-1} - h(X_i, b_{k-1})' \, b]^2 \tag{4.44}$$

Our b_k is then the value that minimizes (4.44) over all possible values of b. But why is that minimization any easier than minimizing (4.41)? To see why, write (4.44) as

$$\sum_{i=1}^{n}[\underbrace{Y_i - g(X_i, b_{k-1}) + h(X_i, b_{k-1})'b_{k-1}} - h(X_i, b_{k-1})' \, b]^2 \tag{4.45}$$

	(4.45)	(2.18)
$Y_i - g(X_i, b_{k-1}) + h(X_i, b_{k-1})' b_{k-1}$		Y_i
	$h(X_i, b_{k-1})$	$\tilde{X}_i$

Table 4.1: Nonlinear/linear correspondences

This should look familiar. It has exactly the same form as (2.18), with the correspondences shown in Table 4.1. In other words, what we have in (4.45) is a *linear* regression problem!

In other words, we can find the minimizing b in (4.45) using **lm()**. There is one small adjustment to be made, though. Recall that in (2.18), the quantity $\tilde{X}_i$ includes a 1 term (Section 2.1), i.e., the first column of A in (2.19) consists of all 1s. That is not the case in Table 4.1 (second row, first column), which we need to indicate in our **lm()** call. We can do this via specifying "-1" in the formula part of the call (Section 2.4.5).

Another issue is the computation of $h()$. Instead of burdening the user with this, it is typical to compute $h()$ using numerical approximation, e.g., using R's **numericDeriv()** function or the **numDeriv** package [57].

4.5.2   Eicker-White Asymptotic Standard Errors

As noted, **nls()** assumes homoscedasticity, which generally is a poor assumption (Section 2.6). It would be nice, then, to somehow apply the Eicker-White method (Section 3.3.3), which is for linear models, to the nonlinear case. Actually, it is remarkably easy to do that adaptation.

The key is to note the linear approximation (4.5.1). One way to look at this is that it has already set things up for us to use the Delta Method, which uses a linear approximation. Thus we can apply Eicker-White to the **lm()** output, say using **vcovHC()**, as in Section 3.3.4.

Below is code along these lines. It requires the user to run **nlsLM()**, an alternate version of **nls()** in the CRAN package **minpack.lm** [45].[8]

---

[8]This version is needed here because it provides the intermediate quantities we need from the computation. However, we will see in Section 4.5.4 that this version has other important advantages as well.

```
library(minpack.lm)
library(sandwich)

# uses output of nlsLM() of the minpack.lm package
# to get an asymptotic covariance matrix without
# assuming homoscedasticity

# arguments:
#
#     nlslmout: return value from nlsLM()
#
# value: approximate covariance matrix for the
#        estimated parameter vector

nlsvcovhc <- function(nlslmout) {
   # notation: g(t,b) is the regression model,
   # where x is the vector of variables for a
   # given observation; b is the estimated parameter
   # vector; x is the matrix of predictor values
   b <- coef(nlslmout)
   m <- nlslmout$m
   # y - g:
   resid <- m$resid()
   # row i of hmat will be deriv of g(x[i,],b)
   # with respect to b
   hmat <- m$gradient()
   # calculate the artificial "x" and "y" of
   # the algorithm
   fakex <- hmat
   fakey <- resid + hmat %*% b
   # -1 means no constant term in the model
   lmout <- lm(fakey ~ fakex - 1)
   vcovHC(lmout)
}
```

In addition to nice convergence behavior, the advantage for us here of **nlsLM()** over **nls()** is that the former gives us access to the quantities we need in (4.45), especially the matrix of $h()$ values. We then apply **lm()** one more time, to get an object of type "lm", needed by **vcovHC()**.

Applying this to the enzyme data, we have

```
> nlsvcovhc(z)
          fakex1      fakex2      fakex3
```

```
fakex1  0.4708209   1.706591    2.410712
fakex2  1.7065910  10.394496   20.314688
fakex3  2.4107117  20.314688   53.086958
```

This is rather startling. Except for the estimated variance of $\widehat{\beta}_1$, the estimated variances and covariances from Eicker-White are much larger than what **nls()** found under the assumption of homoscedasticity.

Of course, with only 60 observations, both of the estimated covariance matrices must be "taken with a grain of salt." So, let's compare the two approaches by performing a simulation. Here

$$E(Y \mid X = t) = \frac{1}{t'\beta} \qquad (4.46)$$

where $t = (t_1, t_2)'$ and $\beta = (\beta_1, \beta_2)'$. We'll take the components of X to be independent and exponentially distributed with mean 1.0, with the heteroscedasticity modeled as being such that the standard deviation of Y given X is proportional to the regression function value at X. We'll use as a check the fact that, 90% of the time, a N(0,1) variable is less than 1.28 Here is the simulation code:

```
sim <- function(n, nreps) {
    b <- 1:2
    res <- replicate(nreps, {
        x <- matrix(rexp(2*n), ncol=2)
        meany <- 1 / (x %*% b)
        y <- meany + (runif(n) - 0.5) * meany
        xy <- data.frame(x, y)
        nlout <- nls(X3 ~ 1 / (b1*X1+b2*X2),
            data=xy, start=list(b1 = 1, b2=1))
        b <- coef(nlout)
        vc <- vcov(nlout)
        vchc <- nlsvcovhc(nlout)
        z1 <- (b[1] - 1) / sqrt(vc[1,1])
        z2 <- (b[1] - 1) / sqrt(vchc[1,1])
        c(z1, z2)
    })
    print(mean(res[1,] < 1.28))
    print(mean(res[2,] < 1.28))
}
```

And here is a run of the code:

```
> sim(250,2500)
[1] 0.6188
[1] 0.9096
```

That's quite a difference! Eicker-White worked well, whereas assuming homoscedasticity fared quite poorly. (Similar results were obtained for $n = 100$.)

4.5.3 Example: Bike Sharing Data

In our bike-sharing data (Section 1.1), there are two kinds of riders, *registered* and *casual*. We may be interested in factors determining the mix, i.e.,

$$\frac{\text{registered}}{\text{registered} + \text{casual}} \tag{4.47}$$

Since the mix proportion is between 0 and 1, we might try the logistic model, introduced in (1.36) in the context of classification. Note, though, that the example here does not involve a classification problem. so we should not reflexively use **glm()** as before. Indeed, that function not only differs from our current situation in that here Y takes on values in [0,1] rather than in {0,1}, but also **glm()** assumes

$$Var(Y \mid X =) = \mu(t)(1 - \mu(t)) \tag{4.48}$$

(as implied by Y being in {0,1}), which we have no basis for assuming here. Thus use of **glm()**, at least in the form we have seen so far, would be inappropriate. Here are the results:

```
> shar <- read.csv("day.csv",header=T)
> shar$temp2 <- shar$temp^2
> shar$summer <- as.integer(shar$season == 3)
> shar$propreg <- shar$reg / (shar$reg+shar$cnt)
> names(shar)[15] <- "reg"
> library(minpack.lm)
> logit <- function(t1,t2,t3,t4,b0,b1,b2,b3,b4)
     1 / (1 + exp(-b0 - b1*t1 -b2*t2 -b3*t3 -b4*t4))
> z <- nlsLM(propreg ~
logit(temp,temp2,workingday,summer,b0,b1,b2,b3,b4),
    data=shar,start=list(b0=1,b1=1,b2=1,b3=1,b4=1))
```

```
> summary(z)
...
Parameters:
     Estimate  Std. Error  t value  Pr(>|t|)
b0  -0.083417   0.020814   -4.008  6.76e-05 ***
b1  -0.876605   0.093773   -9.348   < 2e-16 ***
b2   0.563759   0.100890    5.588  3.25e-08 ***
b3   0.227011   0.006106   37.180   < 2e-16 ***
b4   0.012641   0.009892    1.278     0.202
...
```

As expected, on working days, the proportion of registered riders is higher, as we are dealing with the commute crowd on those days. On the other hand, the proportion doesn't seem to be much different during the summer, even though the vacationers would presumably add to the casual-rider count.

But are those standard errors trustworthy? Let's look at the Eicker-White versions:

```
> sqrt(diag(nlsvcovhc(z)))
      fakex1          fakex2          fakex3          fakex4
0.021936045  0.090544374  0.092647403  0.007766202
      fakex5
0.007798938
```

Again, we see some substantial differences.

4.5.4   The "Elephant in the Room": Convergence Issues

So far we have sidestepped the fact that any iterative method runs the risk of nonconvergence. Or it might converge to some point at which there is only a local minimum, not the global one — worse than nonconvergence, in the sense that the user might be unaware of the situation.

For this reason, it is best to try multiple, diverse sets of starting values. In addition, there are refinements of the Gauss-Newton method that have better convergence behavior, such as the Levenberg-Marquardt method.

Gauss-Newton sometimes has a tendency to "overshoot," producing too large an increment in b from one iteration to the next. Levenberg-Marquardt generates smaller increments. Interestingly it is a forerunner of *ridge re-*

gression that we'll discuss in Chapter 8. It is implemented in the CRAN package **minpack.lm**, which we used earlier in this chapter.

4.6 Further Reading

For more on the Generalized Linear Model, see for instance [47] [3] [41].

Exponential random graph models use logistic regression and similar techniques to analyze relations between nodes in a network, say connections between friends in a group of people [60] [75]. The book by Luke [93] presents various R tools for random graphs, and serves as a short introduction to field.

4.7 Computational Complements

4.7.1 GLM Computation

Though estimation in GLM uses Maximum Likelihood, it can be shown that the actual computation can be done by extending the ideas underlying least-squares models. We use a weight function, which works as follows.

Let's review Section 3.3.2, which discussed weighted least squares in the case of a linear model. Using our usual notation $\mu(t) = E(Y \mid X = t)$ and $\sigma^2(t) = Var(Y \mid X = t)$, the optimal estimator of β is the value of b that minimizes

$$\sum_{i=1}^{n} \frac{1}{\sigma^2(\widetilde{X_i})} (Y_i - \widetilde{X_i}' b)^2 \tag{4.49}$$

Of course, generally $\sigma^2(t)$ is unknown, but it will be estimated.

Now consider the case of Poisson regression. One of the famous properties of the Poisson distribution family is that the variance equals the mean. Thus (4.43) becomes

$$\sum_{i=1}^{n} \frac{1}{g(X_i, b)} [Y_i - g(X_i, b)]^2 \tag{4.50}$$

Then (4.45) bceoms

$$\sum_{i=1}^{n} \frac{1}{g(X_i, b_{k-1})} \underbrace{[Y_i - g(X_i, b_{k-1}) + h(X_i, b_{k-1})'b_{k-1}} -h(X_i, b_{k-1})' \, b]^2$$

(4.51)

and we again solve for the new iterate b_k by calling **lm()**, this time making use of the latter's **weights** argument (Section 3.3.2).

We iterate as before, but now the weights are updated at each iteration too. For that reason, the process is known as *iteratively reweighted least squares*.

4.7.2   R Factors

To explain this R feature, let's look at the famous **iris** dataset included in the R package:

```
> head(iris)
   Sepal.Length  Sepal.Width  Petal.Length  Petal.Width  Species
1           5.1          3.5           1.4          0.2   setosa
2           4.9          3.0           1.4          0.2   setosa
3           4.7          3.2           1.3          0.2   setosa
4           4.6          3.1           1.5          0.2   setosa
5           5.0          3.6           1.4          0.2   setosa
6           5.4          3.9           1.7          0.4   setosa
> is <- iris$Species
> class(is)
[1] "factor"
> str(is)
 Factor w/ 3 levels "setosa","versicolor",..:
    1 1 1 1 1 1 1 1 1 1 ...
> table(is)
is
    setosa versicolor   virginica
        50         50          50
> mode(is)
[1] "numeric"
> levels(is)
[1] "setosa"     "versicolor" "virginica"
```

We see that **is** is basically a numeric vector, with its first few values being 1, 1, 1. But these codes have names, known as *levels*, such as 'setosa' for the code 1.

In some cases, all that machinery actually gets in the way. If so, we can convert to an ordinary vector, e.g.,

```
> s <- as.numeric(is)
```

4.8 Mathematical Complements

4.8.1 Maximum Likelihood Estimation

This is a method of statistical estimation. It is fairly generally applicable to estimation in parametric models, not just in a GLM or regression context, but let's look at the latter for concreteness, making things even more concrete by looking at a very simple example, $n = 2$.

Our model will be Poisson regression, (4.10), and say our data are $(Y_1, X_1) = (3, 20)$ and $(Y_2, X_2) = (8, 16)$. Our estimation will be conditional on the X_i, so they will be treated as constants.

The philosophy of MLE is to ask, "What values of the β_i in (4.10) would maximize the probability of our data occurring?" Using (4.10) and (4.15), and the independence of our data, that probability is

$$\frac{e^{-\lambda_1}\lambda_1^3}{3!} \cdot \frac{e^{-\lambda_2}\lambda_2^8}{8!} \tag{4.52}$$

where

$$\lambda_1 = e^{\beta_0 + \beta_1 \cdot 20}, \quad \lambda_2 = e^{\beta_0 + \beta_1 \cdot 16} \tag{4.53}$$

So, to answer the question of maximizing the probability of our data, we would maximize (4.52) with respect to β_0 and β_1, keeping in mind (4.53). The resulting values are called the Maximum Likelihood Estimators of the β_i.

MLEs form a key part of the theory of mathematical statistics. Under certain conditions, they are optimal [136]. However, this is well beyond the scope of this book.

4.9   Exercises: Data, Code and Math Problems

Data problems:

1. Conduct a negative binomial regression analysis on the Pima data. Compare to the results in Section 4.3.2.

2. Consider the spam example in Section 4.3.6. Find approximate 95% confidence intervals for the effect of the presence of word A.1, if none of the other words is present, and the nonword predictor variables A.49, A.50 and so on are all 0. Then do the same for word A.2. Finally, find a confidence interval for the difference of the two effects.

Mini-CRAN and other computational problems:

3. Though the logit model is plausible for the case $Y = 0, 1$, as noted in Section 4.3.1, we could try modeling $\mu(t)$ as linear. We would then call **lm()** instead of **glm()**, and simply predict Y to be whichever value in $\{0,1\}$ that $\widehat{\mu}(t)$ is closest to.

(a) Write R functions with call forms

```
binlin (indata ,yname)
predict . binlin ( binlinobj , newpts )
```

Here **indata** is a data frame, and **yname** is the name of the variable there that will be taken as the response variable, Y, a vector of 1s and 0s; the predictors will be the remaining columns. The function **binlin()** calls **lm()**, and changes its class to **'binlin'**, a subclass of **'lm'**.

The function **predict.binlin()** then acts on **binlinobj**, an object of class **'binlin'**, predicting on the rows of the data frame **newpts** (which must have the same column names as **indata**). The return value will be a vector of 1s and 0s, computed as in the approach proposed above.

(b) Try this approach on the spam prediction example of Section 4.3.6. Using cross-validation, fit both a logit model and a linear one to the training data, and see which one has better prediction accuracy on the test set.

Math problems:

4. Suppose U_i, $i = 1, 2$ are independent random variables with means λ_i.

Let J be a random variable, independent of the U_i, which takes on the values 1 and 0 with probability q and $1 - q$. Define

$$W = U_J \qquad (4.54)$$

Show that W satisfies the overdispersion condition (4.13). (Hint: Use (1.62).)

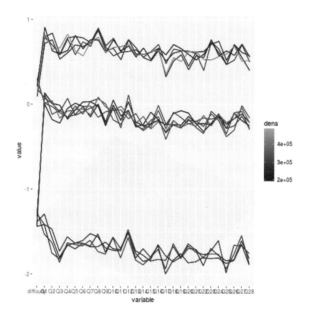

Fig. 1.3, Turkish student evaluations

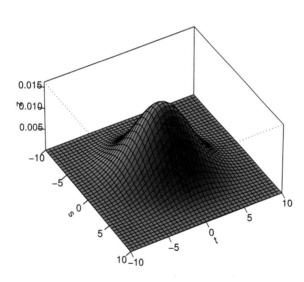

Fig. 2.1, Bivariate normal density, $\rho = 0.2$

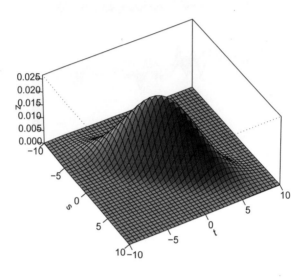

Fig. 2.2, Bivariate normal density, $\rho = 0.8$

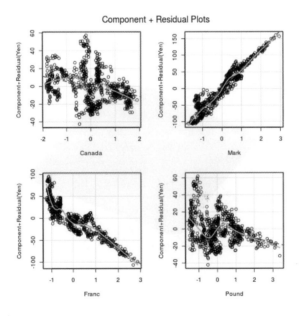

Fig. 6.3, Partial residuals plot, currency data

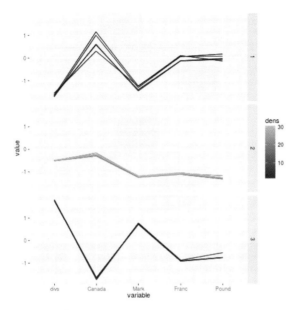

Fig. 6.6, Freqparcoord plot, currency data

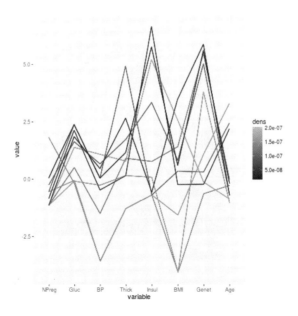

Fig. 6.8, Outlier hunt

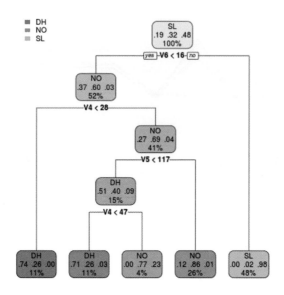

Fig. 10.2, Flow chart for vertebral column data

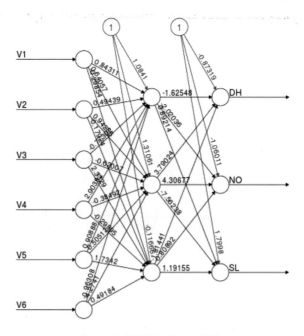

Error: 43.000304 Steps: 1292

Fig. 11.7, Vertebrae NN, 1 hidden layer

Chapter 5

Multiclass Classification Problems

We introduced the classification problem in Section 1.17, and then covered the logistic model for classification in some detail in Section 4.3. But in the classification problems we've discussed so far, we have assumed just two classes. The patient either has the disease in question, or not; the customer chooses to buy a certain item, or not; and so on.

In many applications, we have multiple classes. We may, for instance, be considering several different diseases that a patient might have.[1] In computer vision applications, the number of classes can be quite large, say face recognition with data on a large number of people. Let m denote the number of classes, and label them 0, 1, ..., m - 1.

5.1   Key Notation

The notation is a bit more complex than before, but still quite simple. The reader is urged to read this section carefully in order to acquire a solid grounding for the remaining material.

Say for instance we wish to do machine recognition of handwritten digits, so we have 10 classes, with our variables being various patterns in the

---

[1]For a classification problem, the classes must be mutually exclusive. In this case, there would be the assumption that the patient does not have more than one of the diseases.

pixels, e.g., the number of (approximately) straight line segments. Instead of having a single response variable Y as before, we would now have 10 of them, setting $Y^{(i)}$ to be 1 or 0, according to whether the given digit is i, for $i = 0, 1, ..., 9$. We could run 10 logistic regression models, and then use each one to estimate the probability that our new image represents a certain digit.

In general, as above, let $Y^{(i)}, i = 0, ..., m - 1$ be the indicator variables for the classes, and define the *class probabilities*

$$\pi_i = P(Y^{(i)} = 1), \quad i = 0, 1..., m - 1 \tag{5.1}$$

Of course, we must have

$$\sum_{i=0}^{m-1} \pi_i = 1 \tag{5.2}$$

We will still refer to Y, now meaning the value of i for which $Y^{(i)} = 1$.

Note that in this chapter, we will be concerned primarily with the Prediction goal, rather than Description.

5.2   Key Equations

Equations (4.20) and (4.28), and their generalizations, will play a key role here. Let's relate our new multiclass notation to what we had in the two-class case before. If $m = 2$, then:

- What we called $Y^{(1)}$ above was just called Y in our previous discussion of the two-class case.

- The class probability π_1 here was called simply π previously.

Now, let's review from the earlier material. (Keep in mind that typically X will be vector-valued, i.e., we have more than one predictor variable.) For $m = 2$:

- The quantity of interest is $\mu(t) = P(Y = 1 \mid X = t)$.

- If X has a discrete distribution, then

$$
\begin{aligned}
\mu(t) &= P(Y = 1 \mid X = t) \\
&= \frac{\pi \, P(X = t \mid Y = 1)}{\pi \, P(X = t \mid Y = 1) + (1 - \pi) \, P(X = t \mid Y = 0)}
\end{aligned}
\tag{5.3}
$$

- If X has a continuous distribution, then

$$
\mu(t) = P(Y = 1 \mid X = t) = \frac{\pi \, f_1(t)}{\pi \, f_1(t) + (1 - \pi) \, f_0(t)}
\tag{5.4}
$$

  where the within-class densities of X are f_1 and f_0.[2]

- Sometimes it is more useful to use the following equivalence to (5.4):

$$
\mu(t) = P(Y = 1 \mid X = t) = \frac{1}{1 + \frac{1 - \pi}{\pi} \frac{f_0(t)}{f_1(t)}}
\tag{5.5}
$$

Note that, in keeping with the notion that classification amounts to a regression problem (Section 1.17.1), we have used our regression function notation $\mu(t)$ above.

Things generalize easily to the multiclass case. We are now interested in the quantities

$$
P(Y = i) = \mu_i(t) = P(Y^{(i)} = 1 \mid X = t), \; i = 0, 1, ..., m - 1
\tag{5.6}
$$

For continuous X, (5.4) becomes

$$
P(Y = i) = \mu_i(t) = P(Y^{(i)} = 1 \mid X = t) = \frac{\pi_i \, f_i(t)}{\sum_{j=0}^{m-1} \pi_j \, f_j(t)}
\tag{5.7}
$$

---

[2] Another term for the class probabilities π_i is *prior probabilities*. Readers familar with the debate over *Bayesian* versus *frequentist* approaches to statistics may wonder if we are dealing with (subjective) Bayesian analyses here. Actually, that is not the case; we are not working with "gut feeling" probabilities as in (nonempirical) Bayesian methods. There is some connection, in the sense that (5.3) and (5.4) make use of Bayes' Rule, but the latter is standard for all statisticians, frequentist and Bayesian alike. Note by the way that probabilities like (5.4) are often termed *posterior* probabilities, again sounding Bayesian but again actually Bayesian/frequentist-neutral.

5.3   Estimating the Functions $\mu_i(t)$

We have already estimated $\mu(t)$ in, say, (5.5) using logit models. We can do the same for (5.6), running a logit analysis for each i, and indeed will do so later.

Another possibility would be to take a nonparametric approach. For instance, in the two-class case, one could estimate $\mu(t) = P(Y = 1 | X = t)$ to be the proportion of neighbors of t in our training data that have $Y = 1$. A less direct, but sometimes useful approach is to estimate the $f_i(t)$ in (5.5) and (5.6), and then plug our estimates into (5.5). For instance, one can do this using a k-nearest neighbor method, outlined in Section 5.10.1 of the Mathematical Complements section at the end of this chapter.

5.4   How Do We Use Models for Prediction?

In Section 1.10, we discussed the specifics of predicting new cases, in which we know X but not Y, after fitting a model to training data, in which both X and Y are known (our training data). The parametric and nonparametric cases were slightly different.

Here in the multiclass setting, as noted, we now have multiple functions $\mu_i(t), i = 0, 1, ..., m - 1$ that need to be estimated from our training data, as opposed to just $\mu(t)$ as before. Given a new case for which $X = t$, we guess the class Y to be the value of i for which $\widehat{\mu}_i(t)$ is largest, i.e.,

$$\widehat{Y} = \arg \max_i \widehat{\mu}_i(t) \qquad (5.8)$$

In the two-class case, this reduces to:

$$\text{Given } X. \text{ set } \widehat{Y} = 1 \text{ if and only if } \mu(X) > 0.5 \qquad (5.9)$$

It should be noted, though, that some nonparametric methods do not explicitly estimate $\mu_i(t)$, and instead only estimate "boundaries" involving those functions. These methods will be discussed in Chapter 11.

5.5 One vs. All or All vs. All?

Let's consider the Vertebral Column data from the UC Irvine Machine Learning Repository.[3] Here there are $m = 3$ classes: Normal, Disk Hernia and Spondylolisthesis. The predictors are, as described on the UCI site, "six biomechanical attributes derived from the shape and orientation of the pelvis." Consider two approaches we might take to predicting the status of the vertebral column, based on logistic regression:

- **One vs. All (OVA):** Here we predict each class against all the other classes. In the vertebrae data, we would fit 3 logit models to our training data, predicting each of the 3 classes, one at a time. So, first we would fit a logit model to predict Normal vs. Other, the latter meaning the Disk Hernia and Spondylolisthesis classes combined. Next we would predict Disk Hernia vs. Other, with the latter now being Normal and Spondylolisthesis combined, and the third model would be Spondylolisthesis vs. Other.

  The ith model would regress $Y^{(i)}$ against the 6 predictor variables, yielding $\widehat{\mu}_i(t), i = 0, 1, 2$. To predict Y for $X = t_c$, we would guess Y to be whatever i has the largest value of $\widehat{\mu}_i(t_c)$, i.e., the most likely class, given the predictor values.

- **All vs. All (AVA):** Here we would fit 3 logit models again, but with one model for each possible pair of classes. Our first model would pit class 0 against class 1, meaning that we would restrict our data to only those cases in which the class is 0 or 1, then predict class 0 versus 1 in that restricted data set. Our second logit model would restrict to the classes 0 and 2, and predict 0, while the last model would be for classes 1 and 2, predicting 1. (We would still use our 6 predictor variables in each model.) In each case, we tally which class "wins"; in the case in which we pit class 0 against class 1, our model might predict that the given new data point is of class 1, thus tally a win for that class. Then, whichever class "gets the most votes" in this process is our final predicted class. (If there is a tie, we could employ various tiebreaking procedures.)

Note that it was just coincidence that we have the same number of models in the OVA and AVA approaches here (3 each). In general, with m classes, we will run m logistic models (or k-NN or whatever type of regression

---

[3] *https://archive.ics.uci.edu/ml/datasets/Vertebral+Column*

modeling we like) under OVA, but $C(m, 2) = m(m-1)/2$ models under AVA.[4]

Code for OVA and AVA is given in Section 5.11.1.

5.5.1   Which Is Better?

Clearly, AVA involves a lot of computation, in a more complex form if not increased time. (See the Mathematical Complements section at the end of this chapter for details.) At least at first glance, AVA would not seem to have much to offer to make up for that. Indeed, it may have other problems as well.

For instance, since each of its models uses much less than our full data, the resulting estimated coefficients will likely be less accurate than what we calculate under OVA. And if m is large, we will have so many pairs that at least some will likely be especially inaccurate. (This notion will be discussed in Section 7.6.) And yet some researchers report they find AVA to work better in some settings.

To better understand the situation, let's consider an example and draw upon some intuition.

5.5.2   Example: Vertebrae Data

Here we analyze the vertebrae data first introduced in Section 5.5, applying the OVA and AVA methods to a training set of 225 randomly chosen records, then predicting the remaining records.[5] We'll use the OVA and AVA logistic code from the **regtools** package.[6]

```
> library(regtools)
> vert <- read.table('Vertebrae/column_3C.dat',
      header=FALSE)
> vert$V7 <- as.numeric(vert$V7) - 1a
# for reproducible results
> set.seed(9999)
```

---

[4]Here the notation $C(r, s)$ means the number of combinations one can form from r objects, taking them s at a time.

[5]To avoid clutter, some messages, "glm.fit: fitted probabilities numerically 0 or 1 occurred," have been removed, here and below. The warnings should not present a problem.

[6]Code for them is also shown in the Computational Complements section at the end of this chapter.

```
> trnidxs <- sample(1:310,225)
# args are m and (X,Y) data, Y last
> ovout <- ovalogtrn(3,vert[trnidxs,])
> predy <- ovalogpred(ovout,vert[-trnidxs,1:6])
> mean(predy == vert[-trnidxs,7])
[1] 0.8941176
> avout <- avalogtrn(3,vert[trnidxs,])
> predy <- avalogpred(3,avout,vert[-trnidxs,1:6])
> mean(predy == vert[-trnidxs,7])
[1] 0.8823529
```

Note that **ovalogpred()** requires that Y be coded $0, 1, ..., m-1$, hence the call to **as.numeric()**.

The two correct-classification rates here are, of course, subject to sampling error, but in any case AVA did not seem superior.

5.5.3 Intuition

To put this in context, consider the artificial example in Figure 5.1, adapted from Friedman [51]. Here we have $m = 3$ classes, with $p = 2$ predictors. For each class, the bulk of the distribution of the predictor vectors mass for that group is assumed to lie within one of the circles.

Now suppose a logistic model were used here. It implies that the prediction boundary between our two classes is linear (Section 4.3.7). The figure shows that a logit model would fare well under AVA, because for any pair of classes, there is a straight line (pictured) that separates that pair of classes well. But under OVA, we'd have a problem; though a straight line separates the top circle from the bottom two, there is no straight line that separates the bottom-left circle well from the other two very well; the boundary between that bottom-left circle and the other two would be a curve.

Keep that word *curve* in mind, as it will arise below. The problem here, of course is that the logit, at least in the form implied above, is not a good model in such a situation, so that OVA vs. AVA is not the real issue. In other words, if AVA does do better than OVA on some dataset, it may be due to AVA's helping us overcome model bias. We will explore this in the next section.

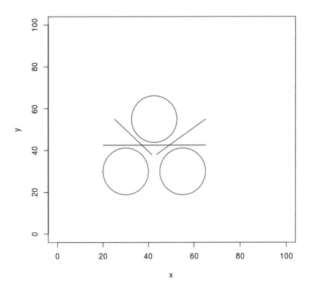

Figure 5.1: Three artificial regression lines

5.5.4   Example: Letter Recognition Data

Following up on the notion at the end of the last section that AVA may work to reduce model bias, i.e., that AVA's value occurs in settings in which our model is not very good, let's look at an example in which we know the model is imperfect.

The UCI Letters Recognition data set[7] uses various summaries of pixel patterns to classify images of capital English letters. A naively applied logistic model may sacrifice some accuracy here, due to the fact that the predictors do not necessarily have monotonic relations with the response variable, the class identity.

Actually, the naive approach doesn't do too poorly:

```
> library(mlbench)
> data(LetterRecognition)
> lr <- LetterRecognition
> lr[,1] <- as.numeric(lr[,1]) - 1
```

---

[7] *https://archive.ics.uci.edu/ml/datasets/Letter+Recognition*; also available in the R package **mlbench** [87].

```
> # training and test sets
> lrtrn <- lr[1:14000,]
> lrtest <- lr[14001:20000,]
> ologout <- ovalogtrn(26,lrtrn[,c(2:17,1)])
> ypred <- ovalogpred(ologout,lrtest[,-1])
> mean(ypred == lrtest[,1])
[1] 0.7193333
```

We will see shortly that one can do considerably better. But for now, we have a live candidate for a "poor model example," on which we can try AVA:

```
> alogout <- avalogtrn(26,lrtrn[,c(2:17,1)])
> ypred <- avalogpred(26,alogout,lrtest[,-1])
> mean(ypred == lrtest[,1])
[1] 0.8355
```

That is quite a difference! So, apparently AVA fixed a poor model. But of course, its better to make a good model in the first place. Based on our previous observation that the boundaries may be better approximated by curves than lines, let's try a quadratic model.

A full quad model will have all squares and interactions among the 16 predictors. But there are $16·15/2+16 = 136$ of them! That risks overfitting, so let's settle for just adding in the squares of the predictors:

```
> for (i in 2:17) lr <- cbind(lr,lr[,i]^2)
> lrtrn <- lr[1:14000,]
> lrtest <- lr[14001:20000,]
> ologout <- ovalogtrn(26,lrtrn[,c(2:33,1)])
> ypred <- ovalogpred(ologout,lrtest[,-1])
> mean(ypred == lrtest[,1])
[1] 0.8086667
```

Ah, much better. Not quite as good as AVA, but the difference is probably commensurate with sampling error, and we didn't even try interaction terms.

5.5.5   Example: k-NN on the Letter Recognition Data

As a basis of comparison to the above analyses of the letter recognition data, let's try k-NN. Note that with a nonparametric method such as k-NN, there is no geometric issue as with Figure 5.1. Also, for convenience we will not do cross-validation here.

Recall that with the **knnest()** function in **regtools**, we need a numeric Y. Our data here has Y as an R factor. So, we'll use the **dummies** package to conveniently generate the $m = 26$ dummy variables for Y.

```
> xd <- preprocessx ( lr [ , −1] ,50)
> library (dummies)
> y <- dummy( lr [ , 1])
> xd <- preprocessx ( lr [ , −1] ,50)
> kout <- knnest (y , xd ,50)
```

Now **kout$regest** has estimated class probabilities. To change those to predictions using (5.8), we'll use R's **apply()** (Section 1.20.2) and **which.max()** functions, and then convert back to characters via R's built-in **LETTERS** vector:

```
> tmp <- apply ( kout$regest ,1 , which.max)
> knnpred <- LETTERS[tmp]
```

How well did we do?

```
> mean( knnpred == lr$lettr )
[1]  0.9121
```

So it appears that neither OVA nor AVA solves the problems of a logit model here.

5.5.6   The Verdict

With proper choice of model, OVA may do as well as AVA, if not better. And a paper supporting OVA, [119] contends that some of the pro-AVA experiments in the research literature were not done properly.

Clearly, though, our letters recognition example shows that AVA is worth considering. We will return to this issue later.

5.6   Fisher Linear Discriminant Analysis

Sir Ronald Fisher (1890–1962) was one of the pioneers of statistics. He called his solution to the multiclass problem *linear discriminant analysis* (LDA), now considered a classic.

It is assumed that within class i, the vector of predictor variables X has a multivariate normal distribution with mean vector μ_i and covariance matrix

Σ (Section 2.6.2). Note that the latter does *not* have a subscript i, i.e., in LDA the covariance matrix for X is assumed the same within each class.

5.6.1 Background

To explain this method, let's review some material from Section 4.3.1.

Let's first temporarily go back to the two-class case, and use our past notation:

$$Y = Y^{(1)}, \quad \pi = \pi_1 \tag{5.10}$$

For convenience, let's reproduce (5.4) here:

$$P(Y = 1 \mid X = t) = \frac{\pi \ f_1(t)}{\pi \ f_1(t) + (1 - \pi) \ f_0(t)} \tag{5.11}$$

5.6.2 Derivation

As noted in Section 4.3.1, after substituting the multivariate normal density for the f_i in (5.11), we find that

$$P(Y = 1 \mid X = t) = \frac{1}{1 + e^{-(\beta_0 + \overline{\beta}' t)}} \tag{5.12}$$

with

$$\beta_0 = \log(1 - \pi) - \log \pi + \frac{1}{2}(\mu_1' \mu_1 - \mu_0' \mu_0) \tag{5.13}$$

and

$$\overline{\beta} = (\mu_0 - \mu_1)' \Sigma^{-1} \tag{5.14}$$

Intuitively, if we observe $X = t$, we should predict Y to be 1 if

$$P(Y = 1 \mid X = t) > 0.5 \tag{5.15}$$

and this was shown in Section 1.17.1 to be the optimal strategy.[8] Combining this with (5.12), we predict Y to be 1 if

$$\frac{1}{1 + e^{-(\beta_0 + \overline{\beta}'t)}} > 0.5 \qquad (5.16)$$

which simplifies to

$$\overline{\beta}'t > -\beta_0 \qquad (5.17)$$

So it turns out that our decision rule is linear in t, hence the term *linear* in *linear discriminant analysis*.[9]

Without the assumption of equal covariance matrices, (5.17) turns out to be quadratic in t, and is called *quadratic discriminant* analysis.

5.6.3   Example: Vertebrae Data

Let's apply this to the vertebrae data, which we analyzed in Section 5.5.2, now using the **lda()** function. The latter is in the **MASS** library that is built-in to R.

5.6.3.1   LDA Code and Results

The **lda()** function assumes that the class variable is an R factor, so we won't convert to numeric codes this time. Here is the code:

```
> library (MASS)
> vert <- read.table ('Vertebrae/column_3C.dat',
    header=FALSE)
> ldaout <- lda (V7 ~ ., data=vert, CV=TRUE)
> mean(ldaout$class == vert$V7)
[1]  0.8096774
```

That **CV** argument tells **lda()** to predict the classes after fitting the model, using (5.4) and the multivariate normal means and covariance matrix that is estimated from the data. Here we find a correct-classification rate of about 81%. This is biased upward, since we didn't bother here to set up

---

[8] Again assuming equal costs of the two types of misclassification.
[9] The word *discriminant* alludes to our trying to distinguish between $Y = 1$ and $Y = 0$.

separate training and test sets, but even then we did not do as well as our earlier logit analysis. Note that in the latter, we didn't assume a common covariance matrix within each class, and that may have made the difference. Of course, we could also try quadratic versions of LDA.

5.7 Multinomial Logistic Model

Within the logit realm, one might also consider *multinomial logistic regression*. This is similar to fitting m separate logit models, as we did in OVA above, with a somewhat different point of view, motivated by the *log-odds ratio* introduced in Section 4.3.3.

5.7.1 Model

The model now is to assume that the log-odds ratio for class i relative to class 0 has a linear form,

$$\log \frac{P(Y = i \mid X = t)}{P(Y = 0 \mid X = t)} = \log \frac{\gamma_i}{\gamma_0} = \beta_{0i} + \beta_{1i}t_1 + ... + \beta_{pi}, \quad i = 1, 2, ..., m - 1 \tag{5.18}$$

(Here i begins at 1 rather than 0, as each of the classes 1 through $m - 1$ is being compared to class 0.)

Note that this is not the same model as we used before, though rather similar in appearance.

The β_{ji} can be estimated via Maximum Likelihood, yielding

$$\log \frac{\widehat{\gamma}_i}{\widehat{\gamma}_0} = \widehat{\beta}_{0i} + \widehat{\beta}_{1i}t_1 + ... + \widehat{\beta}_{pi} \tag{5.19}$$

We then apply **exp()**, yielding the ratios $\widehat{\gamma}_i/\widehat{\gamma}_0$, after which the individual probabilities $\widehat{\gamma}_i$ can be solved algebraically, using the constraint

$$\sum_{i=0}^{m-1} \widehat{\gamma}_i = 1 \tag{5.20}$$

5.7.2   Software

There are several CRAN packages that implement this method. We'll use the **nnet** [121] package here. Though it is primarily for neural networks analysis, it does implement multinomial logit, and has the advantage that it is compatible with R's **stepAIC()** function, used in Chapter 9.

5.7.3   Example: Vertebrae Data

Let's try it out on the vertebrae data. Note that **multinom()** assumes that the response Y is an R factor.

```
> vert <- read.table('column_3C.dat',header=FALSE)
# vert$V7 left as a factor, not numeric
> library(nnet)
> mnout <- multinom(V7 ~ .,data=vert)
...
> cf <- coef(mnout)
> cf
     (Intercept)          V1           V2          V3
NO    -20.23244   -4.584673     4.485069  0.03527065
SL    -21.71458   16.597946   -16.609481  0.02184513
             V4          V5          V6
NO     4.737024  0.13049227  -0.005418782
SL   -16.390098  0.07798643   0.309521250
```

Since $m = 3$, we expect 3-1 = 2 sets of estimated coefficients, which indeed we have above. The class of the return value, **mnout** in our example here, is 'multinom'.

To illustrate how prediction of new cases would then work, let's predict an old case, **vert[1,]**. Remember, in R, many classes have methods for generic functions such as **print()** and **plot()**, and especially in the regression context, **predict()**. There is indeed a **"multinom"** method for the R generic **predict()** function:

```
> vt1 <- vert[1,-7]
> predict(mnout,vt1)
[1] DH
Levels: DH NO SL
```

But it is more informative to determine the estimated conditional class probabilities:

```
> vt1
      V1      V2      V3      V4      V5      V6
1  63.03   22.55   39.61   40.48   98.67   −0.25
> u <- exp(cf %*% c(1,as.numeric(vt1)))
> u
              [,1]
1  0.130388474
2  0.006231212
> c(1,u) / sum(c(1,u))
[1]  0.879801760  0.114716009  0.005482231
```

What happened here? Consider first the computation

cf %*% c(1,as.numeric(vt1))

(The call to **as.numeric()** is needed because **vt1** is an R data frame rather than a vector.)

According to our model (5.18), the above code computes the log-odds ratios. Applying **exp()** gives us the raw odds ratios. Then we use (5.20) to solve for the individual probabilities $P(Y = i|X = t)$, which are 0.879801760 and so on.

In this case, since we are just "predicting" our first observation, we need not go through all that trouble above:

```
> mnout$fitted.values[1,]
          0               1               2
0.879801760  0.114716009  0.005482231
```

But for truly new data, the above sequence of operations will give us the estimated class probabilities, which as mentioned are more informative than merely a predicted class.

5.8 The Issue of "Unbalanced" (and Balanced) Data

Here we will discuss a topic that is usually glossed over in treatments of the classification problem, and indeed is often misunderstood, with much questionable handwringing over "the problem of unbalanced data." This will be explained, and it will be seen below that often the real problem is that the data are *balanced*.

For concreteness and simplicity, consider the two-class problem of predicting whether a customer will purchase a certain item ($Y = 1$) or not ($Y = 0$), based on a single predictor variable, X, the customer's age. Suppose also that most of the customers are older.

5.8.1   Why the Concern Regarding Balance?

Though one typically is interested in the overall rate of incorrect classification, we may also wish to estimate rates of "false positives" and "false negatives," and to also gauge how well we can predict in certain subpopulations. In our customer purchasing example, for instance, we wish to ask, What percentage of the time do we predict that the customer does not purchase the item, among cases in which the purchase actually is made? And how well do we predict among the older customers? One problem is that, although our overall misclassification rate is low, we may do poorly on conditional error rates of this sort. This may occur, for example, if we have unbalanced data, as follows.

Suppose only 1.5% of the customers in the population opt to buy the product.[10] The concern among some machine learning researchers and practitioners is that, with random sampling (note the qualifier), the vast majority of the data in our sample will be from the class $Y = 0$, thus giving us "unbalanced" data. Then our statistical decision rule will likely just predict almost everything to be Class 0, and thus may not predict the other class well. Let's take a closer look at this.

For the time being, assume that we have a random sample from the overall population, rather than separate random samples from the two subclasses, as discussed below.

Say we are using a logit model for $\mu(t)$. If the model is accurate throughout the range of X, unbalanced data is not really a problem. The fact that one class, say $Y = 1$, occurs rarely will likely increase standard errors of the estimated regression coefficients, but it will not be a fundamental issue. We will still have *statistically consistent* estimators (Section 2.7.3) of the coefficients of β.

On the other hand, say we do classification using nonparametric density estimation (Section 5.10.1). Since even among older customers, rather few buy the product, we won't have much data from Class 1, so our estimate

---

[10]As mentioned earlier in the book, in some cases it may be difficult to define a target population, even conceptually. There is not much that can be done about this, unfortunately.

of $\widehat{f}_1$ probably won't be very accurate. Thus Equation (5.5) then suggests we have a problem. We still have statistically consistent estimation, but for finite samples that may not be enough. Nevertheless, short of using a parametric model, there really is no solution to this.

Ironically, a more pressing issue is that we may have data that is *too* balanced. Then we will not even have statistically consistent estimation. This is the subject of our next section.

5.8.2 A Crucial Sampling Issue

In this chapter, we have often dealt with expressions such as $P(Y = 1)$ and $P(Y = 1 \mid X = t)$. These seem straightforward, but actually they may be undefined, due to our sampling design, as we'll see here.

In our customer behavior context, $P(Y = 1)$ is the *un*conditional probability that a customer will buy the given item. If it is equal to 0.12, for example, that means that 12% of all customers purchase this item. By contrast, $P(Y = 1 \mid X = 38)$ is a *conditional* probability, and if it is equal to 0.18, this would mean that *among all people of age 38, 18% of them buy the item.*

The quantities $\pi = P(Y = 1)$ and $1 - \pi = P(Y = 0)$ play a crucial role, as can be seen immediately in (5.3) and (5.4). Let's take a closer look at this. Continuing our customer-age example, X (age) has a continuous distribution, so (5.4) applies. Actually, it will be more useful to look at the equivalent equation, (5.5).

5.8.2.1 It All Depends on How We Sample

Say our training data set consists of records on 1000 customers. Let N_1 and N_0 denote the number of people in our data set who did and did not purchase the item, with $N_1 + N_0 = 1000$. If our data set can be regarded as a statistical random sample from the population of all customers, then we can estimate π from the data. If for instance 141 of the customers in our sample purchased the item, then we would set

$$\widehat{\pi} = \frac{N_1}{1000} = 0.141 \tag{5.21}$$

The trouble is, though, that the expression $P(Y = 1)$ may not even make sense with some data. Consider two sampling plans that may have been followed by whoever assembled our data set.

(a) He sampled 1000 customers from our full customer database.[11]

(b) He opted to sample 500 customers from those who purchased the item, and 500 from those who did not buy it.

Say we are using the density estimation approach to estimate $P(Y \mid X = t)$, in (5.5). In sampling scheme (a), N_1 and N_0 are random variables, and as noted we can estimate π by the quantity $N_1/1000$. But in sampling scheme (b), we have no way to estimate π from our data.

Or, suppose we opt to use a logit model here. It turns out that we will run into similar trouble in sampling scheme (b), as follows. From (4.18) and (5.5), write the population relation

$$\beta_0 + \beta_1 t_1 + \ldots + \beta_p t_p = \ln(\pi/(1-\pi)) + \ln[f_1(t)/f_0(t)] \qquad (5.22)$$

where $t = (t_1, \ldots, t_p)'$. Since this must hold for all t, we see that $\beta_0 = \ln(\pi/(1-\pi))$. So if we follow design (b) above, our estimator, not knowing better, will assume (a), and estimate π to be 0.5. *However*, under design (b), $\beta_i, i > 0$ will not change, because the f_i are *within-class* densities, and their ratio will still be estimated properly. only β_0 changes. In other words, our logit-estimation software will produce the wrong constant term, but be all right on the other coefficients.

In summary:

Under sampling scheme (b), we are obtaining the wrong $\widehat{\beta}_0$, though the other $\widehat{\beta}_i$ are correct.

If our goal is merely Description rather than Prediction, this may not be a concern, since we are usually interested only in the values of $\beta_i, i > 0$. But if Prediction is our goal, as we are assuming in this chapter, we do have a serious problem, since we will need all of the estimated coefficients in order to estimate $P(Y|X = t)$ in (4.18).

A similar problem arises if we use the k-Nearest Neighbor method. Suppose for instance that the true value of π is low, say 0.06, i.e., only 6% of customers buy the product. Consider estimation of $P(Y \mid X = 38)$.

---

[11]Or, this was our entire customer database, which we are treating as a random sample from the population of all customers.

Under the k-NN approach, we would find the k closest observations in our sample data to 38, and estimate $P(Y \mid X = 38)$ to be the proportion of those neighboring observations in which the customer bought the product. The problem is that under sampling scenario (b), there will be many more among those neighbors who bought the product than there "should" be. Our analysis won't be valid.

So, all the focus on unbalanced data in the literature is arguably misplaced. As we saw in Section 5.8.1, it is not so much of an issue in the parametric case, and in any event there really isn't much we can do about it. At least, things do work out as the sample size grows. By contrast, with sampling scheme (b), we have a permanent bias, even as the sample size grows.

Scenario (b) is not uncommon. In the UCI Letters Recognition data set mentioned earlier for instance, there are between 700 and 800 cases for each English capital letter, which does not reflect that wide variation in letter frequencies. The letter 'E', for example, is more than 100 times as frequent as the letter 'Z', according to published data (see below).

Fortunately, there are remedies, as we will now see.

5.8.2.2   Remedies

As noted, use of "unnaturally balanced" data can seriously bias our classification process. In this section, we turn to remedies.

It is assumed here that we have an external data source for the class probabilities π_i. For instance, in the English letters example above, there is much published data, such as at the Web page Practical Cryptography.[12] It turns out that $\pi_A = 0.0855$, $\pi_B = 0.0160$, $\pi_C = 0.0316$ and so on.

So, if we do have external data on the π_i (or possibly want to make some "what if" speculations), how do we adjust our code output to correct the error?

For LDA, R's **lda()** function does the adjustment for us, using its **priors** argument. That code is based on the relation (4.31), which we now see is a special case of (5.22).

The latter equation shows how to deal with the logit case as well: We simply adjust the $\widehat{\beta}_0$ that **glm()** gives us as follows.

---

[12] *http://practicalcryptography.com/cryptanalysis/letter-frequencies-various-languages/english-letter-frequencies/.*

(a) Subtract $\ln(N_1/N_0)$.

(b) Add $\ln[\pi)/(1-\pi)$], where π is the true class probability.

Note that for an OVA m-class setting, we estimate m logistic regression functions, adjusting $\widehat{\beta}_0$ in each case. The function **ovalogtrn()** includes an option for this.

What about nonparametric settings? Equation (5.5) shows us how to make the necessary adjustment, as follows:

(a) Our software has given us an estimate of the left-hand side of that equation for any t.

(b) We know the value that our software has used for its estimate of $(1-\pi)/\pi$, which is N_0/N_1.

(c) Using (a) and (b), we can solve for the estimate of $f_1(t)/f_0(t)$.

(d) Now plug the correct estimate of $(1-\pi)/\pi$, and the result of (c), back into (5.5) to get the proper estimate of the desired conditional probability.

Code for this is straightforward:

```
classadjust <- function(econdprobs, wrongratio,
       trueratio) {
   fratios <- (1 / econdprobs - 1) * (1 / wrongratio)
   1 / (1 + trueratio * fratios)
}
```

Note that if we are taking the approach described in the paragraph labeled "A variation" in Section 1.10.2, we do this adjustment only at the stage in which we fit the training data. No further adjustment at the prediction stage is needed.

5.8.3   Example: Letter Recognition

Let's try the k-NN analysis on the letter data. First, some data prep:

```
> library(mlbench)
> data(LetterRecognition)
> lr <- LetterRecognition
```

```
> # code Y values
> lr[,1] <- as.numeric(lr[,1]) - 1
> # training and test sets
> lrtrn <- lr[1:14000,]
> lrtest <- lr[14001:20000,]
```

As discussed earlier, this data set has approximately equal frequencies for all the letters, which is unrealistic. The **regtools** package contains the correct frequencies [97], obtained from the Practical Cryptography Web site cited before. Let's load those in:

We continue our analysis from Section 5.5.4.

```
> library(mlbench)
> data(LetterRecognition)
> library(regtools)
> tmp <- table(LetterRecognition[,1])
> wrongpriors <- tmp / sum(tmp)
> data(ltrfreqs)
> ltrfreqs <- ltrfreqs[order(ltrfreqs[,1]),]
> truepriors <- ltrfreqs[,2] / 100
```

(Recall from Footnote 2 that the term *priors* refers to class probabilities, and that word is used both by frequentists and Bayesians. It is not "Bayesian" in the sense of subjective probability.)

So, here is the straightforward analysis, taking the letter frequencies as they are, with 50 neighbors:

```
> xdata <- preprocessx(lrtrn[,-1],50)
> trnout <- knntrn(lrtrn[,1],xdata,26,50)
> tmp <- predict(trnout,lrtest[,-1])
> ypred <- apply(as.matrix(tmp),1,which.max) -
> mean(ypred == lrtest[,1])
[1] 0.8641667
```

In light of the fact that we have 26 classes, 86% accuracy is pretty good. But it's misleading: We did take the trouble of separating into training and test sets, but as mentioned, the letter frequencies are unrealistic. How well would our classifier do in the "real world"? To simulate that, let's create a second test set with correct letter frequencies:

```
> newidxs <-
      sample(0:25,6000,replace=T,prob=truepriors)
> lrtest1 <- lrtest[newidxs,]
```

Now we can try our classifier on this more realistic data:

```
> ypred <- knnpred(trnout, lrtest1[,-1])
> mean(ypred == lrtest1[,1])
[1] 0.7543415
```

Only about 75%. But in order to prepare for the real world, we can make use of the **truepriors** argument in **knntrn()**

```
> trnout1 <- knntrn(lrtrn[,1], xdata, 26, 50, truepriors)
> ypred <- predict(trnout1, lrtest1[,-1])
> mean(ypred == lrtest1[,1])
[1] 0.8787988
```

Ah, very nice!

5.9   Going Beyond Using the 0.5 Threshold

As we have seen, the optimal rule (knowing population functions) in the two-class case is given by (5.9). (That formulation assumes that population quantities are known; in practice, of course, we must use estimates, i.e., $\hat{\mu}(t)$ rather than $\mu(t)$.) But here "optimal" referred to minimizing overall misclassification rate, and more detailed analysis may be appropriate.

5.9.1   Unequal Misclassification Costs

Say we are trying to determine whether a patient has a particular disease, based on a vector X of various test results, demographic variables and so on for this patient. Denote the value of X by t_c, and suppose our estimate of $P(Y^{(1)} = 1 \mid X = t_c)$ is 0.02. We estimate that this patient has only a 2% chance of having the disease. This isn't very high, so we might simply stop there.

But in the case of a catastrophic disease, the *misclassification costs* may not be equal. Failing to detect the disease when it's present may be a much more serious error than ordering further medical tests that turn out to be negative. What can be done to address this?

Informal approach:

The physician may have a hunch, based on information not in X and thus not in our sample data, that leads her to suspect that the patient does have the disease. The physician may thus order further tests and so on, in spite of the low estimated probability.

Remember, our estimated $P(Y^{(i)} = 1 \mid X = t_c)$ can be used as just one of several components that may enter into our final decision.

Automatic classification:

In many applications today, our classification process will be automated, done entirely by machine. Consider the example in Section 4.3.1 of classifying subject matter of Twitter tweets, say into financial tweets and all others, a two-class setting. Here again there may be unequal misclassification costs, depending on our goals. If so, the prescription (5.9), i.e.,

$$\text{guess for } Y = \begin{cases} 1, & \text{if } \mu(X) > 0.5 \\ 0, & \text{if } \mu(X) \leq 0.5 \end{cases} \tag{5.23}$$

is not what we want, as it implicitly assumes equal costs. If we wish to automate, we'll probably need to set up a formal cost structure. This will result in the 0.5 threshold above changing to something else. We will use this idea the next section.

5.9.2 Revisiting the Problem of Unbalanced Data

In Section 5.8, it was argued that if our goal is to minimize the overall misclassification rate, the problem of "unbalanced" data is not really a problem in the first place (or if it is a problem, it's insoluble). But as pointed out in Section 5.9.1, we may be more interested in correct prediction for some classes than others, so the overall misclassification rate is not our primary interest.

In the Mathematical Complements section at the end of this chapter, a simple argument shows that we should guess $Y = 1$ if

$$\mu(X) \geq \frac{\ell_0}{\ell_0 + \ell_1} = \frac{1}{1 + l_1/l_0} \tag{5.24}$$

where the l_i are our misclassification costs ("losses"). All that matters is their ratio. For instance, say we consider guessing $Y = 1$ when in fact $Y = 0$ (cost l_0) to be 3 times worse an error than guessing $Y = 0$ when actually $Y = 1$ (cost l_1). Then our new threshold is

$$\frac{1}{1 + \frac{1}{3}} = 0.75 \tag{5.25}$$

So, we set our threshold at 0.75 rather than 0.5. This more stringent criterion for guessing $Y = 1$ means we take such action less often, thus addressing our concern that a false positive is a very serious error.

5.9.3   The Confusion Matrix and the ROC Curve

In an m-class setting, the *confusion matrix* is defined as follows: For $0 \leq i, j \leq m - 1$, the $(i + 1, j + 1)$ element is the number of cases in which the actual class was i and we predicted it to be j. This tells us at a glance on which classes we are predicting well or poorly.

Now consider what would happen, say in the case $m = 2$ for simplicity, if we were to try various values for the threshold in (5.9) instead of just 0.5. Let's call the threshold h, meaning that we guess $Y = 1$ if $\mu(X) > h$.

As we vary h, the confusion matrix would change. As we reduce h starting from near 1.0, we would have more true positive guesses ($Y = \widehat{Y} = 1$) but the number of false positives ($Y = 0$, $\widehat{Y} = 1$) would increae.

The *Receiver Operating Characteristic* (ROC) curve plots the rate of true positives (TPR) against the proportion of false positives (FPR), as h varies. Since we may not have specific misclassification costs in mind, the ROC curve allows us to explore the effects of using different values of h.

Specifically, we plot estimates of

$$P(\text{guess } Y = 1 \text{ when } Y = 1) = P(\mu(X) > h \mid Y = 1) \tag{5.26}$$

versus

$$P(\text{guess } Y = 1 \text{ when } Y = 0) = P(\mu(X) > h \mid Y = 0) \tag{5.27}$$

Look at the disease diagnosis case, for example, where having the disease is coded $Y = 1$. Then TPR is the proportion of time we would guess that the patient has the disease, among those patients who actually have it, versus incorrectly guessing presence of the disease among those who don't have it.

5.9.3.1 Code

The code for plotting ROC is presented in the Computational Complements section at the end of this chapter. It is written from the point of view of clarity of the ROC process, rather than efficiency.

Its call form is simple:

```
roc(x,y,regest,nh)
```

Here **x** is the matrix/data frame of X values, **y** is the vector of Y values, and **regest** is the vector of $\hat{\mu}(X_i)$ values. The optional argument nh is the number of h values to plot, evenly spaced in (0,1). Note that **regest** could come from a logistic model, k-NN analysis and so on.

5.9.3.2 Example: Spam Data

Let's continue the computation we made on the spam data in Section 4.3.6. The object **glmout** was the output from fitting a logit model. The Y variable was in column 58 of the data frame, and all the other columns were used as predictors. So, our call is

```
> roc(spam[,-58],spam[,58],glmout$fitted.values)
```

The plot is shown in Figure 5.2. Note that the quantity h does not appear explicitly. Instead, it is used to generate the (FPR,TPR) pairs, one pair per value of h.

The curve in this case rises steeply for small values of FPR. In general, the steeper the better, because it indicates that we can obtain a good TPR rate at a small "price," i.e., by tolerating just a small amount of false positives. In this case, this does occur, not surprising in light of the high rate of correct classification we found in our earlier analysis of this data.

5.10 Mathematical Complements

5.10.1 Classification via Density Estimation

Since classification amounts to a regression problem, we can use nonparametric regression methods such as k-Nearest Neighbor if we desire a nonparametric approach, as seen above. However, Equations (5.4) and (5.7)

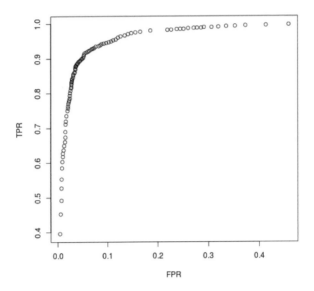

Figure 5.2: ROC curve for spam data

suggest that one approach to the classification problem would be to estimate the within-class densities f_i. Actually, this approach is not commonly used, as it is difficult to get good estimates, especially if the number of predictors is large. However, we will examine it in this section anyway, as it will yield some useful insights.

5.10.1.1   Methods for Density Estimation

Say for simplicity that X is one-dimensional. You are already familiar with one famous nonparametric method for density estimation — the histogram![13] What is more commonly done, though is a variant of that.

Say we are estimating $f()$, the density of X, with corresponding cdf $F(t)$. From introductory calculus we have

$$f(t) \approx \frac{F(t+h) - F(t-h)}{2h} \tag{5.28}$$

---

[13]However, since a density integrates to 1.0, we should scale our histogram accordingly. In R's **hist()** function, we specify this by setting the argument **freq** to FALSE.

for small $h > 0$. Since we can estimate the cdf directly from the data, our estimate is

$$
\begin{aligned}
\widehat{f}(t) &= \frac{\#(t - h, t + h)/n}{2h} & (5.29) \\
&= \frac{\#(t - h, t + h)}{2hn} & (5.30)
\end{aligned}
$$

where # stands for the number of X_i in the given interval and n is the sample size.

For k-NN, with k neighbors, do the following. In the denominator of (5.29), set h equal to the distance from t to the furthest neighbor, and set the numerator to k.

The above is for the case $p = 1$. As noted, things become difficult in higher dimensions, and none of the R packages go past $p = 3$, due to high variances in the estimated values [40].

For an in-depth treatment of density estimation, see [127].

5.10.2 Time Complexity Comparison, OVA vs. AVA

The running time of an algorithm is typically described in terms of "big-O" notation. Consider finding the inverse of a $k \times k$ matrix, for instance. It turns out that this takes about ck^3 time steps, where c is some constant. We usually don't care about the specific value of c, because we are concerned mainly with growth rates. How does the running time grow as k increases? If for exanple the number of rows and columns k is doubled, we see that the time to invert the matrix will increase by a factor of 8, regardless of the value of c.

Let's apply that to the OVA/AVA situation. For fixed number of predictor variables p, here is a rough time estimate. For a logit model, the computation will be proportional to the number of cases n (due to computing various sums over all cases). Say our training data is approximately balanced in terms of sizes of the classes, so that the data corresponding to class i has about n/m cases in it.

Then the computation for one pair of classes in AVA will take $O(n/m)$ time, but there will be $O(m^2)$ pairs, so the total amount of computation will be $O(m^2 \times n/m) = O(mn)$. This is potentially much more than the corresponding $O(n)$ time complexity for OVA.

However, so far we haven't factored p into our analysis. If we do this, then the results become very dependent on the type of classifier we are using — logit, k-NN, SVM (Chapter 11 and so on. Such analysis becomes quite complex, but it can certainly be the case that for some types of estimators the speed advantage of OVA over AVA is not as great.

5.10.3   Optimal Classification Rule for Unequal Error Costs

Let ℓ_0 denote our cost for guessing Y to be 1 when it's actually 0, and define ℓ_1 for the opposite kind of error. Now reason as follows as to what we should guess for Y, knowing that $X = t_c$. For convenience, write

$$p = P(Y = 1 \mid X = t_c) \tag{5.31}$$

Suppose we guess Y to be 1. Then our expected cost is

$$(1 - p)\ell_0 \tag{5.32}$$

If on the other hand we guess Y to be 0, our expected cost is

$$p\ell_1 \tag{5.33}$$

So, our strategy could be to choose our guess to be the one that gives us the smaller of (5.32) and (5.33):

$$\text{guess for } Y = \begin{cases} 1, & \text{if } (1 - p)\ell_0 \le p\ell_1 \\ 0, & \text{if } (1 - p)\ell_0 > p\ell_1 \end{cases} \tag{5.34}$$

Solving for $p = \mu(t_c)$, we have

$$\text{guess for } Y = \begin{cases} 1, & \text{if } (1 - p)\ell_0 \le p\ell_1 \\ 0, & \text{if } (1 - p)\ell_0 > p\ell_1 \end{cases} \tag{5.35}$$

In other words, given X, we guess Y to be 1 if

$$\mu(X) \ge \frac{\ell_0}{\ell_0 + \ell_1} \tag{5.36}$$

5.11 Computational Complements

5.11.1 R Code for OVA and AVA Logit Analysis

To make this concrete, here is code for the two approaches:

```
# One-vs.-All (OVA) and All-vs.All (AVA),
# logit models

# arguments:

#    m:     number of classes
#    trnxy:  X, Y training set; Y in last column;
#            Y coded 0,1,...,m-1 for the m classes
#    predx:  X values from which to predict Y values
#    tstxy:  X, Y test set, same format


##################################################################
# ovalogtrn: generate estimated regression functions
##################################################################

# arguments:

#    m:  as above
#    trnxy:  as above

# value:

#    matrix of the betahat vectors, one per column

ovalogtrn <- function(m, trnxy) {
   p <- ncol(trnxy)
   x <- as.matrix(trnxy[,1:(p-1)])
   y <- trnxy[,p]
   outmat <- NULL
   for (i in 0:(m-1)) {
      ym <- as.integer(y == i)
      betahat <- coef(glm(ym ~ x, family=binomial))
      outmat <- cbind(outmat, betahat)
   }
   outmat
}
```

```
############################################################
# ovalogpred: predict Ys from new Xs
############################################################

# arguments:
#
#    coefmat:   coef. matrix, output from ovalogtrn()
#    predx:   as above
#
# value:
#
#    vector of predicted Y values, in {0,1,...,m-1},
#    one element for each row of predx

ovalogpred <- function(coefmat, predx) {
   # get est reg ftn values for each row of predx
   # and each col of coefmat; vals from
   # coefmat[,] in tmp[,i]
   tmp <- as.matrix(cbind(1, predx)) %*% coefmat
   tmp <- logit(tmp)
   apply(tmp, 1, which.max) - 1
}


############################################################
# avalogtrn: generate estimated regression functions
############################################################

# arguments:

#    m:   as above
#    trnxy:   as above

# value:

#    matrix of the betahat vectors, one per column,
#    in the order of combin()

avalogtrn <- function(m, trnxy) {
   p <- ncol(trnxy)
   n <- nrow(trnxy)
   x <- as.matrix(trnxy[,1:(p-1)])
   y <- trnxy[,p]
   outmat <- NULL
```

```
    ijs <- combn(m, 2)
    doreg <- function(ij) {
        i <- ij[1] - 1
        j <- ij[2] - 1
        tmp <- rep(-1,n)
        tmp[y == i] <- 1
        tmp[y == j] <- 0
        yij <- tmp[tmp != -1]
        xij <- x[tmp != -1,]
        coef(glm(yij ~ xij, family=binomial))
    }
    coefmat <- NULL
    for (k in 1:ncol(ijs)) {
        coefmat <- cbind(coefmat, doreg(ijs[,k]))
    }
    coefmat
}


##################################################
# avalogpred: predict Ys from new Xs
##################################################

# arguments:
#
#    m: as above
#    coefmat: coef. matrix, output from avalogtrn()
#    predx: as above
#
# value:
#
#    vector of predicted Y values, in {0,1,...,m-1},
#    one element for each row of predx

avalogpred <- function(m, coefmat, predx) {
    ijs <- combn(m, 2)   # as in avalogtrn()
    n <- nrow(predx)
    ypred <- vector(length = n)
    for (r in 1:n) {
        # predict the rth new observation
        xrow <- c(1, unlist(predx[r,]))
        # wins[i] tells how many times class i-1
        # has won
        wins <- rep(0,m)
```

```
    for (k in 1:ncol(ijs)) {
        i <- ijs[1,k]    # class i-1
        j <- ijs[2,k]    # class j-1
        bhat <- coefmat[,k]
        mhat <- logit(bhat %*% xrow)
        if (mhat >= 0.5) wins[i] <- wins[i] + 1 else
        wins[j] <- wins[j] + 1
    }
    ypred[r] <- which.max(wins) - 1
  }
  ypred
}

logit <- function(t) 1 / (1+exp(-t))
```

For instance, under OVA, we call **ovalogtrn()** on our training data, yielding a logit coefficient matrix having m columns; the ith column will consist of the estimated coefficients from fitting a logit model predicting $Y^{(i)}$. We then use this matrix as input for predicting Y in all future cases that come our way, by calling **ovalogpred()** whenever we need to do a prediction.

Under AVA, we do the same thing, calling **avalogtrn()** and **avalogpred()**.

5.11.2   ROC Code

Here is the code for ROC plotting:

```
# simple implementation of ROC, meant to show the
# principles rather than be efficient

# arguments:

#     x: matrix/data frame of X values
#     y: vector of Y values (0 or 1)
#     regest: vector of estimated regression function
#              values;  the fitted.values component
#              from glm() and logit
#     nh: number of values of threshold to plot

roc <- function(x,y,regest,nh=100) {
    # find the indices of the
    # Y = 0 and Y = 1 cases
    y0idxs <- which(y == 0)
```

```
ylidxs <- which(y == 1)
# and the estimated values of P(Y = 1 | X)
# for those cases
regest0 <- regest[y0idxs]
regest1 <- regest[y1idxs]
# try various threshold values h
increm <- 1/nh
h <- (1:(nh-1)) / nh
# set vectors for the FPR, TPR values
fprvals <- vector(length = nh-1)
tprvals <- vector(length = nh-1)
# for each possible threshold, find FPR, TPR
for (i in 1:(nh-1)) {
    fprvals[i] <- mean(regest0 > h[i])
    tprvals[i] <- mean(regest1 > h[i])
}
plot(fprvals, tprvals, xlab='FPR', ylab='TPR')
}
```

5.12 Exercises: Data, Code and Math Problems

Data problems:

1. Plot ROC curves for each of the three classes in the Vertebral Column data analyzed in this chapter. Try both logistic and k-NN approaches. (Also see Exercise 4 below.)

2. Consider the OVA vs. AVA comparison, with cross-validation, in Section 5.5.4. Re-run the analysis, recording run time (**system.time()**).

3. Try multinomial logit on the Letter Recognition data, comparing it to the results in Section 5.5.4. Then re-run it after adding squared versions of the predictors, as was also done in that section.

Mini-CRAN and other computational problems:

4. Write a function with call form

```
multiroc(x,y,regestmat,nh=100)
```

to plot multiple ROC curves. Here **x** and **nh** are as in **roc()** (Section **roccode**); **regestmat** is a matrix version of **regest** in **roc()**; and **y** is the class data. The latter could either be an integer taking values in $0, 1, ..., m-1$ or an R factor.

As to the arrangement of these multiple plots, you might show them one at a time, as with **parvsnonparplot()** in Chapter 6, or a grid of plots, say by setting the **mfrow** argument in R's **par()** function.

5. Here we will consider the "OVA and AVA" approaches to the multiclass problem (Section 5.5), using the UCBAdmissions data set that is built in to R. This data set comes in the form of a counts table, which can be viewed in proportion terms via

UCBAdmissions / **sum**(UCBAdmissions)

For the sake of this experiment, let's take those cell proportions to be population values, so that for instance 7.776% of all applicants are male, apply to departmental program F and are rejected. The accuracy of our classification process then is not subject to the issue of variance of estimators of logistic regression coefficients or the like.

(a) Which would work better in this population, OVA or AVA, say in terms of overall misclassification rate?

[Computational hint: First convert the table to an artificial data frame:

ucbd <− **as**.**data**.**frame**(UCBAdmissions)

]

(b) Write a general function

ovaavatbl <− **function**(tbl , yname)

that will perform the computation in part (a) for any table **tbl**, with the class variable having the name **yname**, returning the two misclassification rates. Note that the name can be retrieved via

names(**attr**(tbl , 'dimnames'))

6. Write a function to compute the confusion matrix for the output of **multinom()**, Section 5.7. The call form will be

nmconfmat(mnobj)

where **mnobj** is an object of class **'multinom'**, which is the type returned by **multinom()**. The function will return a matrix, defined as follows.

For $0 \leq i, j \leq m - 1$, the $(i + 1, j + 1)$ element of the matrix will be the proportion of cases in which the actual class is i and the predicted class is j.

Math problems:

7. Extend (5.29) to the case $p = 2$.

8. Section 5.9.2 showed how to handle the problem of unequal costs for the two-class case. Show how to extend this to m classes.

Chapter 6

Model Fit Assessment and Improvement

The famous Box quote from our first chapter is well worth repeating:

> *All models are wrong, but some are useful* — famed statistician George Box

We have quite a bit of powerful machinery to fit parametric models. But are they any good on a given data set? We'll discuss this subject here in this chapter.

6.1 Aims of This Chapter

Most regression books have a chapter on *diagnostics*, methods for assessing model fit and checking assumptions needed for statistical inference (confidence intervals and significance tests).

In this chapter we are concerned only with the model itself. For instance, in a model that is linear in the predictor variables, how accurate is that linearity assumption? Are there extreme or erroneous observations that mar the fit of our model?

We are *not* concerned here with assumptions that only affect inference, as

215

those were treated in Chapter 2.[1]

6.2 Methods

There is a plethora of diagnostic methods! Entire books could and have been written on this topic. Here I will treat only a few such methods here — some classical, some of my own — with the choice of methods presented stemming from these considerations:

- This book generally avoids statistical methods that rely on assuming that our sample data is drawn from a normally distributed population.[2] Accordingly, the material here on unusual observations does not make such an assumption.

- Intuitive clarity of a method is paramount. If a method can't be explained well to, say, a reasonably numerate but nonstatistician client, then I prefer to avoid it.

6.3 Notation

As before, say we have data (X_i, Y_i), $i = 1, ..., n$. Here the X_i are p-component vectors,

$$X_i = (X_i^{(1)}, ..., X_i^{(p)})' \tag{6.1}$$

and the Y_i are scalars (including the case $Y = 0, 1, ..., m-1$ in classification applications). We typically won't worry too much in this chapter whether the n observations are independent. As usual, let

$$\mu(t) = E(Y \mid X = t) \tag{6.2}$$

be our population regression function, and let $\widehat{\mu}(t)$ denote its estimate from our sample data.

---

[1] That chapter showed that the assumption of normal distributions is not very important, and Chapter 3 presented methods for dealing with nonhomogeneous variance. The third assumption, statistical independence, was not covered there, and indeed will not be covered elsewhere in the book, in the sense of methods for assessing independence; there are not many such methods, and typically they depend on their own assumptions, thus "back to Square One."

[2] "Rely on" here means that the method is not robust to the normality assumption.

6.4 Goals of Model Fit-Checking

What do we really mean when we ask whether a model fits a data set well? Our answer ought to be as follows:

> **Possible Fit-Checking Goal:**
>
> Our model fits well if $\widehat{\mu}(t)$ is near $\mu(t)$ for all t.

That criterion is of course only conceptual; we don't know the values of $\mu(t)$, so it's an impossible criterion to truly verify. Nevertheless, it may serve well as a goal, and our various model-checking methods will be aimed at that goal.

Note how the issue of overfitting comes in here. The above Goal could have said, "...if $\widehat{\mu}(t)$ is near $\mu(t)$ for $t = X_i$, $i = 1, 2, ..., n$." As noted before, with $p = 1$, for example, we could fit a polynomial model of degree $n - 1$ and have a "perfect" fit, obviously misleading.

Part of the answer to our goals question goes back to the twin regression goals of Prediction and Description. We'll explore this in the following sections.

6.4.1 Prediction Context

If our regression goal is Prediction and we are doing classification, our above Fit-Checking Goal may be much too stringent. Say for example $m = 2$, just two classes, 0 and 1. Let $(Y_{new}, X'_{new})'$ denote our new observation, with X_{new} known but Y unknown and to be predicted. We will guess $Y = 1$ if $\widehat{\mu}(X_{new}) > 0.5$.

If $\mu(X_{new})$ is 0.9 but $\widehat{\mu}(X_{new}) = 0.62$, we will still make the correct guess, $Y = 1$, even though our regression function estimate is well off the mark. Similarly, if $\mu(X_{new})$ is near 0 (or less than 0.5, actually), we will make the proper guess for Y as long as our estimated value $\widehat{\mu}(X_{new})$ is under 0.5.

Still, other than the classification case, the above Fit-Checking Goal is appropriate. Errors in our estimate of the population regression function will impact our ability to predict.

6.4.2 Description Context

Good model fit is especially important when our regression goal is Description. We really want assurance that the estimated regression coefficients represent the true regression function well. since we will be using them to describe the underlying process.

6.4.3 Center vs. Fringes of the Data Set

Consider a Prediction setting, in the classification case. In Section 6.4.1 above, we saw that we actually can afford a rather poor model fit in regions of the predictor space in which the population regression function is near 1 or 0.

The same reasoning shows, though, that having a good fit in regions where $\mu(t)$ is mid-range, say in $(0.25, 0.75)$ *is* important. If $\widehat{\mu}(t)$ and $\mu(t)$ are on opposite sides of the number 0.5 (i.e., one below 0.5 and the other above), we will make the wrong decision even, though we may still be lucky and guess Y correctly. Say for instance $\mu(t) = 0.6$. The correct decision would be to guess $Y = 1$, but if $\widehat{\mu}(t) = 0.42$, our decision will be to guess $Y = 0$. However, our guess could turn out to be correct anyway.

In classification contexts, the p-variate density of X is often "mound-shaped," if not bell-shaped, within each class. (In fact, many clustering algorithms are aimed at this situation.) For such data, the regions of most sensitivity in the above sense will be the areas near the lines/curves separating the pairs of mounds. (Recall Figure 5.1.) The fringes of the data set, far from these pairwise boundaries, will be regions in which model fit is less important, again assuming a Prediction goal.

In regression contexts (continous Y, count data etc.), the full data set will tend to be mound-shaped. Here good estimation will be important for Prediction and Description throughout the entire region. However, one must keep in mind that model fit will typically be better near the center of the data than at the fringes, i.e., $\widehat{\mu}(t)$ may be approximately linear near the center but depart considerably from linear on the fringes. Moreover, the observations at the fringes typically have the heaviest impact on the estimated coefficients. This latter consideration is of course of great import in the Description case.

We will return to these considerations at various points in this chapter.

6.5   Example: Currency Data

Fong and Ouliaris [48] did an analysis of relations between currency rates for Canada, Germany, France, the UK and Japan. This was in pre-European Union days, with the currency names being the Canadian *dollar*, the German *mark*, the French *franc*, the British *pound* and the Japanese *yen*. The *mark* and *franc* are gone today, of course.

An example question of interest is do the currencies move up and down together? We will assess this by predicting the Japanese *yen* from the others.

The data can be downloaded at *http://qed.econ.queensu.ca/jae/1995-v10.3/ fong-ouliaris/*. The data set does require some wrangling, which we show in Section 6.15.1 in the Computational Complements material at the end of this chapter. In the end, we have a data frame **curr**. Here is the top part of the data frame:

```
> head(curr)
   Canada  Mark  Franc   Pound    Yen
1  0.9770  2.575  4.763  0.41997  301.5
2  0.9768  2.615  4.818  0.42400  302.4
3  0.9776  2.630  4.806  0.42976  303.2
4  0.9882  2.663  4.825  0.43241  301.9
5  0.9864  2.654  4.796  0.43185  302.7
6  0.9876  2.663  4.818  0.43163  302.5
```

This is time series data, and the authors of the above paper do a very sophisticated analysis along those lines. So, the data points, such as for the *pound*, are not independent through time. But since we are just using the data as an example and won't be doing inference (confidence intervals and significance tests), we will not worry about that here.

Let's start with a straightforward linear model:

```
> fout <- lm(Yen ~ ., data=curl)
> summary(fout)
...
Coefficients:
              Estimate  Std. Error  t value  Pr(>|t|)
(Intercept)   102.855     14.663     7.015   5.12e-12  ***
Can           -45.941     11.979    -3.835   0.000136  ***
Mark          147.328      3.325    44.313   < 2e-16   ***
Franc         -21.790      1.463   -14.893   < 2e-16   ***
Pound         -48.771     14.553    -3.351   0.000844  ***
```

...

Multiple **R**-squared: 0.8923, Adjusted **R**-squared: 0.8918

Not surprisingly, this model works well, with an adjusted R-squared value of about 0.89. The signs of the coefficients are interesting, with the *yen* seeming to fluctuate opposite to all of the other currencies except for the German *mark*. Of course, professional financial analysts (*domain experts*, in the data science vernacular) should be consulted as to the reasons for such relations, but here we will proceed without such knowledge.

It may be helpful to scale our data so as to better understand the roles of the predictors, though, so as to make all the predictors commensurate (Section 1.21). Each predictor will be divided by its standard deviation (and have its mean subtracted off first), so all the predictors have standard deviation 1.0:

```
> curr1 <- as.matrix(curr)   # to enable scale()
> curr1[,-5] <- scale(curr1[,-5])
> fout1 <- lm(Yen ~ .,data=curr1)
> summary(fout1)
```
...

Coefficients:

	Estimate	Std. Error	t value	Pr(>|t|)	
(Intercept)	224.9451	0.6197	362.999	< 2e−16	***
Can	−5.6151	1.4641	−3.835	0.000136	***
Mark	57.8886	1.3064	44.313	< 2e−16	***
Franc	−34.7027	2.3301	−14.893	< 2e−16	***
Pound	−5.3316	1.5909	−3.351	0.000844	***

...

So the German and French currencies appear to have the strongest relation to the *yen*. This, and the signs (positive for the *mark*, negative for the *franc*), form a good example of the use of regression analysis for the Description goal.

In the next few sections, we'll use this example to illustrate the basic concepts.

6.6 Overall Measures of Model Fit

We'll look at two broad categories of fit assessment methods. The first will consist of overall measures, while the second will involve relating fit to individual predictor variables.

6.6.1 R-Squared, Revisited

We have already seen one overall measure of model fit, the R-squared value (Section 2.9). As noted before, its cousin, Adjusted R-squared, is considered more useful, as it is aimed at compensating for overfitting.

For the currency data above, the two R-squared values (ordinary and adjusted) were 0.8923 and 0.8918, both rather high. Note that they didn't differ much from each other, as there were well over 700 observations, which should easily handle a model with only 4 predictors (a topic we'll discuss in Chapter 9).

Recall that R-squared, whether a population value or the sample estimate reported by **lm()**, is the squared correlation between Y and its predicted value $\mu(X)$ or $\widehat{\mu}(X)$, respectively. Thus it can be calculated for any method of regression function estimation, not just the linear model. In particular, we can apply the concept to k-Nearest Neighbor methods.

The point of doing this with k-NN is that the latter in principle does not have model-fit issues. Whereas our linear model for the currency data assumes a linear relationship between the *yen* and the other currencies, k-NN makes no assumptions on the form of the regression function. If k-NN were to have a substantially larger R-squared value than that of our linear model, then we may be "leaving money on the table," i.e., not fitting as well as we could with a more sophisticated parametric model.[3]

This indeed seems to be the case:[4]

```
> library(regtools)
> curr2 <- curr1[-762,]
> xdata <- preprocessx(curr2[,-5],25,xval=TRUE)
> kout <- knnest(curr2[,5],xdata,25)
> cor(kout$regest,curr2[,5])^2
          [,1]
[1,]  0.9817131
```

This would indicate that, in spite of a seemingly good fit for our linear model, it does not adequately describe the currency fluctuation process.

So, our linear model, which seemed so nice at first, is missing something. Maybe we can determine why via the methods in the sections below. But

---

[3]We could of course simply use k-NN in the first place. But this would not give us the Description usefulness of the parametric model, and also would give us a higher estimation variance, since the parametric model is pretty good. See Section 1.7.

[4]As noted in the data wrangling, the last row has some NA values, so we will omit it.

first, another measure of fit:

6.6.2 Cross-Validation, Revisited

As discussed before, this involves dividing our n data points into subsets of r and $n - r$ points. We do our data fitting on the first partition, then use the results to predict the second partition. The motivation is to solve the bias problems cited above.

One common variant is *m-fold cross validation*, where we divide the data into m equal-sized subsets, and take $r = n - n/m$. We then perform the above procedure m times. With $m = n$, we have the LOOM technique (Section 2.9.5). This gives us more accuracy than using just one partitioning, at the expense of needing much more computation. When we look at only one partitioning, cross-validation is sometimes called the *holdout* method, as we are "holding out" $n - r$ data points from our fit.

In many discussions of fitting regression models, cross-validation is presented as a panacea. This is certainly not the case, however, and the reader is advised to use it with caution. We will discuss this further in Section 9.3.2.

6.6.3 Plotting Parametric Fit Against a Nonparametric One

Let's plot the $\widehat{\mu}$ values of the linear model against those of k-NN:

```
> library(regtools)
> parvsnonparplot(fout1, kout)
```

The result is shown in Figure 6.1. It suggests that the linear model is overestimating the regression function at times when the *yen* is very low or very high, and possibly underestimating in the moderate range.

We must view this cautiously, though. First, of course, there is the issue of sampling variation; the apparent model bias effects here may just be sampling anomalies.

Second, k-NN itself is subject to some bias at the edges of a data set. This will be discussed in detail in Section 11.1 (and a remedy presented for it), but basically what happens is that k-NN tends to overestimate $\mu(t)$ when the value is low and underestimate when $\mu(t)$ is high. The implication in the currency case, k-NN tends to overestimate for low values of the *yen*,

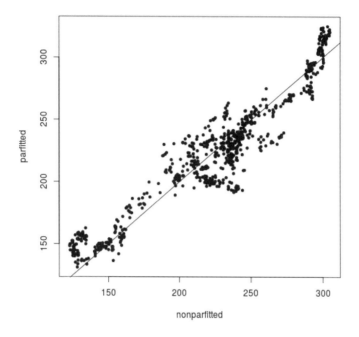

Figure 6.1: Estimation of regression values, two methods

and underestimate at the high end. This can be addressed by doing locally-linear smoothing, an option offered by **knnest()**, but let's not use it for now. And in any event, this k-NN edge bias effect would not entirely explain the patterns we see in the figure.

The "hook shape" at the left end, and a "tail" in the middle suggest odd nonlinear effects, possibly some local nonlinearities, which k-NN is picking up but which the linear model misses.

6.6.4 Residuals vs. Smoothing

In any regression analysis, the quantities

$$r_i = Y_i - \widehat{\mu}(X_i)$$ (6.3)

are traditionally called the *residual values*, or simply the *residuals*. They are of course the prediction errors we obtain when fitting our model and then predicting our Y_i from our X_i. The smaller these values are in absolute value, the better, but also we hope that they may inform us of inaccuracies in our model, say nonlinear relations between Y and our predictor variables.

In the case of a linear model, the residuals are

$$r_i = Y_i - \widehat{\beta}_0 - \widehat{\beta}_1 X_i^{(1)} - \dots - \widehat{\beta}_p X_i^{(p)} \tag{6.4}$$

Many diagnostic methods for checking linear regression models are based on residuals. In turn, their convenient computation typically involves first computing the *hat matrix*, about which there is some material in the Mathematical Complements section at the end of this chapter.

The generic R function **plot()** can be applied to any object of class **"lm"** (including the subclass **"glm"**). Let's do that with **fout1**:

```
> plot(fout1)
Hit <Return> to see next plot:
Hit <Return> to see next plot:
Hit <Return> to see next plot:
Hit <Return> to see next plot:
```

We obtain a series of graphs, displayed sequentially. Most of them involve more intricate concepts than we'll use in this book (recall Section 6.2), but let's look at the first plot, shown in Figure 6.2. The "hook" and "tail" are visible here too.

Arguably the effects are clearer in Figure 6.1. This is due to the fact that the latter figure is plotting smoothed values, not residuals. In other words, residuals and smoothing play complementary roles to each other: Smoothing-based plots can more easily give us "the big picture," but residuals may enable us to spot some fine details.

In any case, it's clear that the linear model does not tell the whole story.

6.7 Diagnostics Related to Individual Predictors

It may be that the relationship with the response variable Y is close to linear for some predictors $X^{(i)}$ but not for others. How might we investigate this?

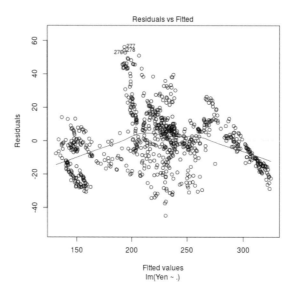

Figure 6.2: Residuals against linear fitted values

6.7.1 Partial Residual Plots

We might approach this question by simply plotting a scatter diagram of Y against each predictor variable. However, the relation of Y with one $X^{(i)}$ may change in the presence of another $X^{(j)}$. A more sophisticated approach may be *partial residual plots*, also known as *component + residual plots*. These would be easy to code on one's own, but the **crPlot()** function in the **car** package [49] does the job nicely for us. Continuing with the currency data, we try

```
> library(car)
> crPlots(fout1)
```

The resulting graph is shown in Figure 6.3. Before discussing these rather bizarre results, let's ask what these plots are depicting.

Here is how the partial-residual method works. The partial residuals for a predictor $X^{(j)}$ are defined to be

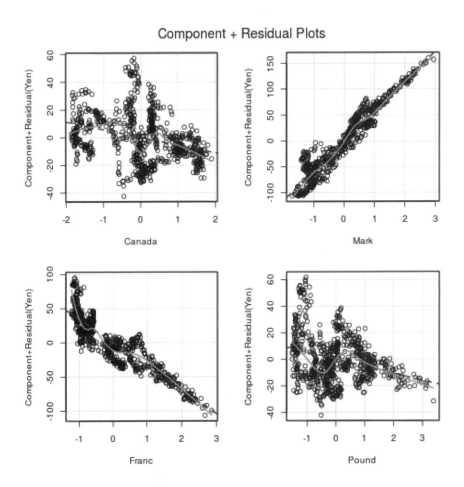

Figure 6.3: Partial residuals plot, currency data

$$\begin{aligned}
p_i &= r_i + \widehat{\beta}_j X_i^{(j)} && (6.5) \\
&= Y_i - \widehat{\beta}_0 - \widehat{\beta}_1 X_i^{(1)} - \ldots - \widehat{\beta}_{j-1} X_i^{(j-1)} - \widehat{\beta}_{j+1} X_i^{(j+1)} - \ldots - \widehat{\beta}_p X_i^{(p)}
\end{aligned}$$

for $i = 1, 2, \ldots, n$.

In other words, we started with the residuals (6.4), but removed the linear term contributed by predictor j, i.e., removed $\widehat{\beta}_j X_i^{(j)}$. We then plot the p_i against predictor j, to try to discern a relation. In effect we are saying,

> Cancel that linear contribution of predictor j. Let's start fresh with this predictor, and see how adding it in a possibly nonlinear form might extend the collective predictive ability of the other predictors.

If the resulting graph looks nonlinear, we may profit from modifying our model to one that reflects a nonlinear relation.

In that light, what might we glean from Figure 6.3? First, we see that the only "clean" relations are the one for the *franc* and the one for the *mark*. No wonder, then, that we found earlier that these two currencies seemed to have the strongest linear relation to the *yen*. There does seem to be some nonlinearity in the case of the *franc*, with a more negative slope for low *franc* values, and this may be worth pursuing, say by adding a quadratic term.

For the Canadian *dollar* and the *pound*, though, the relations don't look "clean" at all. On the contrary, the points in the graphs clump together much more than we typically encounter in scatter plots.

But even the *mark* is not off the hook (pardon the pun), as the "hook" shape noticed earlier is here for that currency, and apparently for the Canadian *dollar* as well. So, whatever odd phenomenon is at work may be related to these two currencies,

6.7.2 Plotting Nonparametric Fit Against Each Predictor

As noted, one approach would be to draw many scatter diagrams, plotting Y individually against each $X^{(i)}$. But scatter diagrams are, well, scattered. A better way is to plot the smoothed nonparametric fit, say using k-NN

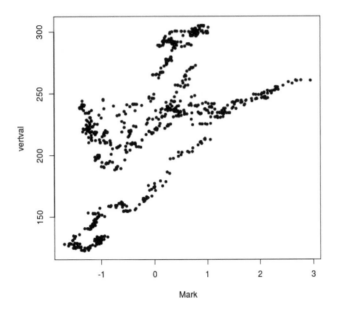

Figure 6.4: Nonparametric fit against the mark

as is done below. against each predictor. The **regtools** function **non-parvsxplot()** does this, plotting one graph for each predictor, presented in succession with user prompts:

```
> nonparvsxplot(kout)
next plot
next plot
next plot
next plot
```

The graph for for the *mark* is shown in Figure 6.4. Oh my gosh! With the partial residual plots, the *mark* and the *franc* seemed to be the only "clean" ones. Now we see that the situation for the *mark* is much more complex. The same is true for the other predictors (not shown here). This is indeed a difficult data set.

Again, note that the use of smoothing has brought these effects into better focus, as discussed in Section 6.6.4.

6.7.3 The freqparcoord Package

Another graphical approach uses the **freqparcoord** package, written by Yingkang Xie and me [104]. To explain this, we must first discuss the notion of *parallel coordinates*, a method for visualizing multidimensional data in a 2-dimensional graph.

6.7.3.1 Parallel Coordinates

The approach dates back to the 1800s, but was first developed in depth in modern times, notably by Alfred Inselberg and Ed Wegman; see [74]. The method is motivated by the problem that scatter plots work fine for displaying a paird of variables, but there is no direct multidimensional analog. The use of parallel coordinates allows us to visualize many variables at once.

The general method of parallel coordinates is quite simple. Here we draw p vertical axes, one for each variable. For each of our n data points, we draw a polygonal line from each axis to the next. The height of the line on axis j is the value of variable j for this data point.

Figure 6.5 shows a very simple example, showing the polygonal lines representing two people. The first person, in the upper line, is 70 inches tall, is 25 years old, and weighs 72.73 kilograms, while the second person has values, 62, 66 and 95.45. By the way, though we have not centered and scaled the data here (Section 1.21), some sort of scaling is typically applied, so that the graph is more balanced in scale.

As we will see, parallel coordinates plots often enable analysts to obtain highly valuable insights in their data, by exposing very telling patterns. Several R functions to create parallel coordinates plots are available, such as **parcoord()** in the **MASS** package included in base R; **parallelplot()** in the **lattice** graphics package [125]; and **ggparcoord** in **GGally**, a **ggplot2**-based graphics package [126].

As Inselberg pointed out, in mathematical terms the plot performs a transformation mapping p-dimensional points to $p - 1$-segment lines. There is an elegant geometric theory arising from this, but for us the practical effect is that we can visualize how our p variables vary together.

6.7.3.2 The freqparcoord Package

One major problem will parallel coordinates is that if the number of data points n is large, our plot will consist of a chaotic jumble of lines, maybe

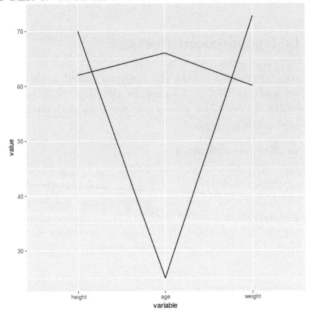

Figure 6.5: Simple parallel coordinates plot

even with the "black screen problem," meaning that so much has been plotted that the graph is mostly black, no defining features.

One solution to that problem is taken by **freqparcoord**. It plots only the most frequently-occurring lines.[5] It starts with all the lines drawn by **ggparcoord()**, but removes most of them, retaining only the most frequently-occurring, i.e., the most representative ones.

6.7.3.3   The regdiag() Function

A function in the **freqparcoord** package, **regdiag()**, applies this to regression diagnostics. The first variable plotted, i.e., the first vertical axis, is what we call the *divergences*, meaning the differences beween the para-

---

[5]There is an issue of what is meant by "most frequent." If our variables take on only integer values, this is clear, but if the variables are continuous, things are a little more involved; it may well be the case that no two lines are exactly the same. Here we might group together lines that are near each other. What **freqparcoord** does is use k-NN for this, estimating the p-dimensional joint density function, then plotting only those points that have the highest values of this function.

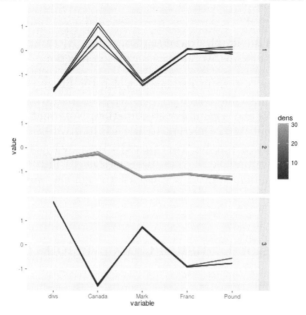

Figure 6.6: Freqparcoord plot, currency data

metric and nonparametric estimates of the population regression function,

$$\widehat{\mu}_{linmod}(X_i) - \widehat{\mu}_{knn}(X_i), \ i = 1, ..., n \tag{6.6}$$

The other axes represent our predictor variables. Vertical measures are numbers of standard deviations from the mean of the given variable.

The code

```
> library(freqparcoord)
> regdiag(fout1)
```

produces the graph in Figure 6.6.

Here we plot five variables, consisting of the divergences and the four currencies. Each variable is centered and scaled.

There are three groups, thus three subgraphs, for the upper 10%, middle 80% and lower 10% of the divergence values, labeled 1, 2 and 3 on the right margin. (The scaling of the variables, including the divergences, is done on the basis of the data as a whole, not within the subgroups.) So for instance the upper subgraph describes data points X_i at which the linear model greatly underestimates the true regression function.

What we see, then, is that in regions in which the linear model underestimates, the Canadian *dollar* tends to be high and the *mark* low, with an opposite relation for the region of overestimation. Note that this is not the same as saying that the correlation between those two currencies is negative; on the contrary, running **cor(curr1)** shows their correlation to be positive and tiny, about 0.01. This suggests that we might try adding a *dollar/mark* interaction term to our model, though the effect here seems mild, with peaks and valleys of only about 1 standard deviation..

6.8 Effects of Unusual Observations on Model Fit

Suppose we are doing a study of rural consumer behavior in a small country C in the developing world. One day, a few billionaires discover the idyllic beauty of rural C and decide to move there. We almost certainly would want to exclude data on these interlopers from our analysis. Second, most data contain errors. Obviously these must be excluded too, or corrected if possible. These two types of observations are sometimes collectively called *outliers*, or simply *unusual*.

Though we may hear that Bill Gates has moved to C, and we can clean the data to remove the obvious errors, such as a human height of 25 feet or a negative weight. other extreme values or errors may not jump out at us. Thus it would be useful to have methods that attempt to find such observations in some organized, mechanical way.

6.8.1 The influence() Function

Base R includes a very handy function, **influence()**. We input an object of type "**lm**" or "**glm**", and it returns an R list. One of the components of that list, **coefficients**, is just what we want: It has a column for each $\widehat{\beta}_j$, and a row for each observation in our data set. Row i, column j tells us how much $\widehat{\beta}_j$ would change if observation i were deleted from the data set.[6] If the change is large, we should take a close look at observation i, to

---

[6]The entire computation does not need to be done from scratch. The Sherman-Morrison-Woodbury formula provides a shortcut. See the Mathematical Complements section at the end of this chapter.

determine whether it is "unusual."

6.8.1.1   Example: Currency Data

Let's take a look, continuing our analysis in Section 6.5:

```
> infcfs <- influence(fout1)$coef
> head(infcfs)
   (Intercept)          Can          Mark         Franc
1  -0.01538183   0.03114368   -0.009961471   -0.07634899
2  -0.02018040   0.03743135   -0.018141461   -0.09488785
3  -0.02196501   0.03583000   -0.024654614   -0.08885551
4  -0.02877846   0.02573926   -0.050914092   -0.08862127
5  -0.02693571   0.02564799   -0.046012297   -0.08261002
6  -0.02827297   0.02524051   -0.050186868   -0.08733993
        Pound
1  0.07751692
2  0.10129967
3  0.10190745
4  0.13201466
5  0.12140985
6  0.13027910
```

So, if we were to remove the second data point, this says that $\widehat{\beta}_0$ would decline by 0.02018040, $\widehat{\beta}_1$ would increase by 0.03743135, and so on. Let's check to be sure:

```
> coef(fout1)
(Intercept)          Can          Mark         Franc
 224.945099     -5.615144     57.888556   -34.702731
        Pound
  -5.331583
> coef(lm(Yen ~ ., data=curr1[-2,]))
(Intercept)          Can          Mark         Franc
 224.965279     -5.652575     57.906698   -34.607843
        Pound
  -5.432882
> -5.652575 + 0.037431
[1]  -5.615144
```

Ah, it checks. Now let's find which points have large influence.

A change in an estimated coefficient should be considered "large" only relative to the standard error, so let's scale accordingly, dividing each change by the standard error of the corresponding coefficient:[7]

```
> se <- sqrt(diag(vcov(fout1)))
> infcfs <- infcfs %*% diag(1/se)
```

So, how big do the changes brought by deletions get in this data? And for which observations does this occur? Let's take a look.[8]

```
> ia <- abs(infcfs)
> max(ia)
[1] 0.1928661
> f15 <- function(rw) any(rw > 0.15)
> ia15 <- apply(ia,1,f15)
> names(ia15) <- NULL
> which(ia15)
 [1] 744 745 747 748 749 750 751 752 753 754 755
[12] 756 757 758 759 760 761
```

Here we (somewhat arbitrarily) decided to identify which deletions of observations would result in an absolute change of *some* coefficient of more than 0.15.

Now this is interesting. There are 761 observations in this data set, and now we find that all of the final 18 (and more) are influential. Let's look more closely:

```
> tail(ia,5)
            [,1]        [,2]        [,3]        [,4]
757 0.05585538 0.1311110 0.1572411 0.1305925
758 0.05851087 0.1294412 0.1563013 0.1259741
759 0.05838813 0.1386614 0.1629851 0.1358875
760 0.05818730 0.1429951 0.1654354 0.1385146
761 0.05626212 0.1316884 0.1534207 0.1211305
            [,5]
757 0.021300177
758 0.015701005
759 0.020431391
760 0.019962502
761 0.006793673
```

---

[7]R's **diag()** function is quite versatile. In our first call here, we are extracting the diagonal elements of the estimated covariance matrix of our coefficients, placing the result in a vector. In the second call, we are creating a diagonal matrix from a vector.

[8]The reader may wish to review Section 1.20.2 before continuing.

So, the influence of these final observations was on the coefficients of the Canadian *dollar*, the *mark* and the *franc* — but not on the one for the *pound*.

Something special was happening in those last time periods. It would be imperative for us to track this down with currency experts.

Each of the observations has something like a 0.15 impact, and intuitively, removing all of these observations should cause quite a change. Let's see:

```
> curr3 <- curr1[-(744:761),]
> lm(Yen ~ .,data=curr3)
...
Coefficients:
(Intercept)        Canada           Mark
    225.780       -10.271         52.926
      Franc         Pound
    -27.126        -6.431

> fout1
...
Coefficients:
(Intercept)        Canada           Mark
    224.945        -5.615         57.889
      Franc         Pound
    -34.703        -5.332
```

These are very substantial changes! The coefficient for the Canadian currency almost doubled, and even the *pound*'s value changed almost 30%. That latter is a dramatic difference, in view of the fact that each individual observation had only about a 2% influence on the *pound*.

A collection of advanced influence measures is provided by another R function, whose name is, not surprisingly, **influence.measures()**.

6.8.2   Use of freqparcoord for Outlier Detection

The versatile **freqparcoord** package can also be used for outlier detection. Here we find the *least*-frequent points, rather than the ones with highest frequemcy as before. Continuing the currency example, we run this code:

```
> freqparcoord(curr1,m=-5,method='maxdens',
     keepidxs=1)$idxs
[1] 547 548 549 551 550
```

The argument *m* indicates how many observations we wish to be reported, i.e., how many polygonal lines to plot. A negative value, combined with an argument **method='maxdens'**, specifies that we want to plot the *least* frequent cases, if **keepidxs** is not NULL.[9] The **idxs** component of the return value gives us the indices of the possible outliers.

Again, a set of consecutive observations turned out to be troubling, And for instance, we might check **curr1[547,]** for possible errors or other undesirable characteristics.

6.9 Automated Outlier Resistance

The term *robust* in statistics generally means the ability of methodology to withstand problematic conditions. Linear regression models, for instance, are said to be "robust to the normality" assumption, meaning that (at least large-sample) inference on the coefficients will work well even though the distribution of Y given X is not normal.

But here we are concerned with robustness of regression models to outliers. Such methods are called *robust regression*. There are many such methods, one of which is *median regression*, to be discussed next.

6.9.1 Median Regression

Suppose we wish to estimate the mean of some variable in some population, but we are concerned about unusual observations. As an alternative, we might consider estimating the median value of the variable, which will be much less sensitive to unusual observations. Recall our hypothetical example earlier, in which we were interested in income distributions. If one very rich person moves into the area, the mean may be affected substantially — but the median would likely not change at all.

We thus say that the median is robust to unusual data points. One can do the same thing to make regression analysis robust in this sense.

Denote the conditional median of Y given $X = t$, i.e., median($Y \mid X = t$), by $\nu(t)$. It turns out (see the Mathematical Complements section at the end of this chapter) that

$$\nu(t) = \mathrm{argmin}_m E(|Y - m| \mid X = t) \tag{6.7}$$

[9]If not NULL, this argument states the variable on which we want the results sorted.

In other words, in contrast to the regression function, i.e., the conditional mean, which minimizes mean squared prediction error, the conditional median minimizes mean *absolute* error.

Remember, as with regression, we are estimating an entire function here, as t varies. A nonparametric approach to this would be to use **knnest()** with **nearf** set to

```
function(predpt, nearxy)
{
    ycol <- ncol(nearxy)
    median(nearxy[, ycol])
}
```

However, in this chapter we are primarily concerned with parametric models. So, we might, in analogy to the linear regression model, make the assumption that (6.7) has the familiar linear form

$$\nu(t) = \beta_0 + \beta_1 t_1 + ... + \beta_p t_p \tag{6.8}$$

Solving this at the sample level is a *linear programming* problem, which has been implemented in the CRAN package **quantreg** [81]. As the package name implies, we can estimate general conditional quantile functions, not just the conditional median. The argument **tau** of the **rq()** function specifies what quantile we want, with the value 0.5, i.e., the median, being the default.

It is important to understand that $\nu(t)$ is *not* the regression function, i.e., not the conditional mean. Thus **rq()** is not estimating the same quantity as is **lm()**. Thus the term *quantile regression*, in this case the term *median regression*, is somewhat misleading here. But we can use $\nu(t)$ as an alternative to $\mu(t)$ in one of two senses:

(a) We may believe that $\nu(t)$ is close to $\mu(t)$. They will be exactly the same, of course, if the conditional distribution of Y given X is symmetric, at least if the unusual observations are excluded.

(b) We may take the point of view that the conditional median is just as meaningful as the conditional mean (no pun intended this time), so why not simply model $\nu(t)$ in the first place?

Sense (a) above will be particularly relevant here.

It should be noted, though, that there are reasons not to use median regression. If estimating the mean of a normally distributed random variable, the sample median's efficiency relative to the sample mean, meaning the ratio of asymptotic variances, can be shown to be only $2/\pi \approx 0.64$. In other words, by using the median regression model rather than a linear one, there is not only the question of whether the models are close to reality, but also that we risk having estimators with large standard errors.

6.9.2 Example: Currency Data

Let's apply **rq()** to the currency data:

```
> qout <- rq(Yen ~ . ,data=curr1)
> qout
...
Coefficients:
(Intercept)          Can           Mark          Franc
 224.517899     -11.038238      53.854005     -27.443584
        Pound
    -5.320035
...
> fout1
...
Coefficients:
(Intercept)          Can           Mark
    224.945        -5.615        57.889
      Franc         Pound
    -34.703        -5.332
```

The results are strikingly similar to what we obtained in Section 6.8.1.1 by calling **lm()** with the bad observations at the end of the data set removed. In other words,

> Median regression can be viewed as an automated method for removing (the effects of) the unusual data points.

6.10 Example: Vocabulary Acquisition

The Wordbank data, *http://wordbank.stanford.edu/*, concerns child vocabulary development, in not only English but also a number of other languages, such as Cantonese and Turkish. These are mainly toddlers, ages from about a year to 2.5 years.

Let's read in the English set:

```
> engl <- read.csv('English.csv')
# take a look
> head(engl)
  data_id age language form birth_order ethnicity
1       1  24  English   WS       First    Asian
2       2  19  English   WS      Second    Black
3       3  24  English   WS       First    Other
4       4  18  English   WS       First    White
5       5  24  English   WS       First    White
6       6  19  English   WS       First    Other
     sex          mom_ed    measure vocab     demo
1 Female        Graduate production   337 All Data
2 Female         College production   384 All Data
3   Male Some Secondary production    76 All Data
4   Male      Secondary production    19 All Data
5 Female      Secondary production   480 All Data
6 Female    Some College production   313 All Data
     n           demo_label
1 5498 All Data (n = 5498)
2 5498 All Data (n = 5498)
3 5498 All Data (n = 5498)
4 5498 All Data (n = 5498)
5 5498 All Data (n = 5498)
6 5498 All Data (n = 5498)
```

We have a number of R factors here, such as birth order, so we need to do some data wrangling here. The details are presented in Section 6.15.2. The result is

```
> head(encc)
  age birth_order mom_ed vocab male asian black
1  24           1     20   337    0     1     0
2  19           2     16   384    0     0     1
3  24           1     10    76    1     0     0
4  18           1     12    19    1     0     0
5  24           1     12   480    0     0     0
6  19           1     14   313    0     0     0
  latino othernonwhite
1      0             0
```

2	0	0
3	0	1
4	0	0
5	0	0
6	0	1

Running **knnest()** (not shown), there seemed to be an approximately linear relation between vocabulary and age (for the age range studied). Let's run a linear regression analysis:

```
> summary(lm(vocab ~ ., data=encc))
...
Coefficients:
```

	Estimate	Std. Error	t value
(Intercept)	−475.8761	23.5059	−20.245
age	33.3968	0.6422	52.005
birth_order	−20.5033	3.0891	−6.637
mom_ed	3.8680	0.9780	3.955
male	−48.9628	5.4842	−8.928
asian	−16.0745	17.2947	−0.929
black	1.1094	10.2321	0.108
latino	−75.5545	13.0172	−5.804
othernonwhite	−53.9844	15.1291	−3.568

	$Pr(>|t|)$	
(Intercept)	< 2e−16	***
age	< 2e−16	***
birth_order	3.84e−11	***
mom_ed	7.85e−05	***
male	< 2e−16	***
asian	0.352741	
black	0.913671	
latino	7.21e−09	***
othernonwhite	0.000366	***

```
...
```
Mult. R-squared: 0.5132, Adj. R-squared: 0.5118

Age is recorded in months. So, the kids seemed to be learning about 33 words per month during the studied age range. Latino kids seem to start slowly, possibly due to speaking Spanish at home. Consistent with the general notion that girls develop faster than boys, the latter have a slower start. Having a lot of older siblings also seems to be related negatively, possibly due to the child being one of several competing for the parents' attention. Having a mother with more education had a modest positive effect, about 19 words for every extra 5 years of schooling.

A word on the standard errors: As each child was measured multiple times as he/she aged, the observations are not independent, and the true standard errors are larger than those given.

Let's try median regression instead:

```
> library(quantreg)
> rq(vocab ~ ., data=encc)
...
Coefficients:
  (Intercept)              age      birth_order
   -581.73077         37.84615        -19.50000
       mom_ed             male            asian
      4.00000        -43.42308        -20.07692
        black           latino  othernonwhite
    -13.73077        -67.65385        -49.65385
```

(There was also a warning, "Solution may be nonunique.")

The robust results here are similar to what we obtained earlier, but with some modest shifts.

It is often worthwhile to investigate other quantiles than the median. Trying that for age only:

```
> plot(c(12,30),c(0,800),type = "n", xlab = "age",
      ylab = "vocab")
> abline(coef(rq(vocab ~ age, data=encc, tau=0.1)))
> abline(coef(rq(vocab ~ age, data=encc, tau=0.5)))
> abline(coef(rq(vocab ~ age, data=encc, tau=0.9)))
```

As seen in Figure 6.7, the middle-level children start out knowing many fewer words than the most voluble ones, but narrow the gap over time. By contrast, the kids with smaller vocabularies start out around the same level as the middle kids, but actually lose ground over time, suggesting that educational interventions may be helpful.

6.11 Classification Settings

Since we treat classification as a special case of regression, we can use the same fit assessment methods, though in some cases some adapting of them is desirable.

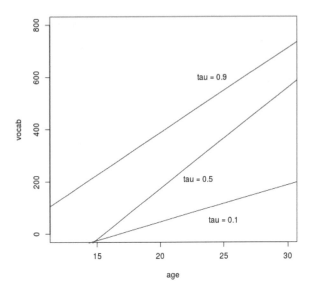

Figure 6.7: Vocabulary vs. age

6.11.1 Example: Pima Diabetes Study

Let's illustrate with the Pima diabetes data set from Section 4.3.2.

```
> pima <- read.csv('pima−indians−diabetes.data')
```

It goes without saying that with any data set, we should first do proper cleaning.[10] This data is actually a very good example. Let's first try the **freqparcoord** package:

```
> library(freqparcoord)
> freqparcoord(pima[,−9],−10)
```

Here we display those 10 data points (predictors only, not response variable) whose estimated joint density is lowest, thus qualifying as "unusual."

The graph is shown in Figure 6.8. Again we see a jumble of lines, but look at the big dips in the variables **BP** and **BMI**, blood pressure and Body Mass Index. They seem unusual. Let's look more closely at blood pressure:

---

[10]And of course we should have done so for the other data earlier in this chapter, but we will keep the first analyses simple.

```
> table(pima$BP)
```

0	24	30	38	40	44	46	48	50	52	54	55
35	1	2	1	1	4	2	5	13	11	11	2
56	58	60	61	62	64	65	66	68	70	72	74
12	21	37	1	34	43	7	30	45	57	44	52
75	76	78	80	82	84	85	86	88	90	92	94
8	39	45	40	30	23	6	21	25	22	8	6
95	96	98	100	102	104	106	108	110	114	122	
1	4	3	3	1	2	3	2	3	1	1	

One cannot have a blood pressure of 0, yet 35 women in our data set are reported as such. The value 24 is suspect too, but the 0s are wrong for sure. What about BMI?

```
> table(pima$BMI)
```

0	18.2	18.4	19.1	19.3	19.4	19.5	19.6	19.9	20
11	3	1	1	1	1	2	3	1	1
20.1	20.4	20.8	21	21.1	21.2	21.7	21.8	21.9	22.1

. . .

Here again, the 0s are clearly wrong. So, at the very least, let's exclude such data points:

```
> pima <- pima[pima$BP > 0 & pima$BMI > 0,]
> dim(pima)
[1] 729    9
```

(We lost 38 cases.)

Now, for our analysis, start with fitting a logit model, then comparing to k-NN. First, what value of k should we use?:

Let's go with 50, and compare the parametric and nonparametric fits:

```
> kout <- knnest(pima$Diab,xdata,50)
> parvsnonparplot(glmout,kout)
```

The results of the plot are shown in Figure 6.10. There does appear to be some overestimation by the logit at very high values of the regression function, indeed all the range past 0.5. This can't be explained by the fact, noted before, that k-NN tends to underestimate at the high end.

Note carefully that if our goal is Prediction, it may not matter much at the high end. Recall the discussion on classification contexts in Section 6.4.1. If

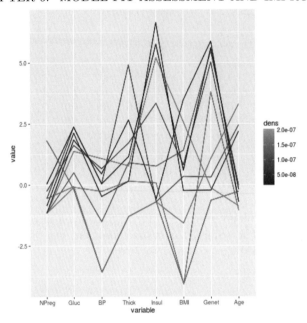

Figure 6.8: Outlier hunt (see color insert)

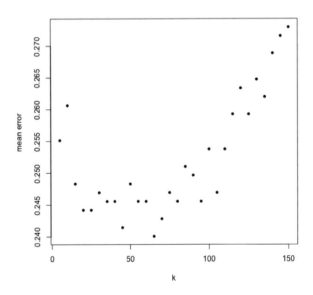

Figure 6.9: Best k for Pima

the true population regression value is 0.8 and we estimate it to be 0.88, we still predict $Y = 1$, which is the same as what we would predict if we knew the true population regression function. Similarly, if the true regression value is 0.25 but we estimate it to be 0.28, we still make the proper guess for Y.

For Description, though, we should consider a richer model. Running **non-parvsxplot()** (not shown) suggests adding quadratic terms for the variables **Gluc**, **Insul**, **Genet** and especially **Age**. Adding these, and rerunning **knnest()** with **nearf = loclin** to deal with k-NN's high-end bias,[11] our new parametric-vs.-nonparametric plot is shown in Figure 6.11.

The reader may ask if we now have *under*estimation by the parametric model at the high end, but we must take into account the fact that with **nearf = loclin**, we can get nonparametric estimates for the regression function that are smaller than 0 or greater than 1, which is impossible in the classification setting. The perceived "underestimation" actually occurs at values at which the nonparametric figures are larger than 1.

In other words, we now seem to have a pretty good model.

6.12 Improving Fit

The currency example seemed so simple at first, with a very nice adjusted R-squared value of 0.89, and with the *yen* seeming to have a clean linear relation with the *franc* and the *mark*. And yet we later encountered some troubling aspects to this data.

First we noticed that the adjusted R-squared value for the k-NN fit was even better, at 0.98. Thus there is more to this data than simple linear relationships. Later we found that the last 18 data points, possibly more, have an inordinate influence on the $\widehat{\beta}_j$. This too could be a reflection of nonlinear relationships between the currencies. The plots exhibited some strange, even grotesque, relations.

So, let's see what we might do to *improve* our parametric model.

6.12.1 Deleting Terms from the Model

Predictors with very little relation to the response variable may actually degrade the fit, and we should consider deleting them. This topic is treated

---

[11]This topic is covered in Section 11.1.

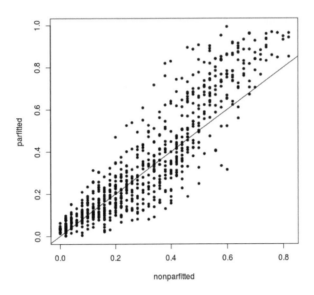

Figure 6.10: Estimation of regression values, two methods (I)

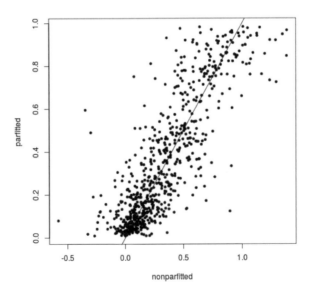

Figure 6.11: Estimation of regression values, two methods (II)

in depth in Chapter 9.

6.12.2 Adding Polynomial Terms

Our current model is linear in the variables. We might add second-degree terms. Note that this means not only squares of the variables, but products of pairs of them. The latter may be important, in view of our comment in Section 6.7.3 that it might be useful to add a Canadian *dollar/mark* interaction term to our model.

6.12.2.1 Example: Currency Data

Let's add squared terms for each variable, and try the interaction term as well. Here's what we get:

```
> curr2 <- curr1
> curr2$C2 <- curr2$Canada^2
> curr2$M2 <- curr2$Mark^2
> curr2$F2 <- curr2$Franc^2
> curr2$P2 <- curr2$Pound^2
> curr2$CM <- curr2$Canada* curr2$Mark
> summary(lm(Yen ~ .,data=curr2))
...
Coefficients:
             Estimate Std. Error  t value
(Intercept) 223.575386   1.270220  176.013
Can          -8.111223   1.540291   -5.266
Mark         50.730731   1.804143   28.119
Franc       -34.082155   2.543639  -13.399
Pound        -3.100987   1.699289   -1.825
C2           -1.514778   0.848240   -1.786
M2           -7.113813   1.175161   -6.053
F2           11.182524   1.734476    6.447
P2           -1.182451   0.977692   -1.209
CM            0.003089   1.432842    0.002
             Pr(>|t|)
(Intercept)  < 2e-16 ***
Can          1.82e-07 ***
Mark         < 2e-16 ***
Franc        < 2e-16 ***
Pound          0.0684 .
```

C2	0.0745 .
M2	2.24e−09 ***
F2	2.04e−10 ***
P2	0.2269
CM	0.9983

―――

. . .

Multiple R-squared: 0.9043, Adj. R-squared: 0.9032

Adjusted R-squared increased only slightly. And this was despite the fact that two of the squared-variable terms were "highly significant," adorned with three asterisks, showing how misleading significance testing can be. The interaction term came out tiny, 0.003089. So, k-NN is still the winner here.

6.12.2.2 Example: Census Data

Let's take another look at the census data on programmers and engineers in Silicon Valley, first introduced in Section 1.16.1.

We run

```
> data(prgeng)
> pe <- prgeng   # see ?knnest
> # dummies for MS, PhD
> pe$ms <- as.integer(pe$educ == 14)
> pe$phd <- as.integer(pe$educ == 16)
> # computer occupations only
> pecs <- pe[pe$occ >= 100 & pe$occ <= 109,]
> pecs1 <- pecs[,c(1,7,9,12,13,8)]
> # predict wage income from age, gender etc.
> # prepare nearest−neighbor data
> xdata <- preprocessx(pecs1[,1:5],150)
> zout <- knnest(pecs1[,6],xdata,5)
> nonparvsxplot(zout)
```

We find that the **age** variable, and possibly **wkswrkd**, seem to have a quadratic relation to **wageinc**, as seen in Figures 6.12 and 6.13. So, let's try adding quadratic terms for those two variables. And, to assess how well this works, let's break the data into training and test sets:

```
> pecs2 <- pecs1
> pecs2$age2 <- pecs1$age^2
```

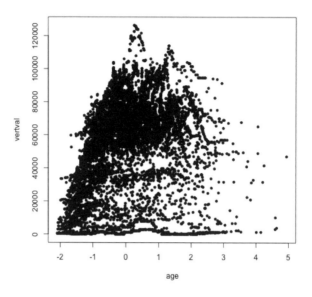

Figure 6.12: Mean wage income vs. age

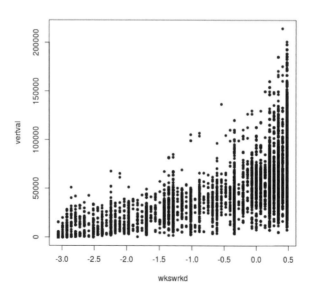

Figure 6.13: Mean wage income vs. weeks worked

```
> pecs2$wks2 <- pecs1$wkswrkd^2
> n <- nrow(pecs1)
> trnidxs <- sample(1:n,12000)
> predidxs <- setdiff(1:n, trnidxs)
> lmout1 <- lm(wageinc ~ ., data=pecs1[trnidxs,])
> lmout2 <- lm(wageinc ~ ., data=pecs2[trnidxs,])
> lmpred1 <- predict(lmout1, pecs1[predidxs,])
> lmpred2 <- predict(lmout2, pecs2[predidxs,])
> ypred <- pecs1$wageinc[predidxs]
> mean(abs(ypred-lmpred1))
[1] 25721.5
> mean(abs(ypred-lmpred2))
[1] 25381.08
```

So, adding the quadratic terms helped slightly, about a 1.3% improvement. From a Prediction point of view, this is at best mild, There was also a slight increase in adjusted R-squared, from 0.22 (not shown) to 0.23 (shown below).

But for Description things are much more useful here:

```
> summary(lmout2)
...
Coefficients:
```

	Estimate	Std. Error	t value
(Intercept)	−63812.415	4471.602	−14.271
age	3795.057	221.615	17.125
sex	−10336.835	841.067	−12.290
wkswrkd	598.969	131.499	4.555
ms	14810.929	928.536	15.951
phd	20557.235	2197.921	9.353
age2	−39.833	2.608	−15.271
wks2	9.874	2.213	4.462

	$Pr(>|t|)$	
(Intercept)	< 2e−16	***
age	< 2e−16	***
sex	< 2e−16	***
wkswrkd	5.29e−06	***
ms	< 2e−16	***
phd	< 2e−16	***
age2	< 2e−16	***
wks2	8.20e−06	***

```
...
```

Multiple R-squared: 0.2385, Adj. R-squared: 0.2381

As usual, we should not make too much of the p-values, especially with a sample size this large (16411 for **pecs1**). So, all those asterisks don't tell us too much. But a confidence interval computed from the standard error shows that the absolute age-squared effect is at least about 34, far from 0, and it does make a difference, say on the first person in the sample:

```
> predict (lmout1 , pecs1 [1 ,])
        1
62406.48
> predict (lmout2 , pecs2 [1 ,])
        1
63471.53
```

The more sophisticated model predicts about an extra $1,000 in wages for this person.

Most important, the negative sign for the age-squared coefficient shows that income tends to level off and even decline with age, something that could be quite interesting in a Description-based analysis.

The positive sign for **wkswrkd** is likely due to the fact that full-time workers tend to have better jobs.

6.12.3   Boosting

One of the techniques that has caused the most excitement in the machine learning community is *boosting*, which in essence is a process of iteratively refining, through reweighting, estimated regression and classification functions (though it has primarily been applied to the latter).

This is a very complex topic, with many variations, and is basically beyond the scope of this book. However, we will present an overview.

6.12.3.1   View from the 30,000-Foot Level

The main idea of boosting is to perform an iterative refitting of the model, adjusting weights of the observations at each iteration. At any given step, the observations predicted most poorly at the last step will now get larger weights.

Berk [16] gives an excellent description of the basic philosophy of boosting, by outlining a procedure that captures the method's spirit. Here is a modified version of his prescription, for the case of the linear model:

(a) Call **lm()** on the data $(X_1, Y_1), ..., (X_n, Y_n)$ as usual, yielding the vector of estimated coefficients, $\widehat{\beta}^{(0)}$.

Repeat steps (b)-(c) for $i = 1, ..., k$:

(b) Form the residuals from the latest beta vector,

$$r_j = Y_j - (\widehat{\beta}_0^{(i-1)} + \widehat{\beta}_1^{(i-1)} X_{i-1,1} + ... \widehat{\beta}_p^{(i-1)} X_{ip-1,}) \tag{6.9}$$

Also compute

$$d_{i-1} = \sum_{j=1}^{n} |r_j| \tag{6.10}$$

our total absolute prediction error.

(c) Run a weighted least squares analysis as in (3.3.2) (**weights** argument in **lm()**), with

$$w_j = |r_j| \tag{6.11}$$

yielding a new vector of estimated coefficients, $\widehat{\beta}^{(i)}$. We are giving the observations with worse prediction errors more weight, in the hope of getting better predictions at those points.

Finally:

(d) Compute d_k as in (6.10), and set the final estimated coefficient vector to

$$\widehat{\beta} = \sum_{s=0}^{k} q_s \widehat{\beta}^{(s)} \tag{6.12}$$

where

$$q_s = \frac{1/d_s}{\sum_t^k 1/d_t} \tag{6.13}$$

In other words, take a weighted average of the estimated coefficient vectors we've computed, with the ones with better total prediction error getting more weight.

The classification case is similar. In the 2-class setting, for instance, we might set

$$r_j = |Y_j - \widehat{Y}_j| \qquad (6.14)$$

where $\widehat{Y}_j$ is our predicted value for Y_j, either 0 or 1. In step (d) we could use "voting," as with AVA in Section 5.5, but with the votes being weighted.

There are many, many variations — AdaBoost, gradient boosting and so on — and their details are beyond the scope of this book. But the above captures the essence of the method.

Why do all this? The key issue (often lost in the technical discussions) is *bias*, in this case model bias, as follows. Putting aside issues of possible heteroscedasticity, ordinary (i.e., unweighted) least squares (OLS) estimation is optimal for homoscedastic linear models: the *Gauss-Markov Theorem* (Section 6.16.4) shows that OLS gives minimum variance among all unbiased estimators. If we are in this setting, why use weights, especially weights that come from such an involved process?

The answer is that the linear model is rarely if ever exactly correct. Thus use of a linear model will result in bias; in some regions of X, the model will overestimate, while in others it will underestimate — no matter how large our sample is. We saw indications of this with our currency data earlier in this chapter. It thus may be profitable to try to reduce bias in regions in which our present predictions are very bad, at the hopefully small sacrifice of some prediction accuracy in places where presently we are doing well. The reweighting process is aimed at achieving a positive tradeoff of that nature.

That tradeoff may be particularly useful in classification settings. As noted in Section 6.4.1, in such settings, we can tolerate large errors in $\widehat{\mu}(t)$ on the fringes of the dataset, so placing more weight in the middle, near the classification boundary, could be a win.

6.12.3.2 Performance

Much has been made of the remark by the late statistician Leo Breiman, that boosting is "the best off-the-shelf classifier in the world" [8], his term *off-the-shelf* meaning that the given method can be used by nonspecialist users without special tweaking. His statement has perhaps been overinterpreted (see Section 1.13), but many analysts have indeed reported that some improvement (though not dramatic) results from the method. On the other hand, it can also perform poorly relative to the nonboosting analysis in some case. And it does seem to require good use of tuning parameters, not "off the shelf" after all.

Note of course that improvement must be measured in terms of accuracy in predicting *new* cases. Since boosting is aimed at reducing errors in individual observations, there is a definite tendency toward overfitting.

6.13 A Tool to Aid Model Selection

In Section 5.5.3, we discussed the effects of nonlinear boundaries between classes, and presented some possible solutions. In light of such situations, a tool that helps us view those boundaries can be helpful. The **regtools** package includes a plotting function, **pwplot()**, motivated by this consideration. As our example, let's again take the Pima diabetes data, Section 4.3.2.

We cannot view things in p dimensions, so **pwplot()** displays two predictors, $X^{(i)}$ and $X^{(j)}$ at a time. Those two variables form the horizontal and vertical axes, and **knnest()**, k-NN estimation, is applied, predicting Y from $X^{i)}$ and $X^{(j)}$. For each point the following computation is done: We compare the estimated values of $P(Y = 1 \mid X^{(i)}, X^{(j)})$ and $P(Y = 1)$, and then determine which is larger. The point here is to ask, Does knowing X^i and X^j make $Y = 1$ more or less likely to occur, relative to having no knowledge of those covariates? A '1' is plotted for each point for which the answer is "more likely," with '0' being plotted otherwise.

One of the function's options is to plot only the points at which the estimated values of $P(Y = 1 \mid X^{(i)}, X^{(j)})$ and $P(Y = 1)$ are close to each other. This produces a "contour" effect, indicating the boundary between $P(Y = 1 \mid X^{(i)}, X^{(j)}) < P(Y = 1)$ and $P(Y = 1 \mid X^{(i)}, X^{(j)}) > P(Y = 1)$. Let's use this option, with our covariates being glucose and body mass:

```
> pwplot(pima$diabetes, pima[,1:8],25,
      pairs=matrix(c(2,6),ncol=1), cex=0.8,band=0.05)
```

The result is shown in Figure 6.14. We see a snake-like contour, suggesting that the boundary is quite nonlinear. If we are fitting a logistic model, we might consider adding quadratic or even higher-order polynomial terms.

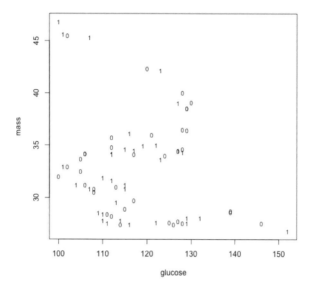

Figure 6.14: Snakelike boundary

6.14 Special Note on the Description Goal

If we are unable to improve the fit of our parametric model in a setting in which k-NN seems to give a substantially better fit, we should be quite wary of placing too much emphasis on the values of the $\widehat{\beta}_j$. As we saw with the currency data, the estimated coefficients can be quite sensitive to unusual observations and so on.

This is not to say the $\widehat{\beta}_j$ are useless in such settings. On the contrary, they may be quite valuable. But they should be used with caution.

6.15 Computational Complements

6.15.1 Data Wrangling for the Currency Dataset

The file **EXC.ASC** has some nondata lines at the end, which need to be removed before running the code below. We then read -it in, and do some wrangling:

```
> curr <- read.table('EXC.ASC', header=FALSE,
     stringsAsFactors=FALSE)
> for (i in 1:ncol(curr)) curr[,i] <-
     as.numeric(curr[,i])
Warning messages:
1: NAs introduced by coercion
2: NAs introduced by coercion
3: NAs introduced by coercion
4: NAs introduced by coercion
> colnames(curr) <-
     c('Canada','Mark','Franc','Pound','Yen')
```

What happened here? The above sequence is a little out of order, in the sense that I ran it with prior knowledge of a certain problem, as follows.

The authors, in compiling this data file, decided to use '.' as their NA code. R, in reading the file, then forced the numeric values, i.e., the bulk of the data, to character strings for consistency. That in turn would have led to each variable, i.e., each currency, being stored as an R factor. Having discovered this earlier (not shown here), I added the argument **stringsAsFactors = FALSE** to my read call.

That still left me with character strings for my numeric values, so I ran **as.numeric()** on each column. Finally, the original data set lacked names for the columns, so I added some.

6.15.2   Data Wrangling for the Word Bank Dataset

There are a number of NA values in the data; let's just look at complete cases.

```
> # get variables of interest
> engl <- engl[,c(2,5:8,10)]
> # exclude cases with NAs
> encc <- engl[complete.cases(engl),]
```

One of the variables is birth order, an R factor. Let's make it numeric:

```
> z <- encc$birth_order
> class(z)
[1] "factor"
> levels(z)
```

```
[1] "Eighth"   "Fifth"    "First"    "Fourth"
[5] "Second"   "Seventh"  "Sixth"    "Third"
> # convert from First, Second, ... to 1, 2, ...
> numcodes <- c(8,5,1,4,2,7,6,3)
> zn <- as.numeric(z)
> encc$birth_order <- numcodes[zn]
```

Similarly, let's convert the variable on the mother's education, another R factor, to a rough number of years of school, e.g., 10 years for the code Some Secondary:

```
> z <- encc$mom_ed
> levels(z)
[1] "College"       "Graduate"
[3] "Primary"       "Secondary"
[5] "Some College"  "Some Graduate"
[7] "Some Secondary"
> numcodes <- c(16,20,4,12,14,18,10)
> encc$mom_ed <- numcodes[zn]
```

Also, create the needed dummy variables, for gender and nonwhite categories:

```
> encc$male <- as.numeric(encc$sex=='Male')
> encc$sex <- NULL
> encc$asian <- as.numeric(encc$ethnicity=='Asian')
> encc$black <- as.numeric(encc$ethnicity=='Black')
> encc$latino <- as.numeric(encc$ethnicity=='Hispanic')
> encc$othernonwhite <- as.numeric(encc$ethnicity=='Other')
> encc$ethnicity <- NULL
```

Note that a column in a data frame (or an element in any R list, of which a data frame is a special case) can be removed by setting it to NULL.

6.16 Mathematical Complements

6.16.1 The Hat Matrix

We'll use the notation of Section 2.4.2 here. The *hat matrix* is defined as the $n \times n$ matrix

$$H = A(A'A)^{-1}A' \qquad (6.15)$$

The name stems from the fact that we use H to obtain "Y-hat," the predicted values for the elements of D,

$$\widehat{D}_i = \widehat{\mu}(\widetilde{X_i}')\,\widehat{\beta} \qquad (6.16)$$

Here is how the hat matrix comes in:

$$\widehat{D} = A\widehat{\beta} = A(A'A)^{-1}A'D = HD \qquad (6.17)$$

Using the famous relation $(VW)' = W'V'$ (Equation (A.14)), it is easily verified that H is a symmetric matrix. Also, some easy algebra shows that H is *idempotent*, i.e.,

$$H^2 = H \qquad (6.18)$$

(The idempotency also follows from the fact that H is a projection operator; once one projects, projecting the result won't change it.)

This leads us directly to the residuals:

$$L = D - A\widehat{\beta} = D - HD = (I - H)D \qquad (6.19)$$

The diagonal elements

$$h_{ii} = H_{ii} \qquad (6.20)$$

are known as the *leverage* values, another measure of influence like those in Section 6.8.1, for the following reason. Looking at (6.17), we see that

$$\widehat{D}_i = h_{ii}D_i \qquad (6.21)$$

This shows us the effect of true value D_i on the fitted value $\widehat{D}_i$:

$$h_{ii} = \frac{\partial \widehat{D}_i}{\partial D_i} \qquad (6.22)$$

So, h_{ii} can be viewed as a measure of how much influence observation i has on its fitted value. A large value might thus raise concern — but how large is "large"?

Let w_i denote row i of H, which is also column i since H is symmetric. Then

$$h_{ii} = H_{ii} = (H^2)_{ii} = w_i' w_i = h_{ii}^2 + \sum_{j=1, j \neq i}^{n} h_{ij}^2 \qquad (6.23)$$

Since the far-right portion of the above equation is a sum of squares, This directly tells us that $h_{ii} \geq 0$. But it also tells us that $h_{ii} \geq h_{ii}^2$, which forces $h_{ii} \leq 1$.

In other words,

$$0 \leq h_{ii} \leq 1 \qquad (6.24)$$

This will help assess whether a particular h_{ii} value is "large."

6.16.2 Matrix Inverse Update

The famous Sherman-Morrison-Woodbury formula says that for an invertible matrix B and vectors u and v

$$(B + uv')^{-1} = B^{-1} - \frac{1}{1 + v'B^{-1}u} B^{-1}uv'B^{-1} \qquad (6.25)$$

In other words, if we have already gone to the trouble of computing a matrix inverse, and the matrix is then updated as above by adding uv', then we do not have to compute the new inverse from scratch; we need only modify the old inverse, as specified above.

Let's apply that to Section 6.8.1, where we discussed the effect of deleting an observation from our data set. Write A in partitioned form as in (2.19), we have that

$$A'A = \sum_{i=1}^{n} \widetilde{X}_i \widetilde{X}_i' \qquad (6.26)$$

Thus the new version of $A'A$ after deleting observation i is

$$(A'A)_{(-i)} = A'A - \widetilde{X}_i \widetilde{X}_i' \qquad (6.27)$$

This is then tailor-made for Sherman-Morrison-Woodbury! We set $B = A'A$. $u = \widetilde{X}_i$ and $v = -\widetilde{X}_i$.

That will be our value for the left-hand side of (6.25). Look what happens to the right-hand side:

$$1 + v'B^{-1}u = 1 - \widetilde{X}'_i(A'A)^{-1}\widetilde{X}_i \tag{6.28}$$

But that subtracted term is just h_{ii}! Now that's convenient.

It should be noted, however, that this approach has poor roundoff error properties if $A'A$ is *ill-conditioned*, meaning that it is nearly singular. This in turn can arise if some of the predictor variables are highly correlated with each other, as we will see in Chapter 8.

6.16.3 The Median Minimizes Mean Absolute Deviation

Let's derive (6.7).

First, suppose a random variable W has a density f_W. What value m minimizes $E[\|W - m\|]$?

$$
\begin{aligned}
E[\|W - m\|] &= \int_{-\infty}^{\infty} |t - m| \, f_W(t)dt \\
&= \int_{-\infty}^{m} (m - t) \, f_W(t)dt + \int_{m}^{\infty} (t - m) \, f_W(t)dt \\
&= mP(W < m) - \int_{-\infty}^{m} t \, f_W(t)dt \\
&\quad + \int_{m}^{\infty} t \, f_W(t)dt - mP(W > m) \tag{6.29}
\end{aligned}
$$

We have

$$mP(W < M) - mP(W > m) = 2mF_W(m) - m \tag{6.30}$$

where F_W is the cdf of W.

Differentiating with respect to m and setting the result to 0, we have

$$0 = 2F_W(m) - 1 \tag{6.31}$$

In other words, $m = \text{median}(W)$.

The extension to conditional median then follows the same argument as in Section 1.19.3.

6.16.4 The Gauss-Markov Theorem

The famous Gauss-Markov Theorem states that under the assumptions of linearity of $\mu(t)$, homoscedasticity and independent observations, the OLS is the Best Linear Unbiaed Estimator (BLUE) [116]. For any linear combination $c'\beta$ of the elements of β, then $c'\widehat{\beta}$ has minimum variance among all unbiased estimators of β that consist of linear functions of the Y_i. Note that the normality assumption for the Y_i is not needed.

Proofs of the theorem tend to take either an algebraic or an optimization approach. We will follow the latter path here.

6.16.4.1 Lagrange Multipliers

A method for optimization under constraints is known as *Lagrange multipliers*. The method will be useful for Gauss-Markov, where our constraint is unbiasedness.

To see how the method works, say we wish to find the maximum value of $x^3 + y$ subject to the constraint $x^2 + y^2 = 1$. We set up the *Lagrangian*,

$$L = x^3 + y + \lambda(x^2 + y^2 - 1) \tag{6.32}$$

We have introduced a number variable, λ, as an artifice to help us meet the constraint. So, we set partial derivatives to 0:

$$0 = \frac{\partial L}{\partial x} = 3x^2 + \lambda \cdot 2x \tag{6.33}$$

$$0 = \frac{\partial L}{\partial y} = 1 + \lambda \cdot 2y \tag{6.34}$$

$$0 = \frac{\partial L}{\partial \lambda} = x^2 + y^2 - 1 \tag{6.35}$$

The second equation gives us $\lambda = -1/(2y)$, so that the first equation becomes

$$0 = 3x^2 - 2x/(2y) \tag{6.36}$$

Setting aside the case $x = 0$ for now, we have $0 = 3x - 1/y$, i.e., $3xy = 1$. Using this and (6.35), we can solve for x and y (and finally dismiss $x = 0$).

6.16.4.2 Proof of Gauss-Markov

Let's start with the case of $p = 1$, with a model with no intercept term. The vector β now consists of a single element, and the matrix A in (2.28) consists of a single column, a. The vector $A'D$ in that equation reduces to $a'Y$.

Consider any linear function of our Y_i, say $u'Y$. Its variance is

$$Var(u'Y) = Var\left(\sum_{i=1}^{n} u_i^2 \sigma^2\right) \tag{6.37}$$

The unbiasedness constraint is

$$\beta = E(u'Y) = u'EY = u'a'\beta \tag{6.38}$$

We are trying to find the BLUE u. Our Lagrangian is

$$L = \sigma^2 \sum_{i=1}^{n} u_i^2 + \lambda(u'a'\beta - \beta) \tag{6.39}$$

Setting derivatives to 0, we have

$$0 = \frac{\partial L}{\partial u_i} = 2\sigma^2 u_i + \lambda a_i \beta \tag{6.40}$$

$$0 = \frac{\partial L}{\partial \lambda} = u'a'\beta - \beta \tag{6.41}$$

From (6.40), we have

$$u_i = -\frac{\lambda a_i \beta}{2\sigma^2} \tag{6.42}$$

Equation (6.41) must hold for any β, so

$$1 = \sum_{j=1}^{n} u_j a_j \tag{6.43}$$

Substituting from (6.42) in (6.43), we have

$$1 = -\frac{\lambda \beta}{2\sigma^2} \sum_{j=1}^{n} a_j^2 \tag{6.44}$$

i.e.,

$$-\frac{\lambda \beta}{2\sigma^2} = \frac{1}{\sum_{j=1}^{n} a_j^2} \tag{6.45}$$

Reusing that in (6.42), we finally have

$$u_i = \frac{a_i}{\sum_{j=1}^{n} a_j^2} \tag{6.46}$$

Our BLUE estimator is then

$$\sum_{i=1}^{n} u_i Y_i \tag{6.47}$$

The reader should check that this is exactly the OLS estimator (2.28): In the latter, for instance,

$$A'A = a'a = \sum_{j=1}^{n} a_j^2 \tag{6.48}$$

For the general case, $p > 1$, one can actually use the same approach (Exercise 12).

6.17   Exercises: Data, Code and Math Problems

Data problems:

1. The contributors of the Forest Fire data set to the UCI Machine Learning Repository, *https://archive.ics.uci.edu/ml/datasets/Forest+Fires*, describe it as "a difficult regression problem." Apply the methods of this chapter to attempt to tame this data set.

2. Using the methods of this chapter, re-evaluate the two competing Poisson-based analyses in Section 4.4.

3. Was logit a good model in the example in Section 4.5.3? Apply the methods of this chapter to check.

4. Add an interaction term for age and gender in the linear model in Section 6.10. Interpret the results.

5. Apply **parvsnonparplot()** to the diabetes data in Section 6.11.1, and discuss possible fit problems.

6. In the discussion of Figure 6.1, it was noted that in investigating possible areas of poor fit for the parametric model, we should keep in mind possible bias of the nonparametric model at the left and right ends of the figure. We saw possible problems with the parametric model at both ends, but one of them may be due in part to the bias issue. State which one, and explain why.

Mini-CRAN and other computational problems:

7. Write an R function analogous to **influence()** for **quantreg** objects.

8. Recall that with Poisson regression we have (4.12), while with overdispersed models we have (4.13). Use **knnest()** to plot variance against mean in the Pima data, Section 4.3.2, in order to partly assess whether a Poisson model works there.

9. The **knnest()** function in the **regtools** package is quite versatile. Its **nearf** argument allows us to apply general functions to Y values in a neighborhood, rather than simply averaging them. Here we will apply that to quantiles.

The γ quantile of a cdf F is defined to be a number d such that $F(d) = \gamma$. For a continuous distribution, that number is unique, and the quantile function is the inverse of the cdf.

But that is not true in the discrete case. The latter is especially problematic in the case of finding sample quantiles. How, for instance, should we define the sample median if the sample size n is an even number? This is such a problem that R's **quantile()** function actually offers the user 9 different definitions of "quantile."

We will be estimating conditional quantiles, defined by

$$q(t, \gamma) = F_t^{-1}(\gamma) \tag{6.49}$$

where $F_t(w)$ is defined to be $P(Y \leq w \mid X = t)$.

(a) Write an R function with call form

quanty (predpt , nearxy ,**gamma**, type)

to be used as **nearf** in calling **knnest()**. Here **type** is as in **quantile()**.

(b) Apply this function to the baseball player data, regression weight against height, and compare to the results of applying the **quantreg** package.

10. Consider m-fold cross-validation, Section 6.6.2, in a k-NN context. Change **preprocessx()** to allow this, so that the new call form will be

preprocessx (x,kmax, xval=FALSE)

Math problems:

11. Say we use parallel coordinates (Section 6.7.3.1) to display some data having $p = 2$. Say some of our points lie on a straight line in $(X^{(1)}, Xi^{(2)})$ space. Show that in the parallel coordinates plot, the lines corresponding to these points will all intersect at a common point. (It might be helpful to generate some data and form their parallel coordinates plot to help your intuition.)

12. Prove the general case of the Gauss-Markov Theorem, showing that the OLS estimator $c'\widehat{\beta}$ is the BLUE of $c'\beta$ for any c. Follow the same pattern as in Section 6.16.4.2, replacing β by $c'\beta$ and so on.

Chapter 7

Disaggregating Regressor Effects

What does the above chapter title mean? It is, admittedly, rather overly abstract, but it does capture the essense of this chapter, as follows.

Recall that a synonym for "predictor variable" is *regressor*. This chapter is almost entirely focused on the Description goal, i.e., analysis of the individual effects of the regressors on the response variable Y. In a linear model, those effects are measured by the β_i. Well, then, what is meant by the term *disaggregating* in our chapter title?

Recall the example in Section 1.11.1, regarding a study of the quality of care given to heart attack patients in a certain hospital chain. The concern was that one of the hospitals served a population with many elderly patients, thus raising the possibility that the analysis would unfairly present this hospital as providing inferior care. We thus want to separate out — *disaggregate*, as the economists say — the age effect, by modeling the probability of survival as a function of hospital branch and age. We could, for instance, use a logistic model, with survival as the binary response variable, and with age and dummy variables for the hospital branches as predictors. The coefficients of the dummies would then assess the quality of care for the various branches, independent of age issues.

Most of this chapter will be concerned with measuring effects of predictor variables, with such disaggregation in mind. We will see, though, that attaining this goal may require some subtle analysis. It will be especially important to bear in mind the following principle:

The Predictor Effects Principle

The sign and magnitude of a regression coefficient (whether sample estimate or population value) for one predictor variable may depend on which other predictors are present.

This in turn is related to correlations among the predictors. In a linear or generalized linear model, the coefficient for predictor i may reflect not only the effect of that predictor, but may also incorporate the effect of some predictor j that is correlated with it.

In a Description context, this may be rather unsettling news, but it clearly can't be ignored. One must indeed be duly cautious in regression applications in which Description is the primary goal. One of the goals of this chapter is to better understand this issue, and thus deal with it in practical settings.

Re-aggregation

In addition, this chapter will also cover what might be termed *re-aggregation* of the predictors. We use this term in the spirit of R's **aggregate()** function, which computes summary statistics for subgroups of the data. As noted above, a straight use of **aggregate()** in the hospital example would be inappropriate; we need to do aggregation in a manner that takes covariates such as age into account, and that will be the focus of Section 7.5.

Our re-aggregation discussion here will also concern statistical methodology known as *small area estimation*, which for example often arises in county data. We may have data on aspects of interest for the larger counties of the state, but have only limited data or none at all for some smaller ones. Small area estimation involves using predictor variables to estimate the missing values.

7.1   A Small Analytical Example

Before analyzing some real data, let's get an overview of the situation, via a small mathematical example. Our model will have two predictors,

$$E(Y|X^{(1)}, X^{(2)}) = \beta_0 + \beta_1 X^{(1} + \beta_2 X^{(2)} \tag{7.1}$$

But what if we use only $X^{(1)}$, i.e., we wish to work with

$$E(Y \mid X^{(1)}) \tag{7.2}$$

Using the Tower Property for conditional expectation (stated in Section 1.19.5.4, proven in Section 7.8.1 in the Mathematical Complements section at the end of this chapter), (7.2) is equal to the conditional expected value of (7.1) with respect to $X^{(1)}$, i.e.,

$$E(Y \mid X^{(1)}) = E\left[E(Y \mid X^{(1)}, X^{(2)}) \mid X^{(1)}\right] \tag{7.3}$$

This is very abstract, of course, but it merely says that the regression function of Y on $X^{(1)}$ and $X^{(2)}$, averaged over the values of $X^{(2)}$ is equal to the regression function of Y on $X^{(1)}$ alone. If we take, say, the mean weight of all people of a given height and age, and then average over the values of age, we obtain mean weight of all people of a given height.

This gives us

$$E(Y \mid X^{(1)}) = \beta_0 + \beta_1 X^{(1)} + \beta_2 E(X^{(2)} \mid X^{(1)}) \tag{7.4}$$

Suppose the regression function of $X^{(2)}$ on $X^{(1)}$ is also linear (which for example will be the case if the three variables have a trivariate normal distribution):

$$E(X^{(2)} \mid X^{(1)}) = \gamma_0 + \gamma_1 X^{(1)} \tag{7.5}$$

so that

$$E(Y \mid X^{(1)}) = \beta_0 + \beta_2 \gamma_0 + (\beta_1 + \beta_2 \gamma_1) X^{(1)} \tag{7.6}$$

Here is the point: Say for convenience that β_1, β_2 and γ_1 are all positive. Comparing (7.6) and (7.1), we see that if we use the two-predictor model (7.1) instead of (7.2), the effect of $X^{(1)}$ on Y shrinks by the amount $\beta_2 \gamma_1$. Putting it more colloquially, adding the predictor $X^{(2)}$ "steals some of $X^{(1)}$'s thunder." So, the effect of $X^{(1)}$ on Y is smaller if we include $X^{(2)}$ in our analysis.

On the other hand, suppose the β_i are positive, but γ_1 is so negative as to make

$$\beta_1 + \beta_2 \gamma_1 < 0 \qquad (7.7)$$

Then the sign of the $X^{(1)}$ effect changes from positive, $\beta_1 > 0$, to negative, (7.7).

Also, observe that if $X^{(1)}$ and $X^{(2)}$ are independent, or at least uncorrelated, then $\gamma_1 = 0$, so that the coefficient of $X^{(1)}$ will be the same with or without $X^{(2)}$ in our analysis.

This little example shows what is occurring in the background in the Predictor Effects Principle.

7.2   Example: Baseball Player Data

In Section 1.9.1.2, we found that the data indicated that older baseball players — of the same height — tend to be heavier, with the difference being about 1 pound gain per year of age. This finding may surprise some, since athletes presumably go to great lengths to keep fit. Ah, so athletes are similar to ordinary people after all.

We may then ask whether a baseball player's weight is also related to the position he plays. So, let's now bring the Position variable in our data into play. First, what is recorded for that variable?

```
> levels(mlb$Position)
[1] "Catcher"            "First_Baseman"
[3] "Outfielder"         "Relief_Pitcher"
[5] "Second_Baseman"     "Shortstop"
[7] "Starting_Pitcher"   "Third_Baseman"
```

So, all the outfield positions have been simply labeled "Outfielder," though pitchers have been separated into starters and relievers.

In order to have a handy basis of comparison below, let's re-run the weight-height-age analysis:

```
> summary(lm(Weight ~ Height + Age, data=nondh))
...
Coefficients:
                Estimate Std. Error t value Pr(>|t|)
```

```
( Intercept )  −187.6382    17.9447   −10.46  < 2e−16
Height            4.9236     0.2344    21.00  < 2e−16
Age               0.9115     0.1257     7.25  8.25e−13

( Intercept )  ***
Height         ***
Age            ***
. . .
Multiple R-squared:    0.318,
Adjusted R-squared:    0.3166
```

Now, for simplicity and also to guard against overfitting, let's consolidate into four kinds of positions: infielders, outfielders, catchers and pitchers. That means we'll need three dummy variables:

```
> pos <− mlb$Position
> infld <− as.integer(pos %in%
     c('First_Baseman','Second_Baseman','Shortstop',
        'Third_Baseman'))
> outfld <− as.integer(pos == 'Outfielder')
> pitcher <− as.integer(pos %in% c('Relief_Pitcher',
     'Starting_Pitcher'))
```

Again, remember that catchers are designated via the other three dummies being 0.

So, let's run the regression:

```
> lmpos <− lm(Weight ∼ Height + Age + infld +
    outfld + pitcher ,data=mlb)
> summary(lmpos)
. . .
Coefficients :
              Estimate Std. Error  t value Pr(>|t|)
( Intercept )  −182.7216   18.3241   −9.972  < 2e−16
Height            4.9858    0.2405   20.729  < 2e−16
Age               0.8628    0.1241    6.952  6.45e−12
infld            −9.2075    1.6836   −5.469  5.71e−08
outfld           −9.2061    1.7856   −5.156  3.04e−07
pitcher         −10.0600    2.2522   −4.467  8.84e−06

( Intercept )  ***
Height         ***
Age            ***
```

```
infld        ***
outfld       ***
pitcher      ***
. . .
```
Mult. **R**-squared: 0.3404, Adj. **R**-squared: 0.3372
. . .

The estimated coefficients for the position variables are all negative. At first, it might look like the town of Lake Wobegon in the radio show *Prairie Home Companion*, "Where all children are above average." Do the above results say that players of all positions are below average?

No, not at all. Look at our model:

$$\text{mean weight} = \tag{7.8}$$

$$\beta_0 + \beta_1 \times \text{ height } + \beta_2 \times \text{ age } + \beta_3 \times \text{ infld } + \beta_4 \times \text{ outfld } + \beta_5 \times \text{ pitcher}$$

Under this model, let's find the difference in mean weight between two subpopulations — 72-inches-tall, 30-year-old pitchers and catchers of the same height and age. Keeping in mind that catchers are coded with 0s in all three dummies, we see that the difference in mean weights is simply β_5!

In other words, β_3, β_4 and β_5 are mean weights *relative to catchers*. Thus for example, the interpretation of the 10.06 figure is that, for a given height and age, pitchers are on average about 10.06 pounds lighter than catchers of the same height and age, while for outfielders the figure is about 9.2 pounds. An approximate 95% confidence interval for the population value of the latter (population mean for outfielders minus population mean for catchers) is

$$-9.2 \pm 2 \times 1.8 = (-12.8, -5.6) \tag{7.9}$$

So, the image of the "beefy" catcher is borne out.

Note that the estimated coefficient for age shrank a little when we added the position variables. In our original analysis, with just height and age as predictors, it had been 0.9115,[1] but now is only 0.8628. The associated confidence interval, (0.61,1.11), still indicates weight increase with age, but the effect is now smaller than before. This is an example of Predictor Effects Principle, mentioned at the outset of this chapter, that the coefficient for one predictor may depend on what other predictors are present.

---

[1] This was the case even after removing the Designated Hitters, not shown here.

It could be that this shrinkage arose because catchers are somewhat older on average. In fact,

```
> mc <- [mlb$PosCategory == 'Catcher',]
> mean(mc$Age)
[1] 29.56368
> mean(mlb$Age)
[1] 28.70835
```

So the catchers "absorbed" some of that age effect, as explained in Section 7.1.

This also suggests that the age effect on weight is not uniform across playing positions. To investigate this, let's add interaction terms:

```
> summary(lm(Weight ~ Height + Age +
    infld + outfld + pitcher +
    Age*infld + Age*outfld + Age*pitcher, data=nondh))
...
Coefficients:
```

	Estimate	Std. Error	t value	Pr(>|t|)
(Intercept)	−168.5453	20.3732	−8.273	4.11e−16
Height	4.9854	0.2407	20.714	< 2e−16
Age	0.3837	0.3335	1.151	0.2501
infld	−22.8916	11.2429	−2.036	0.0420
outfld	−27.9894	11.9201	−2.348	0.0191
pitcher	−31.9341	15.4175	−2.071	0.0386
Age:infld	0.4629	0.3792	1.221	0.2225
Age:outfld	0.6416	0.4033	1.591	0.1120
Age:pitcher	0.7467	0.5232	1.427	0.1539

(Intercept)	***
Height	***
Age	
infld	*
outfld	*
pitcher	*
Age:infld	
Age:outfld	
Age:pitcher	

```
...
Mult. R-squared:  0.3424,    Adj. R-squared:  0.3372
```

(Use of a colon in a regression formula, e.g., **Age:infld** here, means to add an interaction term for the indicated pair of variables, which as we have seen before is just their product. More on this in the Computational Complements section at the end of this chapter.)

This doesn't look helpful. Confidence intervals for the estimated interaction coefficients are near 0,[2] and equally important, are wide. Thus there could be important interaction effects, or they could be tiny; we just don't have a large enough sample to say much.

Note that the coefficients for the position dummies changed quite a bit, but this doesn't mean we now think there is a larger discrepancy between weights of catchers and the other players. For instance, for 30-year-old players, the estimated difference in mean weight between infielders and catchers of a given height is

$$-22.8916 + 30 \times 0.4629 = -9.0046 \tag{7.10}$$

similar to the -9.2075 figure we had before. Indeed, this is another indication that interaction terms are not useful in this case.

7.3   Simpson's Paradox

The famous *Simpson's Paradox* should not be considered a paradox, when viewed in the light of a central point we have been discussing in this chapter, which we will state a little differently here:

> The regression coefficient (sample or population) for a predictor variable may change substantially when another predictor is added. In particular, its sign may change, from positive to negative or *vice versa*.

7.3.1   Example: UCB Admissions Data (Logit)

The most often-cited example, in a tabular context, is that of the UC Berkeley admissions data [20]. The issue at hand was whether the university had been discriminating against women applicants for admission to graduate school.

---

[2]In fact, they include 0, but as discussed before, we should not place any distinction between near 0 and include 0.

On the surface, things looked bad for the school — 44.5% of the male applicants had been admitted, compared to only 30.4% of the women. However,
upon closer inspection it was found that the seemingly-low female rate was
due to the fact that the women tended to apply to more selective academic
departments, compared to the men. After correcting for the Department
variable, it was found that rather than being victims of discrimination, the
women actually were slightly favored over men. There were six departments
in all, labeled A-F.

The data set is actually included in base R. As mentioned, it is stored in
the form of an R table:

```
> ucb <- UCBAdmissions
> class(ucb)
[1] "table"
> ucb
, , Dept = A

          Gender
Admit      Male Female
  Admitted  512     89
  Rejected  313     19

, , Dept = B

          Gender
Admit      Male Female
  Admitted  353     17
  Rejected  207      8
...
```

In R, it is sometimes useful to convert a table to an artificial data frame,
which in this case would have as many rows as there were applicants in the
UCB study, 4526. The **regtools** function **tbltofakedf()** facilitates this:

```
> ucbdf <- tbltofakedf(ucb)
> dim(ucbdf)
[1] 4526    3
> head(ucbdf)
      [,1]        [,2]    [,3]
[1,] "Admitted" "Male"  "A"
[2,] "Admitted" "Male"  "A"
[3,] "Admitted" "Male"  "A"
[4,] "Admitted" "Male"  "A"
```

```
[5,]  "Admitted"  "Male"  "A"
[6,]  "Admitted"  "Male"  "A"
```

The first six rows are the same, and in fact there will be 512 such rows, since, as seen above, there were 512 male applicants who were admitted to Department A.

Let's analyze this data using logistic regression. With such coarsely discrete data, this is not a typical approach,[3] but it will illustrate the dynamics of Simpson's Paradox.

First, convert to usable form, not R factors. It will be convenient to use the **dummies** package [26]:

```
> ucbdf$admit <- as.integer(ucbdf[,1] == 'Admitted')
> ucbdf$male <- as.integer(ucbdf[,2] == 'Male')
# save work by using the 'dummies' package
> library(dummies)
> dept <- ucbdf[,3]
> deptdummies <- dummy(dept)
> head(deptdummies)
     deptA deptB deptC deptD deptE deptF
[1,]    1     0     0     0     0     0
[2,]    1     0     0     0     0     0
[3,]    1     0     0     0     0     0
[4,]    1     0     0     0     0     0
[5,]    1     0     0     0     0     0
[6,]    1     0     0     0     0     0
# only 5 dummies
> ucbdf1 <- cbind(ucbdf,deptdummies[,-6])[,-(1:3)]
> head(ucbdf1)
  admit male deptA deptB deptC deptD deptE
1   1    1     1     1     0     0     0
2   1    1     1     0     0     0     0
3   1    1     1     0     0     0     0
4   1    1     1     0     0     0     0
5   1    1     1     0     0     0     0
6   1    1     1     0     0     0     0
```

Now run the logit, first only with the **male** predictor, then adding the departments:

```
> glm(admit ~ male, data=ucbdf1, family=binomial)
```

---

[3]A popular method for tabular data is *log-linear models* [35] [2].

```
...
Coefficients:
(Intercept)            male
    -0.8305           0.6104
...
> glm( admit ~ ., data=ucbdf1 , family=binomial )
...
Coefficients:
(Intercept)          male           deptA
   -2.62456       -0.09987         3.30648
      deptB          deptC           deptD
    3.26308        2.04388         2.01187
      deptE
    1.56717
...
```

So the sign for the **male** variable switched from positive (men are favored) to slightly negative (women have the advantage). Needless to say, this analysis (again, in the original table form, not logit) caused quite a stir. The evidence against the university had looked so strong, only to find later that an overly simple statistical analysis had led to an invalid conclusion.

By the way, note that the coefficients for all five dummies were positive, which reflects the fact that all the departments A-E had higher admissions rates than department F:

```
> apply( ucb , c ( 1 ,3 ) ,sum )
          Dept
Admit         A    B    C    D    E    F
   Admitted 601  370  322  269  147   46
   Rejected 332  215  596  523  437  668
```

Let's take one more look at this data, this time more explicitly taking the selectivity of departments into account. We'll create a new variable, finding the acceptance rate for each department and then replacing each applicant's department information by the selectivity of that department:

```
> deptsums <- apply( ucb , c ( 1 ,3 ) ,sum )
> deptrates <- deptsums [ 1 ,] / colSums ( deptsums )
> deptsums <- apply( ucb , c ( 1 ,3 ) ,sum )
> deptrates <- deptsums [ 1 ,] / colSums ( deptsums )
> glm( admit ~ male + deptrate , data=ucbdf )
...
Coefficients:
```

```
(Intercept)              male        deptrate
   0.003914          −0.016426       1.015092
```

Consistent with our earlier analysis, the coefficient for the **male** variable is slightly negative. But we can also quantify our notion that the women were applying to the more selective departments:

```
> tapply(ucbdf$deptrate, ucbdf$male, mean)
         0           1
0.2951732   0.4508945
```

The mean admissions rate among departments applied to by women is much lower than the corresponding figure for men.

7.3.2   The Verdict

Simpson's is not really a paradox — let's just call it Simpson's Phenomenon — but is crucially important to keep in mind in applications where Description is the goal. And the solution to the "paradox" is to think twice before deleting any predictor variables.

Ironically, this last point is somewhat at odds with the theme of Chapter 9, in which we try to pare down the number of predictors. When we have correlated variables, such as Gender and Department in the admissions data, it might be tempting to delete one or more of them on the grounds of "redundancy," but we first should check the effects of deletion, e.g., sign change.[4]

On the other hand, this is rather consistent with the method of *ridge regression* in Chapter 8. That approach attempts to ameliorate the effects of correlated predictor variables, rather than resorting to deleting some of them.

Once again, we see that regression and classification methodology does not always offer easy, pat solutions.

7.4   Unobserved Predictor Variables

In Statistical Heaven, we would have data on all the variables having substantial relations to the response variable. Reality is sadly different, and

---

[4]In the admissions data, the correlation, though substantial, would probably not warrant deletion in the first place, but the example does illustrate the dangers.

often we feel that our analyses are hampered for lack of data on crucial variables.

Statisticians have actually developed methodology to deal with this problem, such as the famous Fay-Herriott model. Not surprisingly, the methods have stringent assumptions, and they are hard to verify. This is especially an issue in that many of the models used posit *latent* variables, meaning ones that are unseen. But such methods should be part of the toolkit of any data scientist, either to use where appropriate or at least understand when presented with such analyses done by others. The following sections will provide brief introductions to such methodology.

7.4.1 Instrumental Variables (IVs)

This one is quite controversial. (Or, at least the choice of one's IV often evokes controversy.) It's primarily used by economists, but has become increasingly popular in the social and life sciences. The goal is to solve the problem of not being able to observe data on a variable that ideally we wish to use as a predictor. We find a kind of proxy, known as an *instrument*.

The context is that of a Description goal. Suppose we are interested in the relation between Y and two predictors, $X^{(1)}$ and $X^{(2)}$, with a focus on the former. What we would like to find is the coefficient of $X^{(1)}$ *in the presence of* $X^{(2)}$, i.e., estimate β_1 in

$$E(Y \mid X^{(1)}, X^{(2)}) = \beta_0 + \beta_1 X^{(1)} + \beta_2 X^{(2)} \tag{7.11}$$

But the problem at hand here is that we observe Y and $X^{(1)}$ but not $X^{(2)}$.

We believe that the two population regression functions (one predictor vs. two predictors) are well approximated by linear models:[5]

$$E(Y \mid X^{(1)}) = \beta_{01} + \beta_{11} X^{(1)} \tag{7.12}$$

$$E(Y \mid X^{(1)}, X^{(2)}) = \beta_{02} + \beta_{12} X^{(1)} + \beta_{22} X^{(2)} \tag{7.13}$$

Note that we are doubly-subscripting the β coefficients, since we have two linear models.

---

[5]The second model does not imply the first. What if $X^{(2)}) = (X^{(1)})^2$, for instance?

We are primarily interested in the role of $X^{(1)}$, i.e., the value of β_{12}. However, as has been emphasized so often in this chapter, generally

$$\beta_{11} \neq \beta_{12} \qquad (7.14)$$

Thus an analysis on our data that uses (7.12) — remember, we cannot use (7.13), since we have no data on $X^{(2)}$ — may be seriously misleading.

A commonly offered example concerns a famous economic study regarding the returns to education [31]. Here Y is weekly wage and $X^{(1)}$ is the number of years of schooling. The concern was that this analysis doesn't account for "ability"; highly-able people (however defined) might pursue more years of education, and thus get a good wage due to their ability, rather than the education itself. If a measure of ability were included in our data, we could simply use it as a covariate and fit the model (7.13), but no such measure was included in the data.[6]

The *instrumental variable* (IV) approach involves using another variable, observed, that is intended to remove from $X^{(1)}$ the portion of that variable that involves ability. That variable — the instrument — works as a kind of surrogate for the unobserved variable. If this works — a big "if" — then we will be able to measure the effect of years of schooling without the confounding effect of ability.

In the years-of-schooling example, the instrument proposed is distance from a college. The rationale here is that, if there are no nearby postsecondary institutions, the person will find it difficult to pursue a college education, and may well decide to forego it — even if the person is of high ability. All this will be quantified below, but keep this in mind as a running example.

Note that the study was based on data from 1980, when there were fewer colleges in the U.S. than there are now. Thus this particular instrument may be less useful today, but it was questioned even when first proposed. As noted in the introduction to this section, the IV approach is quite controversial.

Adding to the controversy is that different authors have defined the conditions required for use of IVs differently. Furthermore, in some cases definitions of IV have been insufficiently precise to determine whether they are equivalent to others.

Nevertheless, the wide use of IV in certain disciplines warrants taking a

---

[6]Of course, even with better data, "ability" would be hard to define. Does it mean IQ (of which I am very skeptical), personal drive or what?

closer look, which we do below.

7.4.1.1   The IV Method

Let Z denote our instrument, i.e., an observed variable that we hope will remove the effect of our unobserved variable $X^{(2)}$. The instrument must satisfy two conditions, to be described shortly. In preparation for this, set

$$\epsilon \; = \; Y - E(Y|X^{(1)}, X^{(2)}) \tag{7.15}$$
$$= \; Y - (\beta_{02} + \beta_{12}X^{(1)} + \beta_{22}X^{(2)}) \tag{7.16}$$

Now, letting $\rho(U, V)$ denote the population correlation between and two random variables U and V, the requirements for an IV Z are

(a) $\rho(Z, X^{(1)}) \neq 0$

(b) $\rho(Z, X^{(2)}) = 0$

(c) $\rho(Z, \epsilon) = 0$

In the education example, (a) means that distance is related to years of schooling, as posited above, while (b) means that distance is unrelated to ability. But what about (c)?

Recall that ϵ is called the "error term," and is interpretable here as the collective effect of all variables besides $X^{(1)}$ and $X^{(2)}$ that are related to Y (Section 2.2). These other variables are assumed to have a *causal* relation with Y (another controversy), as opposed to Z whose relation to Y is only via $X^{(1)}$. This motivates (c), which in the schooling example means that not only is distance unrelated to ability, but also distance is unrelated to any other variable having an impact on earnings.

Clearly, (c) is a very strong assumption, and one that is not assessable, i.e., one cannot devise some check for this assumption as we did for various assumptions in Chapter 6.

How are the above conditions used in the mathematics underlying the IV method? Let's perform a (population-level) calculation. Writing

$$Y = \beta_{02} + \beta_{12}X^{(1)} + \beta_{22}X^{(2)} + \epsilon \tag{7.17}$$

and using the linearity of covariance, we have

$$
\begin{aligned}
Cov(Z,Y) &= \beta_{12}Cov(Z,X^{(1)}) + \beta_{22}Cov(Z,X^{(2)}) + Cov(Z,\epsilon) \\
&= \beta_{12}Cov(Z,X^{(1)}) \tag{7.18}
\end{aligned}
$$

and thus

$$
\beta_{12} = \frac{Cov(Z,Y)}{Cov(Z,X^{(1)})} \tag{7.19}
$$

We then take

$$
\widehat{\beta}_{12} = \frac{\widehat{Cov}(Z,Y)}{\widehat{Cov}(Z,X^{(1)})} \tag{7.20}
$$

where the estimated covariances come from the data, e.g., from the R **cov()** function. We can thus estimate the parameter of interest, β_{12} — in spite of not observing $X^{(2)}$.

This is wonderful! Well, wait a minute...is it too good to be true? Well, as noted, the assumptions are crucial, such as:

- We assume the linear models (7.12) and (7.13). The first can be assessed from our data, but the second cannot.

- We assume condition (b) above, i.e., that our instrument is uncorrelated with our unseen variable. Often we are comfortable with that assumption — e.g., that distance from a college is not related to ability — but again, it cannot easily be verified.[7]

- We assume condition (c), which as discussed earlier is quite strong.

- We need the instrument to have a fairly substantial correlation to the observed predictor, i.e., $\rho(Z,X^{(1)})$ should be substantial. If it isn't, then we have a small or even tiny denominator in (7.20), so that the sample variance of the quotient — and thus of our $\widehat{\beta}_{12}$ — will be large, certainly not welcome news.

---

[7] Some tests for this have been developed, but those have their own often-questionable assumptions.

7.4.1.2 Two-Stage Least Squares:

Another way to look at the IV idea is *Two-Stage Least Squares* (2SLS), as follows. Recall the phrasing used above, that the instrument

> is a variable that is intended to remove from $X^{(1)}$ the portion of that variable that involves ability.

That suggests regressing $X^{(1)}$ on Z.[8]

Let's see what happens, again at the population level. Using (7.17), write

$$E(Y \mid Z) = \beta_{02} + \beta_{12}E(X^{(1)}|Z) + \beta_{22}E(X^{(2)}|Z) + E(\epsilon|Z) \qquad (7.21)$$

By assumption, Z and $X^{(2)}$ are uncorrelated, as are Z and ϵ. If these variables also have a multivariate normal distribution, then they are independent. Assuming this, we have

$$E(X^{(2)}|Z) = E[X^{(2)}] \text{ and } E(\epsilon|Z) = E[\epsilon] = 0 \qquad (7.22)$$

In other words,

$$E(Y \mid Z) = c + \beta_{12}E(X^{(1)}|Z) \qquad (7.23)$$

for a constant $c = \beta_{02} + \beta_{22}E[X^{(2)}]$.

Now, remember, our goal is to estimate β_{12}, so its appearing in (7.23) is a welcome sight! Then what does that equation suggest at the sample level?

Well, $E(X^{(1)}|Z)$ is the regression of $X^{(1)}$ on Z. In other words, the process is as follows:

Two-Stage Least Squares:

- First regress $X^{(1)}$ on the instrument Z, to estimate $E(X^{(1)}|Z)$ at our sample values Z_i.
- In view of (7.23), we then treat these estimated regression values as our new "predictor" values — which we use to predict the Y_i.

---

[8] As noted before, the term "regress V on U" means to model the regression function of V given U.

- The resulting estimated slope will be $\widehat{\beta}_{12}$, the estimate we are seeking.

In other words, the IV method can be viewed as an application of 2SLS, with the predictor variable in the first stage being our instrument.

In terms of R, this would mean

```
lmout <- lm(x1 ~ z)
estreg <- lmout$fitted.values
b12hat <- coef(lm(y ~ estreg))[2]
```

The purpose of this section was to explain the notion of IVs. It shows directly where the name "Two-Stage" least squares comes from.

The above just gives us a point estimate of β_{12}. We need a standard error as well. This can be derived using the Delta Method (Section 3.6.1), which we explore in the exercises at the end of this chapter.

However, there are more sophisticated R packages for this, such as **ivmodel**, which give us all this and more, as seen in the next section.

7.4.1.3   Example: Years of Schooling

Data for the schooling example dicussed above is widely available. Here we will use the set **card.data**, available for instance in the **ivmodel** package [79].

There are many variables in the data set. Here we will just follow our earlier example, analyzing the effect of years of schooling on wage (in cents per hour), with nearness to a college as our instrument.

Let's first do the computation "by hand," as above:

```
> library(ivmodel)
> data(card.data)
> sch <- card.data
# without the instrument
> lm(wage ~ educ, data=sch)
...
Coefficients:
(Intercept)            educ
     183.95           29.66
# now with the IV
> stage1 <- lm(educ ~ nearc4, data=sch)
```

```
> lm(sch$wage ~ stage1$fitted.values)
. . .
Coefficients:
          (Intercept)    stage1$fitted.values
             -849.5                     107.6
```

Here is how we can get this from the machinery in **ivmodel**:

```
> ivmodel(Y=sch$wage,D=sch$educ,Z=sch$nearc4)
. . .
Coefficients of k-Class Estimators:

          k Estimate Std. Error  t value
OLS  0.0000  29.6554     1.7075   17.368
. . .
TSLS  1.0000 107.5723    15.3995    6.985
. . .
```

We see our IV result confirmed in the "TSLS" line, with the "OLS" line restating our non-IV result.[9] Note that the standard error increased quite a bit.

Now, what does this tell us? On the one hand, a marginal increase, say 1 year, of schooling seems to pay off much more than the non-IV analysis had indicated, about $1.08 per hour rather than $0.30. However, the original analysis had a much higher estimated intercept term, $1.84 vs. -$8.50. Let's compute predicted wage for 12 years of school, for instance, under both models:

```
> -8.50 + 12*1.07
[1]  4.34
> 1.84 + 12*0.30
[1]  5.44
```

That's quite a discrepancy! We can do much better if we include the other predictor variables available in the data (not shown), but we at least see that blind use of IV models — and of course any models — can lead to poor results.

---

[9]Other lines in the output, not shown here, show the results of applying other IV methods, including those of the authors of the package.

7.4.1.4   The Verdict

In our years-of-schooling example (Section 7.4.1.1), it was mentioned that
the assumption that the distance variable was unreleated to ability was
debatable. For example, we might reason that able children come from
able parents, and able parents believe college is important enough that
they should live near one. This is an example of why the IV approach is so
controversial.

Nevertheless, the possible effect of unseen variables itself can make an analysis controversial. IVs may be used in an attempt to address such problems.
However, extra care is warranted if this method is used.

7.4.2   Random Effects Models

Continuing our theme here in Section 7.4 of approaches to account for
unseen variables, we now turn briefly to *mixed effects models* [77] [52].
Consider a usual linear regression model for one predictor X,

$$E(Y \mid X = t) = \beta_0 + \beta_1 t \qquad (7.24)$$

for unknown constants β_0 and β_1 that we estimated from the data.

Now alter the model so that β_0 is random, each unit (e.g., each person)
having a different value, though all having a common value of β_1. We
might observe people over time, with X representing time and Y being
modeled as having a linear time trend. The slope β_1 of that time trend is
assumed the same for all people, but the starting point β_0 is not.

We might write our new model as

$$E(Y \mid X = t) = \beta_0 + B + \beta_1 t \qquad (7.25)$$

where B is a random variable having mean 0. Each person has a different
value of B, with the intercept for people now being a random variable with
mean β_0 and variance σ_B^2.

It is more common to write

$$Y = \beta_0 + \alpha + \beta_1 X + \epsilon \qquad (7.26)$$

where α and ϵ have mean 0 and variances σ_a^2 and σ_e^2. The population values to be estimated from our data are β_0, β_1, σ_a^2 and σ_e^2. Typically these are estimated via Maximum Likelihood (with the assumptions that α and ϵ have normal distributions, etc.), though the Method of Moments is possible too.

The variables α and ϵ are called *random effects* (they are also called *variance components*), while the $\beta_0 + \beta_1 X$ portion of the model is called a *fixed effecs*. This phrasing is taken from the term *fixed-X regression*, which we saw in Section 2.3; actually, we could view this as a random-X setting, but the point is that even β_1 is fixed. Due to the presence of both fixed and random effects, the term *mixed-effects model* is used.

7.4.2.1 Example: Movie Ratings Data

Consider again the MovieLens data introduced in Section 3.2.4. We'll use the 100,000-rating data here, which includes some demographic variables for the users. The R package **lme4** will be our estimation vehicle [9].

First we need to merge the ratings and demographic data. This entails use of R's **merge()** function, introduced in Section 3.5.1. See Section 7.7.1 for details for our current setting. Our new data frame, after applying the code in that section, is **u**.

We might speculate that older users are more lenient in their ratings. Let's take a look:

```
> z <- lmer(rating ~ age+gender+(1|usernum),data=u)
> summary(z)
...
Random effects:
 Groups     Name         Variance  Std.Dev.
 usernum    (Intercept)  0.175     0.4183
 Residual                1.073     1.0357
Number of obs: 100000, groups:   usernum,  943

Fixed effects:
              Estimate Std. Error  t value
(Intercept)   3.469074   0.048085   72.14
age           0.003525   0.001184    2.98
genderM      -0.002484   0.031795   -0.08

Correlation of Fixed Effects:
        (Intr)  age
```

```
age          -0.829
genderM   -0.461   -0.014
```

First, a word on syntax. Here our regression formula was

rating $\sim$ age + gender + (1|usernum)

Most of this looks the same as what we are accustomed to in **lm()**, but the last term indicates the random effect. In R formulas, '1' is used to denote a constant term in a regression equation (we write '-1' in our formula if we want no such term), and here '(1|usernum)' specifies a random effects intercept term that depends on **usernum** but *is unobserved*.

So, what is the answer to our speculation about age? Blind use of significance testing would mean announcing "Yes, there is a significant positive relation between age and ratings." But the effect is tiny; a 10-year difference in age would mean an average increase of only 0.03525, on a ratings scale of 1 to 5. There doesn't seem to be much difference between men and women either.

The estimated variance of α, 0.175, is much smaller than that for ϵ, 1.073.

Of course, much more sophisticated analyses can be done, adding a variance component for the movies, accounting for the different movie genres and so on.

7.4.3   Multiple Random Effects

Of course, we can have more than one random effect. Consider the movie data again, for instance (for simplicity, without the demographics). We might model a movie rating Y as

$$Y = \mu + \gamma + \nu + \epsilon \tag{7.27}$$

where γ and ν are random effects for the user and for the movie.

The **lme4** package can handle a very wide variety of such models, though speficiation in the call to **lmer()** can become quite complex.

7.4.4   Why Use Random/Mixed Effects Models?

We may be interested in quantities such as σ_α^2 for Description purposes, especially relative to other variances, such as σ_ϵ^2 as in the example above.

This is common in genetics applications, for instance, where we may wish to compare within-family and between-family variation.

Random/mixed effects models may be used in Prediction contexts as well. There is a rich set of methodology for this, concerning Best Linear Unbiased Prediction (BLUP), unfortunately beyond the scope of this book.

7.5 Regression Function Averaging

Recall my old consulting problem from Section 1.11.1:

> Long ago, when I was just finishing my doctoral study, I had my first experience in statistical consulting. A chain of hospitals was interested in comparing the levels of quality of care given to heart attack patients at its various locations. A problem was noticed by the chain regarding straight comparison of raw survival rates: One of the locations served a largely elderly population, and since this demographic presumably has more difficulty surviving a heart attack, this particular hospital may misleadingly appear to be giving inferior care.

How do we deal with such situations? As mentioned in the introduction to this chapter, my approach was to ask the question, How well would Hospital 3 (the one with many elderly) do if it were to serve the populations covered by all the hospitals, not just its own? This led to a method based on computing the average value of an estimated regression function [106] [107]. In subsequent years, this notion, which we'll call *regression function averaging* (RFA), has led not only to methods for handling situations like the hospital example but also methodology for dealing with missing values [34] [113].

Specifically, say we estimate some regression function $\mu(t)$, resulting in the estimate $\widehat{\mu}(t)$. RFA then averages the latter over some random variables $Q_1, ..., Q_r$, forming

$$\frac{1}{r} \sum_{i=1}^{r} \widehat{\mu}(Q_i) \tag{7.28}$$

Let's see what can be done with this.

7.5.1   Estimating the Counterfactual

What an imposing word, *counterfactual*! It simply means, "What would have happened if such-and-such had been different?" In our hospital example above, the counterfactual was

> How well would Hospital 3 (the one with many elderly) do if it were to serve the populations covered by all the hospitals, not just its own?

The problem there, in more specific statistical terms, was that the distribution of one of the predictors, age, was different in Hospital 3 than for the other hospitals. The field of *propensity matching* involves some rather complex machinery designed to equilibrate away differences in predictor variable distributions, by matching similar observations.

Roughly speaking, the idea is to choose a subset of the patients at Hospital 3 who are similar to the patients at other hospitals. We can then fairly compare the survival rate at Hospital 3 with those at the other institutions.

But we can do this more simply with RFA. The basic idea is to estimate the regression function on the Hospital 3 data, then average that function over the predictor variable data on all the hospitals. We would then have an estimate of the overall survival rate if Hospital 3 had treated all the patients. We could do this for each hospital, and thus compare them on a "level playing field" basis.

7.5.1.1   Example: Job Training

The Lalonde data is an oft-used example in the propensity matching literature. It is available in the CRAN **twang** package [29]. The first column is the treatment, 1 for participation in a training program and 0 if not in the program.[10] The outcome of interest is **re78**, earnings in 1978. We wish to compare the training and nontraining groups with respect to earnings.

Let's ask the question, "How much better would the nontraining group have done if they had had the training?" We don't want to simply do a direct comparison of the two sample mean earnings, in the training and nontraining groups, as the predictors likely have different distributions in the two groups. Here is where RFA can be helpful.

---

[10]The term "treatment" is not to be taken literally. It simply indicates a variable on which we wish to compare groups, e.g., male vs. female.

```
> library(twang)
> data(lalonde)
> ll <- lalonde
# separate data frame into
# training and nontraining grps
> trt <- which(ll$treat == 1)
> ll.t <- ll[trt,-1]
> ll.nt <- ll[-trt,-1]
# fit regression on training group
> lmout <- lm(re78 ~ .,data=ll.t)
# find and average the est reg ftn
# values on the nontraining grp
> cfys <- predict(lmout,ll.nt)
> mean(cfys)
[1] 7812.121
# compare to their actual earnings
> mean(ll.nt$re78)
[1] 6984.17
```

So, by this analysis, had those in the nontraining group undergone training, they would have earned about $800 more.

As with formal propensity analysis and many other statistical procedures, there are important assumptions here. The analysis tacitly assumes that the effects of the various predictors are the same for training and nontraining populations, something to ponder seriously in making use of the above figure.

Calculation of standard errors is discussed in Section 7.8.2 of the Mathematical Complements section at the end of this chapter.

7.5.2 Small Area Estimation: "Borrowing from Neighbors"

Mi casa es su casa [my house is your house] — Spanish saying

In this section we apply RFA to *small area estimation*. As noted in the introductory section of this chapter, this term arises from the need to estimate means, totals and other statistics in small subpopulations for which very little data is available [117].

In other words, small-area estimation methods aim to bolster the statistical accuracy of small samples. This is sometimes described as "borrowing" data from other areas to improve estimation in a small area.

Models are typically rather elaborate, and depend heavily on certain population distributions being normal. Moreover, many such models involve latent variables. But we can use RFA to do this more directly and with fewer assumptions.

Let's again look at the year 2000 census data on programmer and engineer salaries, Section 6.12.2.2. Say we are interested in the mean wage of female PhDs under age 35.[11] Let's find them in our sample:

```
> peldoc <- pel[which(pe$phd == 1),]
> z <- peldoc[peldoc$age < 35 & peldoc$sex == 2,]
> nrow(z)
[1] 23
```

Only 23 observations! Somewhat startling, since the full data set has over 20,000 observations, but we are dealing with a very narrow population here. Let's form a confidence interval for the mean wage:

```
> t.test(z$wageinc)
...
95 percent confidence interval:
 43875.90 70419.75
...
```

That is certainly a wide range, though again, not unexpectedly so. So, let's try RFA, as follows. (Justification for the intuitive analysis will be given at the end.) In (7.28), the Q_i are the X_i in the portion of our census data corresponding to our condition (female PhDs under age 35).

So we fit a regression function model to our full data, PhDs and everyone, resulting in fitted values $\hat{\mu}(X_i)$. We then average those values:

$$\frac{1}{N} \sum_{X_i \text{ in } A}^{n} \hat{\mu}(X_i) \tag{7.29}$$

where N is the number of female PhDs under 35.

We can call **predict()** to find the $\hat{\mu}(X_i)$. (Or, we could use **z$fitted.values**; this is left as an exercise for the reader.)

```
> lmout <- lm(wageinc ~ ., data=pel)
```

---

[11]This is a little different from typical SAE applications, where the grouping is geographical, but the principles are the same.

```
> prout <- predict(lmout, z[, -6])
> t.test(prout)
...
95 percent confidence interval:
 58111.78  74948.87
mean of x
 66530.33
```

So, we indeed do obtain a narrower interval, i.e., have a more accurate estimate of the mean wage in this particular subpopulation. Our point estimate (7.29) is $66,530.

Note that we could do this even if we had no Y_i information about the subpopulation in question, i.e., if we did not know the salaries of those femaler PhDs under 35, as long as we had their X_i values. This kind of situation arises often in small-area applications.

It should be mentioned that there is a bit more to that RFA interval than meets the eye. Technically, we have two sources of sampling variation:

- The X_i and Y_i for this subpopulation, i.e., the 23 observations, are considered a random sample from this subpopulation.

- The $\widehat{\beta}_i$ coming from the call to **lm()** are also random, as they are based on the 20,090 observations in our full data set, considered a random sample from the full population.

The confidence interval computed above only takes into account variation in that first bullet, not the second. We could use the methods developed in Section 7.8.2, but actually the situation is simpler here. Owing to the huge discrepancy between the two sample sizes, 23 versus 20090, we can to a good approximation consider the $\widehat{\beta}_i$ to be constants.

But there is more fundamental unfinished business to address. Does RFA even work in this setting? Is it estimating the right thing? For instance, as the sample size grows, does it produce a statistically consistent estimator of the desired population quantity?

To answer those questions, let A denote the subpopulation region of interest, such as female PhDs under age 35 in our example above. We are trying to estimate

$$E(Y \mid X \text{ in } A) \tag{7.30}$$

the population mean wage income given that the person is a female PhD under age 35. Let D denote the dummy variable indicating X in A. Then (7.30) is equal to

$$\frac{E(DY)}{P(X \text{ in } A)} \tag{7.31}$$

This makes intuitive sense. Since D is either 1 or 0, the numerator in (7.31) is the population average value of Y among people for whom X is in A — exactly (7.30) except for the scaling by $P(X \text{ in } A)$.

Now look at $E(DY)$. Using the Law of Total Expectation (1.58), write

$$E(DY) = E[E(DY \mid X)] \tag{7.32}$$

Now assume that D is a function of X. That is the case in our PhD example above, where D is defined in terms of X. Then in the conditional expectation in (7.32), D is a constant and

$$E(DY \mid X) = DE(Y|X) = D\mu(X) \tag{7.33}$$

Combining this with (7.32), we have

$$E(DY) = E[D \, \mu(X)] \tag{7.34}$$

so that (7.31) is

$$\frac{E[D \, \mu(X)]}{P(X \text{ in } A)} \tag{7.35}$$

The numerator here says, "Average the regression function over all people in the population for whom X in A." Now compare that to (7.29), which we will rewrite as

$$\frac{\frac{1}{n}\sum_{X_i \text{ in } A} \widehat{\mu}(X_i)}{(N/n)} \tag{7.36}$$

The numerator here is the sample analog of the numerator in (7.35), and the denominator is the sample estimate of the denominator in (7.35). So, (7.36) is exactly what we need, and it is exactly what we computed in the R code for the PhD example above.

7.5.3 The Verdict

The RFA approaches introduced in the last two sections are appealing in that they require many fewer assumptions than non-regression based methodology for the same settings. Of course, the success of RFA relies on having good predictor variables, at least collecitvely if not individually.

7.6 Multiple Inference

If you beat the data long enough, it will confess — old statistical joke

Events of small probability happen all the time, because there are so many of them — the late Jim Sutton, economics professor and dean

We have already seen many examples of the use of **lm()** and **glm()**, and the use of **summary()** to extract standard errors for the estimated regression coefficients. We can form an approximate 95% confidence interval for each population regression coefficient β_i as

$$\widehat{\beta}_i \pm 1.96 \ s.e.(\widehat{\beta}_i)/\sqrt{n} \tag{7.37}$$

The problem is that the confidence intervals are only *individually* at the 95% level. Let's look into this.

7.6.1 The Frequent Occurence of Extreme Events

Suppose we have a large collection of pennies. Although the minting process results in some random imperfections, resulting in a slight variation in the probability of heads from coin to coin, and though the design of a penny inherently weights one side a bit more likely than the other, we will assume here for simplicity that every penny has heads probability exactly 0.5.

Now suppose we give pennies to 500 people, and have each one toss his coin 100 times. Let's find the probability that at least one of them obtains more than 65 heads.

For a particular person, the probability of obtaining at most 65 heads is

```
> pbinom(65,100,0.5)
[1] 0.999105
```

(See the Computational Complements section at the end of this chapter if you are not familiar with this function.) So, the probability that a particular person gets more than 65 heads is quite small:

> 1 − 0.999105
[1] 0.000895

But the probability that at least one of the 500 people obtains more than 65 heads is not small at all:

> 1 − 0.99910^500
[1] 0.3609039

The point is that if one of the 500 pennies does yield more than 65 heads, we should not say, "Oh, we've found a magic penny." That penny would be the same as the others. But given that we are tossing so many pennies, there is a substantial probability that at least one of them yields a lot of heads.

This issue arises often in statistics, as we will see in the next section. So, let's give it a name:

The Principle of Rare Events

Consider a set of independent events, each with a certain probability of occurrence. Even if the individual even probabilities are small, there may still be a substantial probability that at least one of them occurs.

7.6.2   Relation to Statistical Inference

In many statistical situations, when we form many confidence intervals, we wish to state that they are *collectively* at the 95% level. (The reader may substitute significance testing for confidence intervals in the following discussion.) In the linear regression context, for instance, we wish to form the intervals in such a way that there is a 95% chance that all of the intervals do contain their respective β_i. But that collective probability may be considerably less than 0.95 if each of the intervals is at the 95% level. In the context of the Principle of Frequent Occurrence of Rare Events, the rare events at hand here consist of a confidence interval being incorrect, i.e., failing to contain the parameter being estimated. Each event has only a 0.95 probability, but there is a much larger chance than 0.05 that at least one of the intervals is incorrect.

Why is this a problem? Say one is in some Description application, running regression analysis with a large value of p, i.e., with many predictors. We are interested in seeing which predictors have a substantial impact on Y, and we form a confidence interval for each β_i. If a no-impact situation corresponds to $\beta_i = 0$, we are especially interested in which intervals are substantially away from 0.[12]

The problem is that even if all the true β_i are near 0, with large p the chances are high that at least one interval will accidentally be far from 0. Hence the joke quoted above — "beating the data long enough" alludes to looking at a large number of predictors in hope of finding some with strong impact, and "the data will confess" means at least one of the intervals will (incorrectly) grant our wish.

This goal of attaining an overall 0.95 level for our intervals is known as the *multiple inference* or *simultaneous inference* problem. Many methods have been developed, enough to fill entire books, such as [73] [25]. Many of the methods, though, are either tied exclusively to significance testing or rely heavily on assumptions of normally-distributed populations, both of which are restrictions this book aims to avoid.

Instead, we will focus here on two methods. The first, the *Bonferroni method* is quite famous. The second, the *Scheffe's method*, is well known in its basic form, but little known in the generalized form presented here.

7.6.3 The Bonferroni Inequality

This one is simple. Say we form two 95% confidence intervals.[13] Intuitively, their overall confidence level will be only something like 90%. Formally, let A_i, $i = 1, 2$ denote the events that the intervals fail to cover their respective population quantities. Then

$$P(A_1 \text{ or } A_2) =$$

$$P(A_1) + P(A_2) - P(A_1 \text{ and } A_2) \le P(A_1) + P(A_2) \qquad (7.38)$$

In other words, the probability that at least one of the intervals fails is at most 2 (0.05) = 0.10. If we desire an overall confidence level of 95%,

---

[12]Note that if an interval merely excludes 0 but has both lower and upper bounds near 0, we would probably not judge this predictor as having an important impact on Y.

[13]These typically will be approximate intervals, but for brevity this term will not be repeatedly inserted here.

we can form two intervals of level 97.5%, and be assured that our overall confidence is at least 95%.

Due to the qualifier *at least* above, we say that our intervals are *conservative*.

Using mathematical induction, this can be easily generalized to

$$P(A_1 \text{ or } A_2 \text{ or } ... \text{ or } A_k) \leq \sum_{i=1}^{k} P(A_i) \tag{7.39}$$

Using this for forming simultaneously valid confidence intervals as above, we have:

> If we form k confidence intervals of confidence level $1 - \alpha$, the overall confidence level is at least $1 - k\alpha$.

> Thus, to achieve overall confidence level of at least $1 - \alpha$, we must set the confidence level of each of our intervals to $1 - \alpha/k$.

7.6.4 Scheffe's Method

Consider the setting of Section 2.8.4. There we found that the asymptotic distribution of $\widehat{\beta}$ is multivariate normal with mean β and covariance matrix as in (2.150). The results of Section 7.8.3 imply that the quantity

$$W = s^2(\widehat{\beta} - \beta)'(A'A)^{-1}(\widehat{\beta} - \beta) \tag{7.40}$$

has an asymptotically chi-square distribution with $p+1$ degrees of freedom. That implies that

$$P(W \leq \chi^2_{\alpha,p+1}) \approx 1 - \alpha \tag{7.41}$$

where $\chi^2_{q,k}$ denotes the upper-q quantile of the chi-square distribution with k degrees of freedom.

This in turn implies that the set of all b such that

$$W = s^2(\widehat{\beta} - b)'(A'A)^{-1}(\widehat{\beta} - b) \tag{7.42}$$

forms an approximate $1 - \alpha$ confidence region for the population β in $p + 1$ dimensional space. This in itself is usually not very valuable, but it can be used to show the following highly useful fact:

The confidence intervals

$$c'\widehat{\beta} \pm s\sqrt{\chi^2_{\alpha,p+1}c'(A'A)^{-1}c} \tag{7.43}$$

for the population quantities $c'\beta$ have *overall* confidence level $1 - \alpha$, where c ranges over all $p + 1$ dimensional vectors.

This may seem abstract, but for example consider a vector c consisting of all 0s except for a 1 in position i. Then $c'\widehat{\beta} = \widehat{\beta}_i$ and $c'\beta = \beta_i$. In other words, (7.43) is giving us *simultaneous* confidence intervals for all the coefficients β_i.

Another common usage is to set c to a vector having all 0s except for a 1 at positions i and a -1 at j. This sets up a confidence interval for $\beta_i - \beta_j$, allowing us to compare coefficients.

Recall that the quantity

$$s^2(A'A)^{-1} \tag{7.44}$$

is given by R's **vcov()** function. Denoting this matrix by V, (7.43) becomes

$$c'\widehat{\beta} \pm \sqrt{\chi^2_{\alpha,p+1}c'V^{-1}c} \tag{7.45}$$

The above analysis extends to any asymptotically normal estimator:

Let $\widehat{\theta}$ denote an asymptotically r-variate normal estimator of a population vector θ, with a consistent estimator $\widehat{Cov}(\widehat{\theta})$ of the associated covariance matrix. Then the confidence intervals

$$c'\widehat{\theta} \pm \sqrt{\chi^2_{\alpha,p+1}c'[\widehat{Cov}(\widehat{\theta})]^{-1}c} \tag{7.46}$$

for the population quantities $c'\theta$ have overall confidence level $1 - \alpha$, where c ranges over all r dimensional vectors.

This applies to logistic regression, for instance.

7.6.5   Example: MovieLens Data

Let's predict movie rating from user age, gender and movie genres. The latter are: Unknown, Action, Adventure, Animation, Children's, Comedy, Crime, Documentary, Drama, Fantasy, Film-Noir, Horror, Musical, Mystery, Romance, Sci-Fi, Thriller, War and Western. In order to have independent data points, we will proceed as in Section 3.2.4, where we computed the average rating each user gave to the movies he/she reviewed. But how should we handle genre, which is specific to each movie the user reviews? In order to have independent data, we no longer are looking at the level of each movie rated by the user.

We will handle this by calculating, for each user, the proportions of the various genres in the movies rated by that user. Does a user rate a lot in the Comedy genre, for instance? (Note that these proportions do not sum to 1.0, since a movie can have more than one genre.) In order to perform these calculations, we will use R's **split()** function, which is similar to **tapply()** but more fundamental. (Actually, **tapply()** calls **split()**.) Details are shown in Section 7.7.2, where the final results are placed in a data frame **saout**. Now let's run the regression.

```
> sam <- as.matrix(saout)
> summary(lm(saout[,3] ~ sam[,c(4:24)]))
```

Note that we needed to convert **saout** from a data frame to a matrix, in order to use column numbers in our call to **lm()**. Here is the result:

```
Coefficients:
```

	Estimate	Std. Error	t value
(Intercept)	2.893495	0.325803	8.881
sam[, c(4:24)]age	0.001338	0.001312	1.020
sam[, c(4:24)]gender	−0.008081	0.034095	−0.237
sam[, c(4:24)]GN6	−0.625717	9.531753	−0.066
sam[, c(4:24)]GN7	−0.717794	0.286749	−2.503
sam[, c(4:24)]GN8	1.084081	0.427063	2.538
sam[, c(4:24)]GN9	1.009125	0.660956	1.527
sam[, c(4:24)]GN10	−0.481950	0.448644	−1.074
sam[, c(4:24)]GN11	0.190537	0.278022	0.685
sam[, c(4:24)]GN12	0.083907	0.371962	0.226
sam[, c(4:24)]GN13	2.564606	1.272708	2.015
sam[, c(4:24)]GN14	0.377494	0.291400	1.295

sam [, c (4 : 2 4)] GN15	−0.576265	1.282073	−0.449
sam [, c (4 : 2 4)] GN16	1.268462	0.858390	1.478
sam [, c (4 : 2 4)] GN17	0.274987	0.313803	0.876
sam [, c (4 : 2 4)] GN18	−0.115167	0.577974	−0.199
sam [, c (4 : 2 4)] GN19	−0.578825	0.433863	−1.334
sam [, c (4 : 2 4)] GN20	0.631700	0.222688	2.837
sam [, c (4 : 2 4)] GN21	0.249433	0.345188	0.723
sam [, c (4 : 2 4)] GN22	0.577921	0.296188	1.951
sam [, c (4 : 2 4)] GN23	1.452932	0.356611	4.074
sam [, c (4 : 2 4)] GN24	1.821073	1.057404	1.722

	$\Pr(>|t|)$	
(Intercept)	< 2e−16	***
sam [, c (4 : 2 4)] age	0.30788	
sam [, c (4 : 2 4)] gender	0.81270	
sam [, c (4 : 2 4)] GN6	0.94767	
sam [, c (4 : 2 4)] GN7	0.01248	*
sam [, c (4 : 2 4)] GN8	0.01130	*
sam [, c (4 : 2 4)] GN9	0.12716	
sam [, c (4 : 2 4)] GN10	0.28300	
sam [, c (4 : 2 4)] GN11	0.49331	
sam [, c (4 : 2 4)] GN12	0.82158	
sam [, c (4 : 2 4)] GN13	0.04419	*
sam [, c (4 : 2 4)] GN14	0.19549	
sam [, c (4 : 2 4)] GN15	0.65319	
sam [, c (4 : 2 4)] GN16	0.13982	
sam [, c (4 : 2 4)] GN17	0.38109	
sam [, c (4 : 2 4)] GN18	0.84210	
sam [, c (4 : 2 4)] GN19	0.18249	
sam [, c (4 : 2 4)] GN20	0.00466	**
sam [, c (4 : 2 4)] GN21	0.47011	
sam [, c (4 : 2 4)] GN22	0.05134	.
sam [, c (4 : 2 4)] GN23	5.01e−05	***
sam [, c (4 : 2 4)] GN24	0.08537	.

It is interesting to compare this with the analysis in Section 3.2.4. There age was found "significant" in the statistical sense but not in a practical sense. Its estimated coefficient has now decreased even further, due to the addition of the genre variables (see the "thunder stealing" remark in Section 7.1).

Instead, the relatively big effects here were those associated with genre — but only relatively. One must keep in mind that the genre variables are proportions. Consider for instance the Action genre, with estimated coefficient -0.717794. If we have two users, one that has 20% of her ratings on Action films, and another user for whom that figure is 30%, the estimated impact on mean rating is only -0.07.

Only some of the genre variables came out "significant." One might at first think this odd, since we have 100,000 observations in our data, large enough to pick up even small deviations from $\beta_i = 0$. But that is 100,000 ratings, and since we have collapsed our analysis to the person level, we must take note of the fact that we have only 943 users here.

Now let's apply multiple inference to the genre variables. Since there are 19 of them, in order to achieve a confidence level of (at least) 0.95, we need to set the individual level for each interval at $0.95^{1/19}$. That means that instead of using the standard 1.96 to compute the radius of our interval, we need the value corresponding to this level:

```
> -qnorm((1-0.997304)/2)
[1]  3.000429
```

That means that each of the widths of the confidence intervals will increase by a factor of 3.00/1.96, about 50%. That may be a worthwhile price to pay for the ability to make intervals that hold jointly. But the price is rather dramatic if one does significance testing. This book discourages the use of testing, but it is instructive to look at the effect of multiple inference in a testing context.

Here the "significant" genre variables are those whose entries in the ouptut column labeled "t value" are greater than 3.00 in absolute value. Only one genre, GN24, now qualifies, compared to five if no multiple inference is performed.

What about Scheffe'? Here, instead of 1.96 we will use chi-square quantile in (7.43):

```
> sqrt(qchisq(0.95,19))
[1]  5.490312
```

This is even larger than the Bonferroni value.

So, there is a price to be paid for multiple inference, but it does yield protection that in many contexts is quite important.

7.6.6   The Verdict

As seen in the example above, Scheffe's method is generally quite conservative. It is thus useful only if you wish to form a large number of confidence intervals; otherwise, Bonferroni will be a better choice, among these two methods. It should be noted, though, that both are relying on normal approximations, say of the $\hat{\beta}_i$. These approximations are fine for most uses, but in our call to **qnorm()** above, we are working far out in the right tail of the distribution, where the normal approximation is questionable.

As noted earlier, there are many other methods for multiple inference. Most of them have more restrictions than Bonferroni or Scheffe', such as a population normality assumption or being restricted to hypothesis testing, but many analysts have found them useful. The interested reader should consult the literature, such as the Hsu reference cited earlier.

7.7   Computational Complements

7.7.1   MovieLens Data Wrangling

Here are the details of the merge of the ratings data and demographics data in Section 7.4.2.1.

As in Section 7.7.2, we need to use the R **merge()** function. Things are a little trickier here, because that function relies on having a column of the same name in the two data frames. Thus we need to assign our column names carefully.

```
> ratings <- read.table('u.data')
> names(ratings) <- c('usernum','movienum','rating','
     transID')
> demog <- read.table('u.user',sep='|')
> names(demog) <- c('usernum','age','gender','occ',
     'ZIP')
> u.big <- merge(ratings,demog,by.x=1,by.y=1)
> u <- u.big[,c(1,3,5,6)]
```

7.7.2   More Data Wrangling in the MovieLens Example

While **tapply()** partitions a data frame into groups and then applies some summary function, **split()** does only the partitioning, returning an R list, with each list element being the rows of the data frame corresponding to one group. In our case here, we will group the user data by user ID.

```
> # read data, form column names
> ud <- read.table('u.data',header=F,sep='\t')
> uu <- read.table('u.user',header=F,sep='|')
> ui <- read.table('u.item',header=F,sep='|')
> ud <- ud[,-4]    # remove timestamp
> uu <- uu[,1:3]    # user, age, gender
> ui <- ui[,c(1,6:24)]    # item num, genres
> names(ud) <- c('user','item','rating')
> names(uu) <- c('user','age','gender')
> names(ui)[1] <- 'item'
> names(ui)[-1] <-
    gsub('V','GN',names(ui)[-1])  # genres
> uu$gender <- as.integer(uu$gender == 'M')
> # merge the 3 dfs
> uall <- merge(ud,uu)
> uall <- merge(uall,ui)
> # uall now is ud+uu+ui; now split by user
> users <- split(uall,uall$user)
```

At this point, for instance, **users[[1]]** will consist of all the rows in **uall** for user ID 1:

```
> head(users[[1]],3)
```

	item	user	rating	age	gender	GN6	GN7	GN8	GN9
1	1	1	5	24	1	0	0	0	1
510	2	1	3	24	1	0	1	1	0
637	3	1	4	24	1	0	0	0	0

	GN10	GN11	GN12	GN13	GN14	GN15	GN16	GN17	GN18
1	1	1	0	0	0	0	0	0	0
510	0	0	0	0	0	0	0	0	0
637	0	0	0	0	0	0	0	0	0

	GN19	GN20	GN21	GN22	GN23	GN24
1	0	0	0	0	0	0
510	0	0	0	1	0	0
637	0	0	0	1	0	0

Now, for each user, we need to find the mean rating and the proportions in each genre. Since the genre variables are dummies, i.e., 0,1-valued, those proportions are just their means. Here is the code:

```
> saout <- sapply(users,colMeans)
> saout <- t(saout)
```

The R function **sapply()** applies a given function to each element of an R
list, producing a new list, and then attempting to collapse the latter into
a matrix or vector. In our case here, since each element of **users** is a data
frame, applying **colMeans()** to it produces a vector, after which **sapply()**
combines those vectors into a matrix.

As is typical with the output of **sapply()**, the row and colum arrangement
is not convenient, so we take the transpose. Here is the top of the final
matrix:

```
> head(saout,3)
        item  user     rating  age gender         GN6
1 136.5000       1 3.610294    24      1 0.003676471
2 249.5000       2 3.709677    53      0 0.000000000
3 318.8148       3 2.796296    23      1 0.000000000
        GN7          GN8           GN9          GN10
1 0.2757353 0.15441176  0.04411765  0.09191176
2 0.1612903 0.04838710  0.01612903  0.06451613
3 0.2592593 0.07407407  0.00000000  0.00000000
        GN11         GN12          GN13         GN14
1 0.3345588 0.09191176  0.01838235  0.3897059
2 0.2580645 0.14516129  0.00000000  0.5645161
3 0.2222222 0.18518519  0.01851852  0.4074074
        GN15           GN16          GN17          GN18
1 0.007352941 0.003676471  0.04779412  0.04779412
2 0.016129032 0.032258065  0.03225806  0.01612903
3 0.000000000 0.037037037  0.09259259  0.03703704
        GN19          GN20          GN21          GN22
1 0.01838235 0.16176471  0.15808824  0.1911765
2 0.06451613 0.25806452  0.06451613  0.1935484
3 0.20370370 0.09259259  0.14814815  0.3888889
        GN23         GN24
1 0.09191176 0.02205882
2 0.04838710 0.00000000
3 0.09259259 0.00000000
```

Look at row 1, for instance. User 1 had a mean rating of 3.610294 in the
movies he rated. 27.57353% of the movies were of genre GN7 (Action), and
so on. Note that within each element of the list **users**, columns such as
age and gender are constant, so we have simply taken the average of those
constants, a bit wasteful but no problem.

7.8   Mathematical Complements

7.8.1   Iterated Projections

The Tower Property of conditional expectations states that

$$E\left[E(V|U_1,U_2)\mid U_1\right] = E(V\mid U_1) \tag{7.47}$$

This can be shown quite simply and elegantly using the vector space methods of Sections 1.19.5.5 and 2.12.8, as follows. (Readers may wish to review Section 2.12.8 before continuing.) Our only assumption is that the variables involved have finite variances, so that the vector space in 2.12.8 exists.

Let $\mathcal{A}$ and $\mathcal{B}$ denote the subspaces spanned by all functions of (U_1, U_2) and U_1, respectively. Denote the full space as $\mathcal{C}$.

Write

$$u = V \tag{7.48}$$

$$v = E(V|U_1,U_2) \tag{7.49}$$

$$w = E\left[E(V|U_1,U_2)\mid U_1\right] \tag{7.50}$$

Note that $\mathcal{B}$ is a subspace of $\mathcal{A}$. Indeed, w, a vector in $\mathcal{B}$, is the projection of v, a vector in $\mathcal{A}$, onto $\mathcal{B}$.

What we need to show is that w is also the projection of u onto $\mathcal{B}$. Since projections are unique, we will accomplish this if we show that

$$(w, u - w) = 0 \tag{7.51}$$

Write

$$u - w = (u - v) + (v - w) \tag{7.52}$$

Now consider each of the two terms on the right, which we will show are orthogonal to w. First,

$$(w, v - w) = 0 \tag{7.53}$$

because w is the projection of v onto $\mathcal{B}$. Second, since v is the projection of u onto $\mathcal{A}$, we know that

$$(q, u - v) = 0 \tag{7.54}$$

for any $q \in \mathcal{A}$, including the case $q = w$. Thus we have

$$(w, u - w) = 0 \tag{7.55}$$

as desired, completing the proof.

7.8.2   Standard Errors for RFA

Recall that RFA is defined as

$$\widehat{\nu} = \frac{1}{r} \sum_{i=1}^{r} \widehat{\mu}(Q_i) \tag{7.56}$$

We will assume that $\mu(t)$ is linear in t, with the notation of Section 2.4.2. And for convenience, assume a model with no intercept term (Section 2.4.5). Then (7.56) becomes

$$\widehat{\nu} = \frac{1}{r} \sum_{i=1}^{r} Q_i' \widehat{\beta} = \overline{Q}' \widehat{\beta} \tag{7.57}$$

where $\overline{Q}$ is the (vector-valued) sample mean of the Q_i.

As in Section 2.8, our standard errors will be conditional on the X_j. We will also assume that the Q_i are i.i.d. and independent of the (Y_i, X_i) from which $\widehat{\beta}$ is computed.

To derive an estimate of $Var(\widehat{\nu})$, we will use the Delta Method (Section 3.6.1),[14] as follows.

Let Q denote a generic random variable having the distribution of the Q_i. From (7.57), our function $f()$ in (3.26) is simply

$$f(s,t) = st \tag{7.58}$$

with $a = (EQ, \beta)'$. Using the reasoning in Section 3.6.1,

$$\widehat{\nu} \approx EQ'\,\widehat{\beta} + \overline{Q}'\beta = EQ'\,\widehat{\beta} + \beta'\overline{Q} \tag{7.59}$$

so that

$$AVar(\widehat{\nu}) = EQ'\,Cov(\widehat{\beta})\,EQ + \beta'Cov(Q)\beta \tag{7.60}$$

To get standard errors, we then replace EQ by $\overline{Q}$, $Cov(\widehat{\beta})$ by (2.57), β by $\widehat{\beta}$ and $Cov(Q)$ by the sample covariance matrix of the Q_i,

$$\frac{1}{r}\sum_{i=1}^{r}(Q_i - \overline{Q})(Q_i - \overline{Q})' \tag{7.61}$$

Here is code for the computation:

```
rfase <- function(q, lmout) {
    quadf <- function(u, v)  (t(u) %*% v %*% u)[1]
    qbar <- colMeans(q)
    qbarcov <- cov(q) / nrow(q)
    bhat <- coef(lmout)
    bhatcov <- vcov(lmout)
    sqrt(quadf(qbar, bhatcov) + quadf(bhat, qbarcov))
}
```

7.8.3   Asymptotic Chi-Square Distributions

The *chi-square* family of distributions has a single parameter, m, called its *degrees of freedom*. The family is defined as follows. Let $Z_1, ..., Z_m$ be

---

[14]In that section, we had $n \to \infty$. Here we need $r \to \infty$ too. For convenience take r/n to be constant.

independent random variables, each having a N(0,1) distribution, and set

$$Y = Z_1^2 + \ldots + Z_m^2 \tag{7.62}$$

The chi-square distribution with m degrees of freedom is defined to be the distribution of Y.

More generally, suppose V has an r-variate normal distribution with mean vector ν and invertible covariance matrix Q. Then the diagonalizability of Q can be used to show that the quantity

$$W = (V - \nu)'Q^{-1}(V - \nu) \tag{7.63}$$

has a chi-square distribution with r degrees of freedom (Exercise 6).

Now consider any asymptotically multivariate normal estimator $\widehat{\theta}$ of an r-dimensional population vector θ. Let C denote the associated covariance matrix, and let $\widehat{C}$ denote a consistent estimate of C. Then the quantity

$$(\widehat{\theta} - \theta)'\widehat{C}^{-1}(\widehat{\theta} - \theta) \tag{7.64}$$

is then asymptotically chi-square distributed with r degrees of freedom.

7.9 Exercises: Data, Code and Math Problems

Data problems:

1. Extend the analysis of the schooling data, Section 7.4.1.3, by incorporating other (observed) predictors besides **educ**, years of school. Pay particular attention to fit, in view of the finding in that section that the Two-Stage Least Squares fit was quite different from the OLS fit at the point we tried, 12 years of education.

2. Form simultaneous confidence intervals for the 48 word variables in Section 4.3.6. Try both the Bonferroni and Scheffe' approaches.

3. In Chapter 1, we found that baseball players gain about a pound of weight per year. We might ask whether there is a team effect. Add team membership as a predictor, and investigate this possibility.

Mini-CRAN and other computational problems:

4. Write an R function of call form

```
simulci(lmout, singles=NULL, diffs=NULL,
    alph=0.05,method=c('bonf','scheffe'))
```

that will form simultaneous confidence intervals, using the Bonferroni or Scheffe' method, of overall level **1 - alph**, from the output **lmout** from **lm()** or **glm()**. The intervals will be for the coefficients named in **singles** and/or differences of coefficients named in **diffs**. The latter is a two-column **character** matrix, one row per difference.

5. In the RFA computation in Section 7.5.2, it was suggested that **fitted.values** be used insead of **predict()**, thereby saving a computational step. Show how to do this. Hint: Use **rownames()**.

Math problems:

6. Show that under the given assumptions, W in (7.63) has a chi-square distribution as claimed.

7. Derive a closed-form expression for the Two-Stage Least Squares estimator in the case in which the instrument Z is binary, i.e., a dummy variable. Extend this to the case of any categorical variable (i.e. R factor) Z.

8. Use the Delta Method (Section 3.6.1) to derive the standard error for $\widehat{\beta}_{12}$ in (7.20).

Chapter 8

Shrinkage Estimators

Suppose we are estimating a vector mean. Consider for instance the baseball player example, Section 1.6. Let the vector (H,W,A) denote the height, weight and age of a player. Say we are interested in the population mean vector

$$\mu = (EH, EW, EA) \qquad (8.1)$$

The natural estimator would be the vector of sample means,

$$\widehat{\mu} = (\overline{H}, \overline{W}, \overline{A}) \qquad (8.2)$$

where $\overline{H}$ is the mean height of all players in our sample, and so on. Yet it turns out that this "natural" estimator is not optimal in a certain theoretical sense. Instead, the theory suggests that it is actually better to "shrink" (8.2) to a smaller size.

Theory is of course not a focus of this book. However, this particular theoretical finding regarding "shrunken" estimators has had a major impact on some of the current applied methodology in regression and classification. Among the various methods developed for multivariate analysis in recent years, many employ something called *regularization*. This technique shrinks estimators, or equivalently, keeps them from getting too large. In particular, the *LASSO* has become popular in machine learning circles and elsewhere. The theoretical findings can help guide our intuition in practical settings, so we will begin with a brief summary of the theory.

Shrinkage estimators will be the subject of this chapter, and will influence Chapter 9 as well.

8.1 Relevance of James-Stein to Regression Estimation

Recall that the very definition of the regression function is the conditional mean,

$$\mu(t) = E(Y \mid X = t) \tag{8.3}$$

Since James-Stein theory says estimates of means should be shrunken, it is not surprising that the same is true for estimates of regression functions, at least say in linear models. In the latter context, μ becomes $\mu(t)$ and the shrinking is applied to estimates of the vector β that parameterizes $\mu(t)$.

This illustrates the bias-variance tradeoff introduced in Section 1.11.1. In the regression context, shrinking will likely produce bias (definitely so in the case of the linear model), but by making the estimate smaller it should reduce variance. If a good amount of shrinkage can be determined, it may be a "win."

Define the quantity

$$v = \frac{\sigma}{||\widehat{\beta}||} \tag{8.4}$$

where σ is as in (2.48). Then the discussion in Section 8.11.1.3, carried over to our regression context, suggests the following about shrinking $\widehat{\beta}$:

- For fixed p and v, the smaller our sample size n is, the more we need to shrink.

- For fixed n and v, the more predictor variables we have, the more we should shrink.

- For fixed n, p and β, the larger σ is, the more we need to shrink.

Again, it is questionable whether the assumptions of James-Stein theory are directly relevant to real life. Nevertheless, the qualitative message of that theory is highly relevant to actual practice.

For instance, let's consider what James-Stein might say about situations in which our predictors are highly correlated with each other. Recall (2.54):

$$Cov(\widehat{\beta}) = \sigma^2 (A'A)^{-1} \tag{8.5}$$

This quantity might be "large" (loosely defined) if the matrix inverse $(A'A)^{-1}$ is large, even if σ is small. The inverse will be large if $A'A$ is "small." Again, the latter term is loosely defined for now, but the main point that will emerge below is that all this will occur if our predictors are *multicollinear*, meaning that they are highly correlated.

Thus another bullet should be added to the rough guidelines above:

- For fixed values of n, p and v, the stronger the degree of multicollinearity, the more the need for shrinkage.

What exactly is multicollinearity? We take this up next.

8.2 Multicollinearity

As we have been discussing in this chapter, the coefficient for a predictor variable may be greatly affected by the presence or absence of other predictors. This becomes a particular concern when there are strong correlations among the predictors.

8.2.1 What's All the Fuss About?

8.2.2 A Simple Guiding Model

To see the problem, first recall the R function **scale()**, introduced in Section 1.21. For each variable in the data it is applied to, this function centers and scales the variable, i.e., subtracts the variable's mean and divides by its standard deviation. The resulting new versions of the variables now have mean 0 and variance 1.

Suppose we apply **scale()** to our predictor variables. Also rewrite (2.28) as

$$\widehat{\beta} = (\frac{1}{n} A'A)^{-1} \cdot \frac{1}{n} A'D \tag{8.6}$$

This won't change the resulting $\widehat{\beta}$, but it will prove to be quite useful.

Due to the centering and scaling, we are in essence fitting a model without a constant term β_0 (Section 1.21), thus no 1s column in the matrix A. The matrix $\frac{1}{n}A'A$ in (2.28) then becomes the correlation matrix of the predictors (Exercise 4).

If we have just two predictors, for instance, we will have

$$\frac{1}{n}A'A = \begin{pmatrix} 1 & c \\ c & 1 \end{pmatrix} \tag{8.7}$$

where c is the correlation between the two predictors. Then to get the estimated regression coefficients, we will compute the inverse of $\frac{1}{n}A'A$ in (2.28). That inverse, from (A.18), is

$$\frac{1}{1-c^2} \begin{pmatrix} 1 & -c \\ -c & 1 \end{pmatrix} \tag{8.8}$$

Now we can see the problem arising if the two predictor variables are highly correlated with each other: The value of c will be near 1, so the quantity $1/(1-c^2)$ may be huge. In that light, (2.54) tells us that our estimated coefficients will have very large standard errors. This is bad — confidence intervals will be very wide, and significance tests will have low power.

8.2.3   Checking for Multicollinearity

The previous section shows how to check for multicollinearity in the case of just two predictors. How do we do this in general?

8.2.3.1   The Variance Inflation Factor

Noting the squared correlation in the demoninator of (8.8), a natural generalization would be to look at multiple R^2. For predictor variable $X^{(i)}$, we would run **lm()**, predicting $X^{(i)}$ from all the other $X^{(j)}$; denoting the resulting multiple R^2 by R_i^2. To complete the analogy to (8.8), we define the *variance inflation factor* (VIF) for $X^{(i)}$ by

$$VIF_i = \frac{1}{1 - R_i^2} \tag{8.9}$$

A rule of thumb used by some analysts is to treat VIF values over 10 as cause for concern, though setting this threshhold is up to each analyst.

We could calculate the VIF values ourselves, by repeated calls to **lm()**, but the **vif()** function in the **car** package does this for us [49].

8.2.3.2 Example: Currency Data

Recall the currency data analysis in Section 6.8.1.1. We had regressed the *yen* against the Canadian *dollar*, the *mark*, the *franc* and the *pound*, assigning the output to **fout1**. We can input the latter object to **vif()**:

```
> vif(fout1)
      Can        Mark      Franc      Pound
  5.575051   4.438270  14.119749   6.582210
```

Let's check the Can case:

```
> tmp <- summary(lm(Can ~Mark+Franc+Pound,data=curr1))
> 1 / (1 - tmp$r.squared)
[1]  5.575051
```

By the way, various definitions for VIF for the generalized linear model, i.e., **glm()**, have been proposed. The **vif()** function in **car** applies one of them.

8.2.4 What Can/Should One Do?

Once one has determined that the data has serious multicollinearity issues — and note again, this is up to the individual analyst to decide whether it is a problem — what are possible remedies?

8.2.4.1 Do Nothing

Multicollinearity can produce counterintuitive results in the signs of the $\widehat{\beta}_j$. However, the analyst should be extremely reluctant to resort to anti-multicollinearity measures shown below (or any others) such as ridge regression and the LASSO, simply because he/she does not like the signs. On the contrary, they may be the "real" signs in various senses, as discussed in Chapter 7.

If our application involves the Description goal of regression analysis, we may really want to know the effects of all the predictor variables. The currency data analyzed above might be such a case. In this situation, we might simply accept the fact that the standard errors of the coefficients are large. Indeed, if the sample is sizable, then even those "large" standard errors might be quite acceptable.

8.2.4.2   Eliminate Some Predictors

In principle, a variable with a high VIF is essentially redundant, and thus not likely to add much predictive power. One might then remove it from one's set of predictor variables. Elimination of predictor variables is a major issue in regression and classification analysis, and all of Chapter 9 will be devoted to it.

8.2.4.3   Employ a Shrinkage Method

We turn to this very topic now.

8.3   Ridge Regression

Equation (8.8) suggested that multicollinearity results in $(\frac{1}{n}A'A)^{-1}$ becoming "too large," causing our estimated coefficient vector $\widehat{\beta}$ to be too large. This fits in with the "too large, too often" intuitive motivation for James-Stein theory, and thus the James-Stein crowd would react to the multicollinearity problem by saying "Shrink it!" Two common methods for this are *ridge regression* and the *LASSO*. The latter is more widely used, but we can get better insight by considering the former first.

The approach involves adding a positive constant λ to the diagonal of $A'A$, to make the matrix "larger," resulting in a "smaller" $(\frac{1}{n}A'A)^{-1}$. Here λ is a tuning parameter, to be chosen by the analyst.

Equation (2.28) now becomes

$$\widehat{\beta}_{ridge} = (A'A + \lambda I)^{-1} \cdot A'D \qquad (8.10)$$

In fact, shrinking does indeed occur, as we will see shortly.

A generalization would be to add different constants to different elements of the diagonal, say because the predictors are of widely different scales. But it is easier just to apply the R function **scale()** to the predictors, as we saw earlier in this chapter.

To avoid confusion, let us denote the ordinary $\widehat{\beta}$ by $\widehat{\beta}_{OLS}$ (Section 3.3.)

8.3.1 Alternate Definitions

There are other ways of defining ridge regression. These are not just mathematical parlor tricks; we will see later that they will be quite useful.

The first alternative definition is that for fixed λ, the ridge estimator is the value of b that minimizes

$$\sum_{i=1}^{n}(Y_i - \widetilde{X}_i b)^2 + \lambda ||b||_2^2 \tag{8.11}$$

The idea here is that on the one hand we want to make the first term (the sum of squares) small, so that b gives a good fit to our data, but on the other hand we don't want b to be too large. The second term acts like a "governor" to try to prevent having too large a vector of estimated regression coefficients; the larger we set λ, the more we penalize large values of b.

Let's see why this is consistent with (8.10). First, rewrite (8.11) as

$$(D - Ab)'(D - Ab) + \lambda b'b \tag{8.12}$$

in the notation of Section 2.4.2. Since we are discussing minimization of (8.11), let's take the derivative with respect to b, which, again following Section 2.4.2, is

$$-2A'(D - Ab) + 2\lambda b \tag{8.13}$$

Setting this to 0, and writing $b = Ib$, we have

$$(A'A + \lambda I)b = A'D \tag{8.14}$$

just as in (8.10).

This leads to our second alternative definition of ridge: It can be shown that minimizing (8.12) is equivalent to minimizing

$$(D - Ab)'(D - Ab) \qquad (8.15)$$

subject to a constraint of the form

$$||b||_2^2 \leq \gamma \qquad (8.16)$$

(γ and λ are not numerically equal, but are functions of each other.)

This really makes the point that we want to avoid letting $\widehat{\beta}$ get too large. (See also Section 8.11.2 in the Mathematical Complements portion of this chapter.)

8.3.2   Choosing the Value of λ

Well, then. Faced with actual data, what is the practioner to do? Using ridge regression might solve the practioner's multicollinearity problem, but how is he to choose the value of λ?

One way to choose λ is visual: For each predictor variable, we draw a graph, the *ridge trace*, that plots the associated estimated coefficient against the value of λ. We choose the latter to be at the "knee" of the curve. The function **lm.ridge()** in the **MASS** package (part of the base R distribution) can be used for this, but here we will use the function **ridgelm()** from **regtools**, due to its approach to scaling. Here is why:

As is standard, the **ridgelm()** function calls **scale()** on the predictors and *centers* the response variable. But **ridgelm()** goes a little further, also applying the $1/n$ scaling we used in Section 8.2.1. Equation (8.6) becomes

$$\widehat{\beta} = (\frac{1}{n}A'A + \lambda I)^{-1} \cdot \frac{1}{n}A'D \qquad (8.17)$$

Here is the point: As noted earlier, $\frac{1}{n}A'A$ is the correlation matrix of our predictor data. Thus this matrix has 1s on the diagonal, with the off-diagonal elements being smaller than 1 in absolute value. *This makes it much easier to choose a range of values to try for λ, more or less independent of what data we have.* Thus the default value for the λ argument in **ridgelm()** is $(0.01, 0.02, ..., 1.00)$, meaning that the function will try these values of λ.

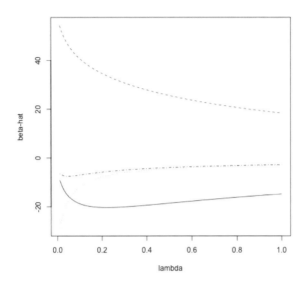

Figure 8.1: Ridge analysis of the currency data

The code for **ridgeln()** includes some noteworthy computational aspects, and is discussed in the Computational Complements at the end of the chapter.

Another approach to choosing λ is cross validation. We then choose λ to be the value that predicts new cases best. The function **ridge.cv()** in the **parcor** package does this [82].

8.3.3 Example: Currency Data

We continue from Section 8.2.3.2:

```
> library(regtools)
> plot(ridgelm(curr1))
```

The results are displayed in Figure 8.1. From top to bottom along the left edge, the curves show the values of $\widehat{\beta}_{Mark}$, $\widehat{\beta}_{Pound}$, $\widehat{\beta}_{Can}$ and $\widehat{\beta}_{Franc}$. The "knee" is visible for the *franc* and Canadian *dollar* at about 0.15 or 0.20, though interestingly the curve for the *mark* continues to decline significantly after that.

Ridge regression is thus rather subjective, without much "science" behind it. Perhaps the most scientific approach is to use cross validation. Thus let's try **ridge.cv()**:

```
> ridge.cv(as.matrix(curr1[,−5]),curr1[,5])
$intercept

224.9451

$coefficients
    XCanada      XMark      XFranc      XPound
  −5.786784   57.713978   −34.399449   −5.394038

$lambda.opt
[1]  0.3168208
```

The recomended λ value here is about 0.32, rather larger than what we might have chosen using the "knee" method. On the other hand, this larger value makes sense in light of our earlier observation concerning the *mark*.

Shrinkage did occur. Here are the OLS estimates:

```
> lm(Yen ~ ., data=curr1)
...
Coefficients:
(Intercept)        Canada          Mark          Franc
    224.945        −5.615        57.889        −34.703
        Pound
      −5.332
```

Ridge slightly reduced the absolute values of most of the coefficients, Canada being the exception. The fact that the reductions were only slight should not surprise us, given the rough guidelines in Section 8.11.1.3. The n/p ratio is pretty large, and even the multicollinearity was mild according to the generally used rule of thumb (Section 8.2.3.1).

8.4   The LASSO

Much of our material on the LASSO will appear in Chapters 9 and 12 but we introduce it in this chapter due to its status as a shrinkage estimator. To motivate this method, recall first that shrinkage estimators form another example of the bias-variance tradeoff. With ridge regression, for instance, by shrinking $\widehat{\beta}$, we are reducing its variance (actually, its covariance matrix), at the expense of introducing some bias. If we can choose a good value of λ, we can find a "sweet spot" in that tradeoff, and hopefully improve predictive ability. This of course is the motivation for using cross-validation to choose λ.

The *Least Absolute Shrinkage and Selection Operator* — the LASSO — takes another approach to shrinking. As with ridge regression, the LASSO actually has two equivalent formulations, which in rough terms are:

- Penalize large values of $\widehat{\beta}$.

- Place an explicit limit to the size of $\widehat{\beta}$.

We will begin with the first of these.

8.4.1 Definition

As noted in Section 8.2.4.2 and in earlier chapters, a more traditional way than shrinkage to improve prediction error is *subset selection*, meaning to pare down the set of predictor variables into a smaller but representative set. As discussed earlier, this reduces variance, though again increasing bias. One advantage of this approach is that it is appealing to deal with just a small number of predictors, often termed a *parsimonious* model.

The LASSO was invented with the goal of combining the best aspects of ridge regression on the one hand, and subset selection on the other. It involves shrinkage, like ridge regression, but often results in a roundabout way of doing subset selection.

So, how does the LASSO accomplish all this? The answer is remarkably simple: In (8.11), simply replace $||b||_2^2$ by $||b||_1$ (see (A.2)). In other words, the LASSO estimator is defined to be the value of b that minimizes

$$\sum_{i=1}^{n}(Y_i - \widetilde{X}_i b)^2 + \lambda||b||_1 \tag{8.18}$$

Similar to the ridge case, one can show that an equivalent definition is that the LASSO estimator is chosen to minimize

$$\sum_{i=1}^{n}(Y_i - \widetilde{X}_i b)^2 \tag{8.19}$$

subject to a constraint of the form

$$||b||_1 \leq \gamma \tag{8.20}$$

Using the argument in Section 8.11.2, we see that the LASSO does produce a shrinkage estimator. But it is designed so that typically many of the estimated coefficients turn out to be 0, thus effecting subset selection, which we will see in Section 9.7.7.1.

8.4.2   The lars Package

We'll use the R package **lars** [63]. It starts with no predictors in the model, then adds them (in some cases changing its mind and deleting some) one at a time. At each step, the action is taken that is deemed to best improve the model, as with *forward stepwise regression*, to be discussed in Chapter 9. At each step, the LASSO is applied, with λ determined by cross-validation.

The **lars** package is quite versatile. Only its basic capabilities will be shown here.

8.4.3   Example: Currency Data

As noted, the LASSO is commonly used as a method for variable selection, the topic of Chapter 9. Since we have only $p = 4$ predictors, and more than 700 observations, variable selection is not really an issue. But in this chapter's context of multicollinearity, it is of interest to see how much the software decides to shrink.

```
> lassout <- lars(as.matrix(curr1[,-5]),curr1[,5])
> lassout
...
R-squared: 0.892
Sequence of LASSO moves:
      Canada Mark Pound Franc
Var        1    2     4     3
```

Step   1   2   3   4

Note that **lars** requires the predictor values to be given as a matrix.

At Step 0, there are no predictors; it is a regression model with just a constant term, so we are just predicting Y from its unconditional mean. We see that at Step 1, **lars()** brought in the Canada predictor, then the *mark*, then the *pound* and lastly, the *franc*.

Let's take a closer look:

```
> summary( lassout )
LARS/LASSO
Call: lars (x = as.matrix(curr1 [, -5]), y = curr1 [, 5])
   Df      Rss         Cp
0   1  2052191  6263.50
1   2  2041869  6230.18
2   3   392264   587.31
3   4   377574   539.04
4   5   220927     5.00
```

The C_p criterion is similar to adjusted-R^2, and will be discussed in full in Chapter 9. The user may choose to use the C_p value as a guide as to which model to use. In this case, that approach would choose the full model, with all predictors, not surprising in this context of $p << n$.

Since the LASSO is mainly used for subset selection, the actual values of the estimated coefficients are rather secondary, and not presented in the output of **summary()**. But they are indeed accessible:

```
> lassout$beta
          Canada        Mark       Franc       Pound
0    0.0000000    0.00000     0.00000    0.000000
1   -0.2042481    0.00000     0.00000    0.000000
2  -28.6567963   28.45255     0.00000    0.000000
3  -28.1081479   29.61350     0.00000   -1.393401
4   -5.6151436   57.88856   -34.70273   -5.331583
. . .
```

Again, this is presented in terms of the values at each step, and the 0s show which predictors were *not* in the model as of that step. In our multi-collinearity context in this chapter, we are interested in the final values, at Step 4. They are seen to provide shrinkage similar to the mild amount we saw in Section 8.3.3.

8.4.4   The Elastic Net

Although ridge regression had been around for many years, its popularity was rather limited. But the introduction of the LASSO in the 1990s (and some related methods proposed slightly prior to it) revived interest in shrinkage estimators for regression contexts. A cottage industry among statistics/machine learning researchers has thrived ever since then, including various refinements of the LASSO idea.

One of those refinements is the *elastic net*, defined to be the value of b minimizing

$$\sum_{i=1}^{n}(Y_i - \widetilde{X}_i b)^2 + \lambda_1 b_1 + \lambda_2 ||b||_2^2 \qquad (8.21)$$

The idea behind this is that one might not be sure whether ridge or LASSO would have better predictive power in a given context, so one can "hedge one's bets" by using both at once! Again, one could then rely on cross-validation to choose the values of the λ_i.

8.5   Cases of Exact Multicollinearity, Including $p > n$

These days it is common to have more predictors than observations, i.e., to have $p > n$. Such a situation used to be dismissed as impossible, since the matrix A in (2.28) would necessarily be of less than full rank, so that $A'A$ would not be invertible.

However, today, people are more adventurous (some might say recklessly so), and they hope to do regression and classification analysis in such situations. Shrinkage estimators provide a possible solution.

8.5.1   Why It May Work

This may be easily seen in the case of ridge regression. The key point is that even if $A'A$ is not invertible, $A'A + \lambda I$ *will* be invertible for any $\lambda > 0$). (This follows from the analysis of Section 8.11.3 and the fact that the rank of a matrix is the number of nonzero eigenvalues.) So, to many people, there is hope for the case $p > n$!

8.5.2 Example: R mtcars Data

This is one of the famous built-in data sets in R, with $n = 32$ and $p = 11$. So this is NOT an example of the $p > n$ situation, but as will be seen, we can use this data to show how ridge can resolve a situation in which $A'A$ is not invertible.

The **cyl** column in this data set shows the number of cylinders in the car's engine, 4, 6 or 8. Let's create dummy variables for each of these three types:

```
> library (dummies)
> dmy <- dummy( mtcars$cyl )
> mtcars <- cbind ( mtcars , dmy )
```

Let's predict **mpg**, gas mileage, just from the number of cylinders. Since there are three categories of engine, we should only use two of those dummy variables.[1] But let's see what happens if we retain all three dummies.

In the matrix A of predictor variables, our first column will consist of all 1s as usual, but consider what happens when we add the vectors in columns 2, 3 and 4, where our dummies are: Their sum will be an all-1s vector, i.e., column 1! Thus one column of A will be equal to a linear combination of some other columns (Section A.4), so A will be less than full rank. That makes $A'A$ noninvertible.

Let's use all three anyway:

```
> library (dummies)
> d <- dummy( mtcars$cyl )
> mtcars <- cbind ( mtcars , d )
> lm(mpg ~ cyl4+cyl6+cyl8 , data=mtcars )
...
Coefficients :
(Intercept )          cyl4          cyl6          cyl8
     15.100         11.564         4.643            NA
```

R was smart enough to notice the column dependency, so it simply ignored the **cyl8** column. Indeed, if we omit that column ourselves, we get the same result:

---

[1]Readers familiar with *Analysis of Variance* (ANOVA) will recognize this as one-way ANOVA. There the model is $EY_{ij} = \mu + \alpha_i + \epsilon_{ij}$, where in our case i would equal 1,2,3. There again would be a redundancy, but it is handled with the constraint $\sum_{i=1}^{3} \alpha_i = 0$.

```
> lm(mpg ~ cyl4+cyl6 , data=mtcars)
. . .
Coefficients:
(Intercept)              cyl4              cyl6
      15.100            11.564             4.643
```

We of course should have omitted that column in the first place. But
let's see how ridge regression could serve as an alternate solution to the
dependency problem:

```
> head(t(ridgelm(mtcars[,c(12:13,1)])$bhats))
           cyl4        cyl6
[1,]   11.40739   4.527374
[2,]   11.25571   4.416195
[3,]   11.10838   4.309103
[4,]   10.96522   4.205893
[5,]   10.82605   4.106375
[6,]   10.69068   4.010370
```

Recall that each row here is for a different value of λ. We see that shrinking
occurs, as anticipated, thus producing bias over repeated sampling, but use
of ridge indeed allowed us to overcome the column dependency problem.

8.5.2.1   Additional Motivation for the Elastic Net

In many senses, the elastic net was developed with the case $p \gg n$ in mind.
This occurs, for instance, in many genomics studies, with there being one
predictor for each gene under consideration. In this setting, Description,
not Prediction, is the main goal, as we wish to determine which genes affect
certain traits.

Though the LASSO would seem to have potential in such settings, it is
limited to finding p nonzero regression coefficients. This may be fine for
Prediction but problematic in, say, in genomics settings.

Similarly, the LASSO tends to weed out a predictor if it is highly correlated
with some other predictors. And this of course is exactly the issue of
concern, multicollinearity, in the early part of this chapter, and again, it
is fine for Prediction. But in the genomics setting, if a group of genes is
correlated but influential, we may wish to know about all of them.

The elastic net is motivated largely by these perceived drawbacks of the
LASSO. It is implemented in, for instance, the R package **elasticvnet**.

8.6   Bias, Standard Errors and Signficance Tests

Due to their bias, shrinkage estimators generally do not yield straightforward calculation of standard errors or p-values. Though some methods have been proposed, mainly for p-values [27], they tend to have restrictive assumptions and are not widely used and notably, software libraries for shrinkage methods do not report standard errors.

Some analysts view the inclusion of a predictor by LASSO as akin to finding that the predictor is "statistically significant." This interpretation of LASSO output is debatable, for the same reason as is the basic notion of significance testing of coefficients: For large n (but fixed p), the LASSO will include all the predictors.

8.7   Principal Components Analysis

The Variance Inflation Factor makes sense as a gauge of multicollinearity, because it measures the possible deleterious impact of that problem. But we might investigate multicollinearity in our data directly: Is one predictor nearly equal to some linear combination of the others? The material in Section A.6 is perfect for answering this question.

For convenience, suppose our predictor variables are centered, i.e., have their means subtracted from them, so they have mean 0.

Here is our strategy: We will find p new predictor variables that are linear combinations of our p original ones. Some of them will have very small variance, which essentially makes them approximately constant. Since the original predictors have mean 0, so will the new predictors, and thus in the preceding sentence "approximately constant" will mean approximately 0 — i.e., one of the original predictors will approximately be a linear combination of the others.

In this way, we will identify multicollinearities. Note the plural in that latter word, by the way. More than one of the new predictors could be approximately 0.

The key equation is (A.21), which, to avoid a clash of symbols, we will rewrite as

$$U'BU = G \qquad (8.22)$$

As in Section 2.4.2, let A denote the matrix of our predictor variables (without a 1s column, due to centering of our data). In (8.22) take

$$B = A'A \qquad (8.23)$$

i.e.,

$$G = U'A'AU = (AU)'AU \qquad (8.24)$$

from (A.14).

Column j of A is our data on predictor variable j. Now, in the product AU, consider *its* first column. It will be A times the first column of U, and will thus be a linear combination of the columns of A (the coefficients in that linear combination are the elements of the first column of U).

In other words, the first column of AU is a new variable, formed as a linear combination of our original variables. This new variable is called a *principal component* of A. The second column of AU will be a different linear combination of the columns of A, forming another principal component, and so on.

Also, Equation (2.79) implies[2] that the covariance matrix of the new variables is G. Since G is diagonal, this means the new variables are uncorrelated. That will be useful later, but for now, the implication is that the diagonal elements of G are the variances of the principal components.

And that is the point: If one of those diagonal elements is small, it corresponds to a linear combination with small variance. And in general, a random variable with a small variance is close to constant. In other words, multicollinearity!

"The bottom line," is that we can identify multicollinearity by inspecting the elements of G, which are the eigenvalues of $A'A$. This can be done using the R function **prcomp()**.[3]

---

[2] Again, beware of the clash of symbols.

[3] We could also use **svd()**, which computes the *singular value decomposition* of A. It would probably be faster.

This will be especially useful for generalized linear models, as VIF does not directly apply. We will discuss this below.

8.8 Generalized Linear Models

Even though multicollinearity is typically discussed in the context of linear models, it is certainly a concern from generalized linear models. The latter "contain" linear models in their specification, and use them in their computation (Section 4.5). The shrinkage parameter λ is used as a counterweight to extreme values of the log-likelihood function.

The R package **glmnet** can be used to compute shrinkage estimators in the generalized linear models sense. Both the LASSO and the elastic net are available as options in this package..

8.8.1 Example: Vertebrae Data

Consider the Vertebral Column data in Section 5.5.2. First, let's check this data for multicollinearity using diagonalization as discussed earlier:

```
> vx <- as.matrix(vert[,-7])
> prcomp(vx)
Standard deviations:
[1]  42.201442532  18.582872132  13.739609112  10.296340128
[5]   9.413356950   0.003085624
```

That's quite a spread, with the standard deviation of one of the linear combinations being especially small. That would suggest removing one of the predictor variables. But let's see what **glmnet()** does.

At first, let's take the simple 2-class case, with classes DH and not-DH. Continuing from Section 5.5, we have

```
> vy <- as.integer(vert$V7 == 'DH')
> coef(glmnet(vx,vy,family='binomial'))
V1  .    .                    .                   .
V2  .    .                    .                   .
V3  .    .             -0.0001184646       -0.002835562
V4  . -0.008013506     -0.0156299299       -0.020591892
V5  .    .                    .                   .
V6  .    .                    .                   .
```

```
V1    .              .           .
V2    .              .           .
V3  −0.005369667  −0.00775579  −0.01001947
V4  −0.025365930  −0.02996199  −0.03438807
V5    .              .           .
V6    .              .           .
. . .
```

i

The coefficients estimates from the first six iterations are shown, one column per λ. The values of the latter were

```
> vout$lambda
 [1]  0.1836231872  0.1673106093  0.1524471959
 [4]  0.1389042072  0.1265643403  0.1153207131
. . .
```

There is as usual the question of which λ value is best. Unfortunately, it is even less clear than the corresponding question for a linear model, and we will not pursue it here. The analyst may wish to use the old standby, cross-validation here.

8.9   Other Terminology

The notion of placing restraints on classical methods such as least-squares is often called *regularization*. Ridge regression is called *Tykhanov regularization* in the numerical analysis community. Another common term is *penalized methods*, motivated by the fact that, for instance, the second term in (8.18) "penalizes" us if we make b too large.

8.10   Further Reading

The classic paper on ridge regression is [68]. It's well worth reading for the rationale for the method.

A general treatment of the LASSO is [65].

8.11 Mathematical Complements

8.11.1 James-Stein Theory

Consider further the height-weight-age example at the beginning of this chapter, but with some other variables in there too, a total of p in all. We have a random sample of size n, and are interested in the mean vector.

Assume that each of these variables has a normal distribution with variance σ^2, with the latter value being known; these assumptions are rather unrealistic, but our goal is to gain insight from the resulting theory, which we will apply by extension to the regression case.

8.11.1.1 Definition

The *James-Stein estimator [136] is defined to be*

$$\left(1 - \frac{(p-2)\sigma^2/n}{||\widehat{\mu}||^2}\right)\widehat{\mu} \tag{8.25}$$

Here $|| \; ||$ denotes the Euclidean norm, the square root of the sums of squares of the vector's elements (Equation (A.1)).

8.11.1.2 Theoretical Properties

It can be shown that the James-Stein estimator outperforms $\widehat{\mu}$ in the sense of Mean Squared Error,[4] as long as $p \geq 3$. This is pretty remarkable! In the ballplayer example above, it says that $\widehat{\mu}$ is nonoptimal in the height-weight-age case, but NOT if we are simply estimating mean height and weight. Something changes when we go from two dimensions to three or more.[5]

---

[4]Defined in the vector case as $E(||\widehat{\mu} - \mu||^2)$.

[5]Oddly, there is also a fundamental difference between one and two dimensions versus three or more for *random walk*. It can be shown that a walker stepping in random directions will definitely return to his starting point if he is wandering in one or two dimensions, but not in three or more. In the latter case, there is a nonzero probability that he will never return.

A key aspect of the theory is that if

$$\frac{(p-2)\sigma^2/n}{||\widehat{\mu}||^2} < 1 \qquad (8.26)$$

then James-Stein *shrinks* $\widehat{\mu}$ toward 0.

8.11.1.3   When Might Shrunken Estimators Be Helpful?

Looking at (8.25), we can see the circumstances in which the James-Stein (J-S) estimator will be substantially different from the ordinary one. We can see immediately, for instance, that for large n, *ceteris paribus*, J-S will essentially be the same as our ordinary sample-means vector $\widehat{\mu}$. J-S is useful only for small samples. However, the factor $p-2$ tells us that here the definition of "small" depends on dimension; for larger values of p, J-S may be helpful even in somewhat larger samples.

On a more subtle level, recall the quantity known as the *coefficient of variation* from statistical practice, defined to be the ratio of a variable's standard deviation to its mean. We don't know whether to consider a standard deviation of say, 2.8, small or large, so we compare that 2.8 value to the mean of the variable. In (8.25), the quantity $\sigma/||\widehat{\mu}||$, so J-S won't be very useful if σ is small relative to $||\widehat{\mu}||$.

8.11.2   Yes, It Is Smaller

So, does ridge regression actually shrink $\widehat{\beta}$, as advertised? The answer is yes, i.e.,

$$||\widehat{\beta}_{ridge}||_2 \leq ||\widehat{\beta}_{OLS}||_2 \qquad (8.27)$$

(In fact, the inequality above will be strict in any practical situation.) But why?

There are intuitive answers to this question. We remarked earlier that adding λI to $A'A$ would intuitively seem to make that matrix "larger," thus making $\widehat{\beta}$ smaller. And the constraint (8.16) would seem to imply shrinkage, but such an argument is not airtight. What if, say, $\widehat{\beta}$ would have satisified (8.16) anyway, without applying the ridge procedure?

Instead, there is actually a very easy way to see that shrinking does indeed occur. Let's look again at (8.12), giving names to the various expressions:

$$f(b) = (D - Ab)'(D - Ab) + \lambda b'b = g(b) + h(b) \qquad (8.28)$$

In minimizing that expression with respect to b, we have a tradeoff between the two terms, $g(b)$ and $h(b)$. Since setting $b = \widehat{\beta}_{OLS}$ minimizes $g(b)$, taking $b = \widehat{\beta}_{ridge}$ means accepting a larger value of $g(b)$. But that must be accompanied by a reduction in $h(b)$; otherwise $b = \widehat{\beta}_{OLS}$ would give us a smaller $f(b)$ than would $b = \widehat{\beta}_{ridge}$, a contradiction, since the latter is the b that minimizes $f(b)$. So, $h(b)$ does get smaller when we go from OLS to ridge, and thus the ridge estimator is indeed shrunken.

More formally,

$$
\begin{array}{rll}
g(\widehat{\beta}_{ridge}) + h(\widehat{\beta}_{ridge}) & = & f(\widehat{\beta}_{ridge}) \qquad (8.29) \\
& \leq & f(\widehat{\beta}_{OLS}) \qquad (8.30) \\
& = & g(\widehat{\beta}_{OLS}) + h(\widehat{\beta}_{OLS}) \qquad (8.31) \\
& \leq & g(\widehat{\beta}_{ridge}) + h(\widehat{\beta}_{OLS}) \qquad (8.32)
\end{array}
$$

In other words, it must be that

$$h(\widehat{\beta}_{ridge}) \leq h(\widehat{\beta}_{OLS}) \qquad (8.33)$$

i.e.,

$$||\widehat{\beta}_{ridge}||_2 \leq ||\widehat{\beta}_{OLS}||_2 \qquad (8.34)$$

Not only is this the simplest way to demonstrate mathematically that ridge estimators are shrinkage estimators, the same argument above shows that shrinkage occurs for any vector norm, not just l_2. The LASSO, to be discussed later in this chapter, uses the l_1 norm, so we immediately see that the LASSO shrinks too.

8.11.3   Ridge Action Increases Eigenvalues

Let $M = A'A$ in the context of (2.19). It is nonnegative definite (Section A.8).

Under ridge regression, a value λ is added to the diagonal of M, i.e., M is replaced by

$$M + \lambda I \qquad (8.35)$$

Now suppose ν is an eigenvalue of M, with eigenvector x. Then

$$(M + \lambda I)x = Mx + \lambda x = \nu x + \lambda x \qquad (8.36)$$

So, $\nu + \lambda$ is an eigenvalue of the "ridge-ized" matrix $M + \nu I$, again with eigenvector x. Since λ is positive, applying the ridge action to M has increased its eigenvalues.

8.12   Computational Complements

8.12.1   Code for ridgelm()

There are several noteworthy aspects. Here first is the code:

```
ridgelm <- function(xy, lambda=seq(0.01, 1.00, 0.01),
    mapback=TRUE) {
    p <- ncol(xy) - 1;  n <- nrow(xy)
    x <- xy[,1:p]
    y <- xy[,p+1]
    x <- scale(x);  y <- y - mean(y)
    tx <- t(x)
    xpx <- tx %*% x / n
    xpy <- tx %*% y / n
    mapftn <- function(lambval)
        qr.solve(xpx + lambval*diag(p),xpy)
    tmp <- Map(mapftn,lambda)
    tmp <- Reduce(cbind,tmp)
    if (mapback) {
        sds <- attr(x,'scaled:scale')
        for (i in 1:p) tmp[i,] <- tmp[i,] / sds[i]
    }
    result <- list(bhats=tmp,lambda=lambda)
    class(result) <- 'rlm'
    result
}
```

```
plot.rlm <- function(ridgelm.out) {
    lamb <- ridgelm.out$lambda
    bhs <- t(ridgelm.out$bhats)
    matplot(lamb,bhs,type='l',pch='.',
        xlab='lambda',ylab='beta-hat')
}

print.rlm <-
    function(ridgelm.out) print(t(ridgelm.out$bhats))
```

Note the use of the R functions **Map()** and **Reduce()**, borrowed from functional languages such as LISP. The line

```
tmp <- Map(mapftn,lambda)
```

results in **mapftn()** being applied to each element of the vector **lambda**, yielding a vector of $\widehat{\beta}_i$ for each of those elements. Those vectors are returned in an R list, and we wish to combine them into a matrix. The call

```
tmp <- Reduce(cbind,tmp)
```

accomplishes this, by applying **cbind()** to all those vectors.[6]

Now look at this code excerpt:

```
    x <- scale(x); y <- y - mean(y)
    ...
    if (mapback) {
        sds <- attr(x,'scaled:scale')
        for (i in 1:p) tmp[i,] <- tmp[i,] / sds[i]
    }
```

When we called **scale()** on **x**, R did as asked, and divided each of the columns of **x**, i.e., each of the predictor variables, by its standard deviation. That results in a corresponding increase in each $\widehat{\beta}_i$ by the same factor. Thus, if we wish to have the $\widehat{\beta}_i$ on the original scale, we need to divide by those standard deviations. Fortunately, they were saved for us — in **x**! Calling **scale()** bestowed an R *attribute* on **x**, with information about the original means and standard deviations of the columns of **x**. As seen above, we can then retrieve the standard deviations via a call to **attr()**.

---

[6]The R function **do.call()** could also have been used here.

Finally, recall our brief discussion of R's S3 classes in Section 1.20.4.[7] How does one actually create an object of a certain S3 class? This is illustrated above. We first form an R list, containing the elements of our intended class object, then set its class:

```
result <- list(bhats=tmp,lambda=lambda)
class(result) <- 'rlm'
```

The object **result** is now an object of S3 class **'rlm'**.

We can then do function dispatch on objects of class **'rlm'**, just as in Section 1.20.4. In particular, calling the generic function **plot()** on such objects will result in the call being dispatched to **plot.rlm()**. The case of another generic function, **print()**, is similar.

Note by the way the use of the R graphing function **matplot()**, which plots multiple curves based on columns of matrix.

8.13   Exercises: Data, Code and Math Problems

Data problems:

1. As discussed at various points in this book, one may improve a parametric model by adding squared and interaction terms (Section 1.16). In principle, one could continue in that vein, adding cubic terms, quartic terms and so on. However, in doing so, we are likely to quickly run into multicollinearity issues. Indeed, **lm()** may detect that the $A'A$ matrix of Section 2.4.2 is so close to singular that the function will refuse to do the computation.

Try this scheme on the baseball player data in Section 1.6, predicting weight from height. Keep adding higher-degree powers of height until **lm()** complains, or until at least one of the coefficients returned by the function is the NA value.

Mini-CRAN and other computational problems:

2. A simple alternative approach to shrinkage estimators in the regression context would be to do shrinking directly, as follows, in the LOOM context:

---

[7]R also features two other types of classes, S4 and *reference classes*.

We choose γ to minimize

$$\sum_{i=1}^{n}(Y_i - \gamma \tilde{X}_i \widehat{\beta}_{-i})^2 \tag{8.37}$$

where $\widehat{\beta}_{-i}$ is the $\widehat{\beta}$ vector resulting from fitting the model to all observations but the ith. Write an R function to do this.

3. The output of **summary.lars()** shows C_p values. Alter that code so that it also shows adjusted-R^2. You may wish to review Section 1.20.4 and the fact that R treats functions as objects, i.e., they are mutable.

Math problems:

4. In Section 8.2.2, it was stated that if our predictor variables are centered and scaled, then $A'A$ will become the correlation matrix of our predictors. Derive this fact.

Note: Take the sample correlation between vectors U and V of length n to be

$$\frac{\frac{1}{n}\sum_{i=1}^{n}(U_i - \overline{U})(V_i - \overline{V})}{s_U s_V} \tag{8.38}$$

using the sample means and standard deviations.

5. Consider a fixed-X setting (Section 2.3), with an orthogonal design (Section 2.3. Find a simple, closed-form expression for $\widehat{\beta}_r$, $r = 1, ..., p$ under the ridge method for a specified λ.

Chapter 9

Variable Selection and Dimension Reduction

[Mathematical models] should be made as simple as possible, but not simpler. — Albert Einstein

If I had more time, I would have written a shorter letter — 18th century mathematician Blaise Pascal

For every complex problem there is an answer that is clear, simple, and wrong — H.L. Mencken

In the long and continuing history of regression and classification methodology, the burning question has always been, How should one select predictor variables? We may have data on many potential predictors, but somehow feel compelled to "thin them out." Motivations generally include:

- Avoidance of overfitting, thus hoping for better estimation or prediction.

- *Parsimony* — simple models are more appealing.

- Determination of the "important" predictors.

Denote the number of predictors used by p. Since we are working in p-dimensional space (or $p + 1$, counting the response variable), some refer to the variable selection process as *dimension reduction*. The term is especially apt in view of approaches to the problem that are explicitly based on vector subspace searches, to be covered in Section 9.9.

In spite of the long quest for the Dimension Reduction Holy Grail, in many settings there are good reasons NOT to delete any potential predictor variable, such as:

- If one's goal is Description, then variable selection generally renders statistical inference techniques invalid. A nominal 95% confidence interval no longer has that confidence level, because the distribution of the data changes, once one takes into account the fact that that distribution is now conditional on the variable-selection process. In addition, there is the multiple inference problem, discussed in Section 7.6. (More on this in Section 9.8.)

- Again in the Description setting, a parsimonious model may be misleading. We have already seen, in Chapter 7, that omission of some predictors can substantially change the estimated coefficients of the remaining predictors in a parametric model. This goes to the Mencken quote at the top of this chapter.

- In the Prediction setting, an omitted variable, though not generally very useful, might have good predictive power in some key subregion of the predictor space.

On the other hand, with many modern data sets, the dimension reduction issue cannot be ignored. In particular, it is now common for p to be much larger than the number of observations n, a setting in which, for instance, a classic linear parametric model cannot be fit. In such a situation, our hand is forced; we have very few options other than to do variable selection.[1]

Much theoretical (and empirical) work has been done on this subject. This is not a theoretical book, but at some points in this chapter we will discuss in nontheoretical terms what the implications of the research work are for the real-world practice of regression and classification analysis.

The aim of this chapter, then, is to investigate the dimension reduction issue, both from *why* and *how* points of view: *Why* might it be desirable in some situations, and how can one do it? The layout of the chapter is:

- Sections 9.1 through 9.2 will cover the *why*.

- Section 9.3 through 9.9 will handle the *how*.

---

[1]Ridge regression is a possibility, but questionable if p is much larger than n.

Note that polynomial and interaction terms (Section 1.16) are also predictors. so, deciding on a model may also entail which quadratic terms to include, for instance.

By the way, we will treat the case $p >> n$ separately, in Chapter 12.

9.1 A Closer Look at Under/Overfitting

We introduced the notion of under/overfitting in Section 1.11, and will go into more detail here, as it has strong implications for the methods to be discussed in this chapter. (The reader may wish to review Section 1.11 before coninuing.)

Recall that our estimated regression function $\widehat{\mu}$ plays two roles:

- The quantity $\widehat{\mu}(t)$ is an estimate of $\mu(t)$, and in parametric models it is thus an estimate of the coefficients. These are of interest in Description contexts.

- In Prediction contexts, the quantity $\widehat{\mu}(X_{new})$ is our predicted Y value for a new data point with predictor vector X_{new}.

The point is that, whether our interest is in Description or Prediction, $\widehat{\mu}$ is central. But as an estimator, it is subject to both variance and bias considerations. In particular, as noted in Section 1.11, we have that for any t,

$$\text{MSE}(\widehat{\mu}(t)) = Var[\widehat{\mu}(t)] + [\text{bias}(\widehat{\mu}(t))]^2 \tag{9.1}$$

So, again we see the famous *variance-bias tradeoff*: Richer models, i.e., those with more predictors, have smaller bias but larger variance.

Equation (9.1) concerns estimation of the regression function at a single point t. For the Prediction goal, we need to see how well we are doing over a range of t. In Random-X contexts (Section 2.3), we typically have t range according to the distribution of X, yielding the Mean Squared Prediction Error (MSPE),

$$E\left(Var[\widehat{\mu}(X)]\right) + E\left([\text{bias}(\widehat{\mu}(X))]^2\right) \tag{9.2}$$

9.1.1   A Simple Guiding Example

Here we present a simple toy example to help guide our intuition in under-standing (9.1). Suppose we have the samples of boys' and girls' heights at some age, say 10, $X_1, ..., X_n$ and $Y_1, ..., Y_n$. Assume for simplicity that the variance of height is the same for each gender, σ^2. The means of the two populations are designated by μ_1 and μ_2.

Say we wish to estimate μ_1 and μ_2. The "obvious" estimators are

$$\widehat{\mu}_1 = \frac{1}{n} \sum_{i=1}^{n} X_i \tag{9.3}$$

and

$$\widehat{\mu}_2 = \frac{1}{n} \sum_{i=1}^{n} Y_i \tag{9.4}$$

But at age 10, boys and girls tend to be about the same height. So if n is small, we may wish to make the simplifying assumption that $\mu_1 = \mu_2$, and then just use the overall mean as our estimate of μ_1 and μ_2:

$$\check{\mu}_i = \frac{1}{2}(\overline{X} + \overline{Y}), \quad i = 1, 2 \tag{9.5}$$

Each estimate is now based on $2n$ observations instead of just n, thus reducing variance. On the other hand, if $\mu_1 - \mu_2$ is large — we don't know this, which is why we would be doing the estimation in the first place, but we can still imagine the consequences of a large $\mu_1 - \mu_2$ — then we will have introduced substantial bias, thus possibly negating the advantage of using $2n$ data points for each estimator. Let's investigate this precisely.

We'll take as our criterion total MSE,

$$MSE(\widehat{\mu}_1) + MSE(\widehat{\mu}_2) \tag{9.6}$$

and

$$MSE(\check{\mu}_1) + MSE(\check{\mu}_2) \tag{9.7}$$

The computations in the Mathematical Complements portion of this chapter, Section 9.13.1, yield that (9.6) has the value

$$2(\frac{\sigma^2}{n} + 0^2) = \frac{2\sigma^2}{n} \tag{9.8}$$

where the 0 quantity is the bias.

The computations for (9.7) yield

$$\frac{\sigma^2}{n} + \frac{1}{2}(\mu_1 - \mu_2)^2 \tag{9.9}$$

So, let's call our model

$$\mu_1 = \mu_2 \tag{9.10}$$

the Lower-Dimensional Model, and call $\mu_1 \neq \mu_2$ the Higher-Dimensional Model, reflecting the fact that the latter has two parameters, μ_1 and μ_2, while the former has only one, the presumed common value of μ_1 and μ_2.

Comparing the two MSE values above, we see that using the Lower-Dimensional Model will pay off if and only if

$$(\mu_1 - \mu_2)^2 < 2\sigma^2/n \tag{9.11}$$

Again, we don't know the values of the μ_i and σ^2, but we can ask "what if" questions: The Lower-Dimensional Model will be a "win" if

- $(\mu_1 - \mu_2)^2$ is small, i.e., (9.10) is approximately true,

- n is small, i.e., we just don't have enough data to estimate two separate means, or

- σ^2 is large, which again amounts to not having enough data.

In such cases, the Higher-Dimensional Model is overfitting, i.e., is too rich a model for our situation.

This, in a nutshell, is the essence of the notion of overfitting. We know that technically (9.10) is incorrect, but if it is approximately correct, or if our data are meager, it is helpful to make that assumption.

The above example is trivially a regression problem. The Higher Dimensional Model has $p = 1$, with our predictor X being a single dummy variable for say, male. Our Lower Dimensional Model has $p = 0$, no predictors at all, just our intercept term. (The latter is the common value of the μ_i.)

If we have some k-category variable, that would mean $p = k - 1$ predictors. In this case, an assumption like, say,

$$\mu_1 = ..., = \mu_k \qquad\qquad (9.12)$$

would be a huge dimension reduction, and of course there may be groupings that might be considered for more moderate reductions.

The general regression/classification context is similar. Consider parametric models with p predictors on n observations. If we do not use all of our p predictors, we have fewer β_i to estimate, i.e., we have a lower-dimensional model. If p is large and/or n is small, it may be desirable to use that simplfiied model, say by omitting the predictors for which β_i seems to be small.

9.2   How Many Is Too Many?

To avoid overfitting, we should not have too many predictor variables p, relative to our sample size n. So, how many is too many?

In Chapter 1, we set a rough rule of thumb, due to Tukey, that one should have no more than $\sqrt{n}$ predictors. Later work by Portnoy [115] produced a similar result, and we will use this rule informally here. We will examine it more closely in Chapter 12.

9.3   Fit Criteria

Let's say we have m predictor variables in all, and we wish to choose a good subset of them. Let p denote the size of the subset we ultimately choose. Keep in mind that we typically do not choose p ourselves, but decide upon a value based on one of the processes to be described shortly in this chapter, or on some similar process.

9.3.1 Some Common Measures

In choosing a subset of predictor variables, we need a criterion to describe how good a particular subset is. The criteria below (other than the first) will be easier to explain if we assume a multivariate normal setting (Section 2.6.2). Let m denote the total number of variables we have available to serve as predictors, so we are assuming that (X, Y) has an (m+1) variate normal distribution.

Suppose also that we have some algorithm by which to add predictors one at a time (we will see such algorithms below), So we will only end up looking at $m + 1$ sets of predictors (including the case of having none), not the 2^m possible sets we could examine.

In the multivariate normal setting, the regression function of Y against any set of our predictors will be linear. Let σ_p^2 denote the variance of Y given our predictor set, $p = 0, 1, ..., m$.

Here are some of the most common criteria (the first two of which have been covered earlier in this book):

- **Re-predicting the Original Data**

  R^2 is an example of this. Recall (Section 2.9.2) that it is the squared sample correlation between the predicted and actual Y values, computed by "predicting" Y in the data in which we did our model fitting. In classification problems, we can compute the overall rate of correct classification in the same manner.

  As discussed earlier, e.g., in Section 6.6.1, this approach has a bias problem. Since our fit procedures find the best fit in the particular sample data at hand, and since the sample data will diverge somewhat from the population distribution, the fitted model will probably not be the best fit to the population. Thus R^2 and the rate of correct classification are generally too optimistic, i.e., are biased upward.

- **Adjusted R^2:** This was introduced in Section 2.9.4. Let's take a closer look. The quantity is defined to be

$$R_{adj}^2 = 1 - \frac{\frac{1}{n-p-1} \sum_{i=1}^{n} (Y_i - \widetilde{X}_i' \widehat{\beta})^2}{\frac{1}{n-1} \sum_{i=1}^{n} (Y_i - \overline{Y})^2} \tag{9.13}$$

  Recall here that $\widetilde{X}_i$ is X_i with a 1 prepended.

  Note that there will be a different $\widehat{\beta}$ value for each subset of predictors being used.

Recall that the numerator and denominator here are unbiased estimates of the values of σ_p^2 and σ_0^2. Recall too from Section 2.9.4 that both R^2 and adjusted R^2, in their sample-based forms, are interpreted as the estimated reduction in mean squared prediction error obtained when one uses the p predictor variables, versus just predicting by a constant.

As we add more and more predictors, R^2 is guaranteed to increase, since with more variables our minimization of (2.18) will be made over a wider set of choices. By contrast, the adjusted R^2 could go up or down when we add a predictor.

We could take adjusted R^2 as our *stopping criterion*: In adding our first few predictors, adjusted R^2 may increase, but eventually it might start to come back down. We might choose our predictor set to be the one that yields maximum adjusted R^2.

- **Mallows' C_p:** This amounts to an alternative way to adjust R^2:

$$C_p = \frac{SSE_p}{\frac{1}{n-m+1}SSE_m} - n + 2p \tag{9.14}$$

where SSE_p is

$$\sum_{i=1}^{n}(Y_i - \tilde{X}_i'\widehat{\beta})^2 \tag{9.15}$$

for the model with p predictors, $p = 0, 1, ..., m$.

Let's see what motivates this definition. The reader should keep in mind, by the way, that this is just a heuristic, so don't worry much about "leaps of faith" that may seem to occur.

Think of what would happen if a set of p predictors told the whole story, i.e., were as good as using all m predictors. Then we would have

$$\sigma_p^2 = \sigma_{p+1}^2 = ... = \sigma_m^2 \tag{9.16}$$

As noted, the numerator in (2.71) is an unbiased estimate of σ_p^2, so in (9.14), SSE_p is approximately $(n - p + 1)\sigma_p^2$. The denominator is approximately σ_m^2, no matter whether the set of p predictors tells the whole story, but under our assumption it is equal to σ_p^2. Then (9.14) is approximately

$$\frac{(n - p - 1)\sigma_p^2}{\sigma_p^2} - n + 2p = p - 1 \tag{9.17}$$

In other words, our stopping criterion might be to use the value of p that achieves approximate equality in (9.17), with estimated values for the σ_i^2.

For typical data sets, the graph of C_p against p is more or less convex (concave-upward), possibly with some small deviations from that shape. The reason for the convexity is that in (9.14), SSE_p decreases with p while of course $-n + 2p$ is an increasing function of p (n is fixed, as we are on the same data set). Moreover, if there is value of p such that (9.16) holds, then in (9.14), SSE_j will be approximately constant for $j = p, p + 1, ..., m$, making it even more likely that the graph will be increasing after $j = p$.

The bottom line, then, is that we might take as our stopping criterion the model with minimum C_p.

- **Akaike's Information Criterion:** This is basically

$$-2 \log \text{ likelihood} + 2 \text{ p} \qquad (9.18)$$

In the linear regression model with the classic normality assumption, this boils down to (except for unimportant constants)

$$n \ \log(s^2) + 2p \qquad (9.19)$$

where s^2 is as in (2.55).

Note carefully that AIC assumes that the conditional distribution of Y given X is known, in this case normal. Divergence from this assumption has unknown impacts of the use of AIC for variable selection.

AIC can also be computed for the logit model, though, and there we are on safe ground, since by definition the conditional distribution in question is Bernoulli/binomial.

Like C_p, AIC reflects a tradeoff between within-sample fit and number of predictors. Using this criterion in choosing among a sequence of models, we might choose the one with smallest AIC value.

- **Cross-validation:** As discussed earlier, here we split our data into a *training set* and a *test set*. To compare several different predictor sets, we fit each one to the training set, and see how well the resulting model predicts the test set.

Consider a single partitioning of our data into r and $n - r$ data points, and consider the context mentioned above, in which we add predictors to our model one at a time, with p denoting the number of predictors at any given step. For fixed n and fixed population distribution, there is an optimal value of p, which we will call $p_{opt}(n)$. Its value is unknown, of course, but it does exist, and our model selection process is, *inter alia*, estimating $p_{opt}(n)$. Now consider the ramifications of our choice for r:

- If we make r too small, we are estimating $p_{opt}(r)$ instead of $p_{opt}(n)$, potentially a problem if r is much smaller than n.

- If we make r too large, our test set, consisting of $n - r$ observations, will be small, and thus the measure of prediction accuracy it gives us will have high variability.

This is the motivation for m-fold cross-validation: We randomly divide our data into m subsets of about equal size, setting r to about n/m. For rather small m, r will be close to n, solving the problem in the first bullet above. But since we will look at m partitionings, the variability problem in the second bullet may be ameliorated with large m. But keep in mind that this is only a rough analysis, since for large m, our estimates from each subset will have higher variance.

9.3.2   There Is No Panacea!

Choosing a subset of predictors on the basis of cross-validation, AIC and so on is not foolproof by any means. Due to the Principle of Frequent Occurrence of Rare Events (Section 7.6.1), some subsets may look very good yet actually be artifacts.

On a theoretical basis, it has been shown that LOOM is not statistically consistent [128]. That paper does find conditions under which one gets consistency by leaving w out rather than 1, providing $w/n \to 1$, but it is not clear what the practical implications are. Empirical doubt on cross-validation was cast in [118].

9.4   Variable Selection Methods

Alan Miller wrote a comprehensive account of the state of the art for variable selection in 1990, and published a second edition in 2002 [110]. His comment speaks volumes:

> What has happened to the field since the first edition was first
> published in 1990? The short answer is that there has been very
> little progress.

The same statement applies today. The general issue of good variable
selection is still an unsolved problem. Nevertheless, a number of methods
have been developed that enjoy wide usage, and that many analysts have
found effective. We present some of them in the next few sections.

9.5 Simple Use of p-Values: Pitfalls

In applications in which Description is the primary goal, this method is the
most widespread, at least among nonstatisticians. The analyst simply reads
the output of, say, **lm()**, and decides that the predictors with asterisks are
the "important" ones, discarding the rest.

This book has discouraged such thinking (Section 2.10). The approach can
be very misleading, as shown in the MovieLens example in Section 3.2.4.
Moreover, if Prediction is the goal, it was disccovered long ago [14] that the
standard 0.05 cutoff for p-values may not be best, with values such as 0.25
and 0.35 giving better performance.[2]

9.6 Asking "What If" Questions

An alternative is to form confidence intervals for the various β_i, and gauge
the effects of the $X^{(i)}$ by comparing the locations of the intervals to the
general size of the Y variable. Examples of this were presented in Sections
2.10 and 3.2.4. Let's review a finding in the latter example, involving the
possible impact of user age on movie ratings:

> A 10-year difference in age only has an impact of about 0.03 on
> ratings, which are on the scale 1 to 5.

Clearly, user age is not an important variable for predicting ratings, or for
analyzing the underlying process that affect ratings.

On the other hand, consider the baseball player example, results of which
are discussed in Section 1.9.1.2. The estimated regression coefficient for the

---

[2]This is in the stepwise context to be discussed below.

age variable was 0.9115, indicating that players do gain weight as they age, almost a pound a year, in spite of needing to keep fit. Of course, one would need to form a confidence interval for this, as it is only an estimate, but the result is useful.

In other words, we would select predictors "by hand," using common sense and our knowledge of the setting in question. This is in contrast to using an automated process such as selection by p-values or the stepwise methods to be presented shortly. The drawback — to some people — is that one must work harder this way, no automated system to make our decisions for us. Nonetheless, this method has direct practical meaning, which the p-value and other approaches do not.

In the case of something like a logit model, as noted in Section 4.3.3, the β_i are a little more difficult to interpret than in linear models. After forming confidence intervals for the coefficients, how do we decide if they are "large" or "small"? One quick way would be to compare them to the intercept term, $\widehat{\beta}_0$.

For example, consider the example of diabetes in indigenous Americans, Section 4.3.2. The p-value for NPreg was tiny, generally considered "very highly significant." Yet the estimated effect of one extra pregnancy is only 0.1232, quite small compared to the intercept term, -8.4047. If our goal is Description, the finding that having more pregancies puts a women at greater risk for developing diabetes would be of interest, small but meaningful. On the other hand, in a Prediction context, it is clear that NPreg would not be of major help.

Note that we may not have a good domain-expert knowledge of the predictors. For instance we know, in the diabetes example above, what is common for the number of pregancies a woman has had. But we may not have similar knowledge of the triceps skin fold thickness variable. By looking at that variable's mean and standard deviation, we can attain at least a first-pass understanding of its scale of variation, and thus gauge the size of the coefficient in proper context. It gives us an idea of how much of a change in that variable is typical, and we can multiply that value by the coefficient to see how much $\widehat{\mu}(t)$ changes.

For this reason, the analyst may find it useful, for instance, to routinely run **colMeans()** on the predictor matrix/data frame after running **lm()** or **glm()**. Similarly, we should run

apply(dataname , 2 , **sd**)

where **dataname** is our matrix/data frame. Of course, a more direct way would be to apply **scale()** to the predictors before running

On the other hand, with a small sample (especially relative to the size of p), we have the opposite danger: There may be some good predictors that a simple p-value analysis may overlook. It may pay to take a good look at the "non-signicant" variables, especially if we are armed with domain expertise.

Again, this approach will not satisfy those who "want the computer to make their decisions for them." But placing the decision-making into the hands of the actual user may yield a better result in the end. (Recall Section 1.18.)

9.7 Stepwise Selection

When the use of computers first became common in statistical applications, stepwise methods were the rage. The idea is quite straightforward.

9.7.1 Basic Notion

We start with no predictors in the model, and test the hypothesis

$$H_0 : \beta_i = 0 \tag{9.20}$$

for each predictor $X^{(i)}$. Whichever one yields the smallest p-value, that predictor is added to the model. We then determine which predictor to add to the model next in the same manner, via (9.20). We stop when we no longer have any p-values below a cutoff value, which classically has been the usual 0.05 for stepwise methods.[3]

This is called *forward* selection; *backward* selection begins with all the predictors in the model, and removes them one by one, again using (9.20).

The original forms of these methods are now considered out of date by many, but the notion of stepwise adding/deleting predictors is still quite in favor.

---

[3]As noted earlier, research has shown that a much larger cutoff tends to work better. An interesting related fact is that including a predictor if the Z-score in Equation (2.56) is greater than 1.0 in absolute value — a p-value of about 0.32 — is equivalent to including the predictor if it raises the Adjusted R-squared value [58].

9.7.2   Forward vs. Backward Selection

Forward selection seems especially appealing. Since our goal is to find a parsimonious model, it is natural to start with the most parsimonious model of all — the one without any predictors at all — and then slowly add more predictors. One setting in which forward selection is virtually forced upon us is that in which $p > n$, making linear or other parametric models impossible to fit (without resorting to ridge regression or the like).

On the other hand, many analysts believe that forward selection may under-fit, in the sense that it may miss a group of predictors in which each one is not a strong predictor but collectively the group has substantial predictive power.

Consider the MovieLens data (Section 3.2.4). After calculating the T_i as our Y variable here, we added age and gender predictors. We might also add dummy variables for the individual movies. Presumably each one will contribute only a negligible amount to prediction, but collectively they could be rather powerful. We will investigate this in Chapter 12.

We can also try *both* forward *and* backward stepping. In the forward case, for instance, after adding a few predictors it may be that one of the ones added early in the process is now not very helpful.

9.7.3   R Functions for Stepwise Regression

There are many such functions. Here, we will use **stepAIC()** for the linear and logit models. It is part of the **MASS** package included in the base R distribution. It uses AIC for its fit criterion, so that for instance in forward selection, the predictor that is added will be the one that brings about the largest drop in AIC. The argument **direction** allows the user to specify forward or backward selection, or even both. In the latter case, which is the default value, backward elimination is used but variables can be re-added to the model at various stages of the process.

By the way, **lars()**, which we will use for LASSO below, also does offer other options, including one for stepwise selection.

9.7.4   Example: Bodyfat Data

This data set was introduced in Section 1.2, with $n = 252$ observations on $p = 13$ predictor variables. The role of prediction for that data was explained:

> Body fat is expensive and unwieldy to measure directly, as it
> involves underwater weighing. Thus it would be highly desirable
> to "predict" that quantity from easily measurable variables such
> as height, age, weight, abdomen circumference and so on.

The first column of the data set is the case number. If the numbering is
sequential in time, this might be useful for investigating time trends, but
we will omit it.

The next three columns all involve the underwater weighing, something
we are trying to avoid. We wish to see how various body circumference
methods can predict body fat, so we will use only one of the available
measures, say the first:

```
> library(mfp)
> data(bodyfat)
> bodyfat <- bodyfat[,-c(1,3:4)]
```

Now let's run **lm()**:

```
> lmall <- lm(brozek ~ ., data=bodyfat)
> summary(lmall)
...
```
Coefficients:

	Estimate	Std. Error	t value	Pr(>|t|)	
(Intercept)	−15.29255	16.06992	−0.952	0.34225	
age	0.05679	0.02996	1.895	0.05929	.
weight	−0.08031	0.04958	−1.620	0.10660	
height	−0.06460	0.08893	−0.726	0.46830	
neck	−0.43754	0.21533	−2.032	0.04327	*
chest	−0.02360	0.09184	−0.257	0.79740	
abdomen	0.88543	0.08008	11.057	< 2e−16	***
hip	−0.19842	0.13516	−1.468	0.14341	
thigh	0.23190	0.13372	1.734	0.08418	.
knee	−0.01168	0.22414	−0.052	0.95850	
ankle	0.16354	0.20514	0.797	0.42614	
biceps	0.15280	0.15851	0.964	0.33605	
forearm	0.43049	0.18445	2.334	0.02044	*
wrist	−1.47654	0.49552	−2.980	0.00318	**

```
...
```
Multiple R-squared: 0.749, Adj. R-squared: 0.7353

This is rather good fit, with a straightforward analysis with all predictor variables present: Adjusted R^2 was 0.7353. Why, then, should we do variable selection?

Again, the answer is expense. The point of predicting bodyfat in the first place was to save on cost, and since collecting data on the predictor variables entails labor and thus costs, it would be nice if we found a parsimonious subset. Let's see what **stepAIC()** does with it.

The function **stepAIC()** needs the full model to be fit first, using it to acquire information about the data, model and so on. We have already done that, so now we run the analysis, showing the output of the first step here:

```
> library (MASS)
> stepout <− stepAIC (lmall)
Start:   AIC=710.77
brozek ~ age + weight + height + neck + chest +
      abdomen + hip + thigh + knee + ankle + biceps +
      forearm + wrist
```

	Df	Sum of Sq	RSS	AIC
− knee	1	0.04	3785.2	708.77
− chest	1	1.05	3786.2	708.84
− height	1	8.39	3793.5	709.33
− ankle	1	10.11	3795.2	709.44
− biceps	1	14.78	3799.9	709.75
<none>			3785.1	710.77
− hip	1	34.28	3819.4	711.04
− weight	1	41.73	3826.9	711.53
− thigh	1	47.83	3833.0	711.94
− age	1	57.12	3842.3	712.55
− neck	1	65.66	3850.8	713.10
− forearm	1	86.63	3871.8	714.47
− wrist	1	141.21	3926.3	718.00
− abdomen	1	1944.46	5729.6	813.24

The initial AIC value, i.e., with all predictors present, was 710.77. The function then entertained removal of the various predictors, one by one. Removing the knee measurement, for instance, would reduce AIC to 708.77, while removing chest would achieve 708.84, and so on.

By contrast, removing the hip measurement would be worse than doing nothing, actually increasing AIC to 711.04. (The <none> line separates the variables whose removal decreases AIC from those that increase it.

The difference between the knee and chest variables is negligible, but the function decides to remove the knee variable, as seen in the model after one step,

```
Step:   AIC=708.77
brozek ~ age + weight + height + neck + chest +
    abdomen + hip + thigh + ankle + biceps +
    forearm + wrist
```

At the second step, the algorithm was faced with the following choices:

	Df	Sum of Sq	RSS	AIC
− ankle	1	11.20	3805.1	704.09
− biceps	1	16.21	3810.1	704.43
− hip	1	28.16	3822.0	705.22
<none>			3793.9	705.35
− thigh	1	63.66	3857.5	707.55
− neck	1	65.45	3859.3	707.66
− age	1	66.23	3860.1	707.71
− forearm	1	88.14	3882.0	709.14
− weight	1	102.94	3896.8	710.10
− wrist	1	151.52	3945.4	713.22
− abdomen	1	2737.19	6531.1	840.23

It then chose to remove the ankle variable.

Eventually, all variables fall below the <none> line,

```
Step:   AIC=703.08
brozek ~ age + weight + neck +
    abdomen + hip + thigh + forearm + wrist
```

	Df	Sum of Sq	RSS	AIC
<none>			3820.0	703.08
− hip	1	33.23	3853.2	703.26
− neck	1	67.79	3887.8	705.51
− age	1	67.88	3887.9	705.52
− weight	1	81.50	3901.5	706.40
− thigh	1	90.34	3910.3	706.97
− forearm	1	122.99	3943.0	709.07
− wrist	1	139.46	3959.4	710.12
− abdomen	1	2726.49	6546.5	836.83

and the final model is chosen:

```
Coefficients:
(Intercept)                age              weight
  -20.06213            0.05922            -0.08414
        neck            abdomen               hip
   -0.43189            0.87721            -0.18641
       thigh           forearm              wrist
    0.28644            0.48255            -1.40487
```

So, how does this reduced model fare in terms of predictive ability?

> **summary**(stepout)
...
```
Coefficients:
               Estimate Std. Error  t value  Pr(>|t|)
(Intercept)   -20.06213   10.84654   -1.850   0.06558
age             0.05922    0.02850    2.078   0.03876
weight         -0.08414    0.03695   -2.277   0.02366
neck           -0.43189    0.20799   -2.077   0.03889
abdomen         0.87721    0.06661   13.170   < 2e-16
hip            -0.18641    0.12821   -1.454   0.14727
thigh           0.28644    0.11949    2.397   0.01727
forearm         0.48255    0.17251    2.797   0.00557
wrist          -1.40487    0.47167   -2.978   0.00319

(Intercept)   .
age           *
weight        *
neck          *
abdomen       ***
hip
thigh         *
forearm       **
wrist         **
...
Multiple R-squared:   0.7467,
    Adjusted R-squared:   0.7383
```

The two R-squared values are quite close to those of the full model. In other words, our pared-down predictor set has about the same predictive power as the full model, but at a much lower data collection cost.

Note, though, that the variable selection process changes all the distributions. The p-values are overly optimistic, as is adjusted R-squared. Nevertheless, use of the more restrictive predictor set, with the attendant cost savings, does seem to be a safe bet. As is often the case, it would be wise to check this with a domain expert.

9.7.5 Classification Settings

Dimension reduction is of course an important issue in classification settings. Continuing our use of **stepAIC()** as our vehicle for stepwise predictor selection, let's take a look at what can be done.

9.7.5.1 Example: Bank Marketing Data

This data set is one of the best known in the UCI collection. It consists of data on a marketing campaign by a Portuguese bank, with the goal of predicting whether a customer would open a new term deposit account. The latter is indicated by the **y** column in the data frame:

```
# note the ';' separator symbol, not commas
> bank <- read.csv('bank.csv',sep=';')
> head(bank)
  age         job marital education default
1  30  unemployed married   primary      no
2  33    services married secondary      no
3  35  management  single  tertiary      no
4  30  management married  tertiary      no
5  59 blue-collar married secondary      no
6  35  management  single  tertiary      no
  balance housing loan  contact day month
1    1787      no   no cellular  19   oct
2    4789     yes  yes cellular  11   may
3    1350     yes   no cellular  16   apr
4    1476     yes  yes  unknown   3   jun
5       0     yes   no  unknown   5   may
6     747      no   no cellular  23   feb
  duration campaign pdays previous poutcome  y
1       79        1    -1        0  unknown no
2      220        1   339        4  failure no
3      185        1   330        1  failure no
4      199        4    -1        0  unknown no
5      226        1    -1        0  unknown no
6      141        2   176        3  failure no
> dim(bank)
```

[1] 4521   17

So we have $n = 4521$ and a nominal value $p = 16$. The true value of p is much larger, due to the presence of R factors, which correspond to dummy variables; more on this below. Another point to keep in mind is that this is actually the smaller data set in this bank data package.

Now, what about those dummy variables? In R, a popular way to store a *categorical* variable, say eye color, is as a *factor*, essentially a numeric vector that is normally dealt with by the names of the values. See the Computational Complements section on factors at the end of this chapter for details, but for now the point is that with factors we can avoid forming dummy variables; R does this work for us:

```
> glout <- glm(y ~ ., data=bank, family=binomial)
> summary(glout)
Coefficients:
```

	Estimate	Std. Error	z value
(Intercept)	−2.462e+00	6.038e−01	−4.077
age	−4.232e−03	7.125e−03	−0.594
jobblue−collar	−3.924e−01	2.420e−01	−1.621
jobentrepreneur	−2.498e−01	3.811e−01	−0.655
jobhousemaid	−3.530e−01	4.176e−01	−0.845
jobmanagement	−7.302e−02	2.407e−01	−0.303
jobretired	6.315e−01	3.112e−01	2.029
jobself−employed	−1.812e−01	3.533e−01	−0.513
jobservices	−1.457e−01	2.729e−01	−0.534
jobstudent	3.784e−01	3.750e−01	1.009
jobtechnician	−1.926e−01	2.301e−01	−0.837
jobunemployed	−6.395e−01	4.214e−01	−1.518
jobunknown	5.207e−01	5.853e−01	0.890
maritalmarried	−4.696e−01	1.743e−01	−2.694
maritalsingle	−3.051e−01	2.038e−01	−1.497

. . .

	Pr(>|z|)	
(Intercept)	4.55e−05	***
age	0.552537	
jobblue−collar	0.104937	
jobentrepreneur	0.512199	
jobhousemaid	0.398000	
jobmanagement	0.761602	
jobretired	0.042454	*
jobself−employed	0.608167	
jobservices	0.593542	

```
jobstudent          0.312958
jobtechnician       0.402496
jobunemployed       0.129138
jobunknown          0.373669
maritalmarried      0.007058  **
maritalsingle       0.134354
```

Notice that R not only created dummies, but gave them names according to the levels; for instance, R noticed that one of the levels of **job** was **blue-collar**, and thus named the dummy **jobblue-collar**. If we want to know which levels R chose for constructing the dummies, we can examine the **xlevels** component:

```
glout$xlevels
$job
 [1] "admin."         "blue−collar"
 [3] "entrepreneur"   "housemaid"
 [5] "management"     "retired"
 [7] "self−employed"  "services"
 [9] "student"        "technician"
[11] "unemployed"     "unknown"

$marital
[1] "divorced"  "married"    "single"

$education
[1] "primary"    "secondary"  "tertiary"
[4] "unknown"
...
```

We can see that the **marital** column in the data frame has levels "divorced," "married" and "single," yet **lm()** only produced coefficients for the latter two. So those coefficients are relative to the "divorced" level. Since they are both negative, it seems that the divorced customers are more likely to open the new account.

It's important to know what the real value of p is (as opposed to the nominal value 16 we mentioned earlier), as the larger p is, the more we risk overfitting. As already discussed, there is no magic rule for determining if p is too large, but we should at least know what value of p we have:

```
> length(coef(glout)) − 1   # exclude intercept
[1] 42
```

Let's apply the stepwise selection process:

```
> stepout <- stepAIC(glout)
> stepout
...
```
Coefficients:

(Intercept)	maritalmarried
−2.731855	−0.510829
maritalsingle	educationsecondary
−0.318135	0.126807
educationtertiary	educationunknown
0.381100	−0.283123
housingyes	loanyes
−0.343417	−0.643090
contacttelephone	contactunknown
−0.007710	−1.457777
day	monthaug
0.016295	−0.318045
monthdec	monthfeb
0.202759	0.139813
monthjan	monthjul
−1.085164	−0.731248
monthjun	monthmar
0.560045	1.626269
monthmay	monthnov
−0.473499	−0.861876
monthoct	monthsep
1.410274	0.755895
duration	campaign
0.004184	−0.072216
poutcomeother	poutcomesuccess
0.481385	2.414545
poutcomeunknown	
−0.093110	

How much was trimmed?

```
> length(coef(stepout))
[1] 27
```

So, 15 of the original 42 predictors were removed.

Let's see whether this resulted in much compromise in predictive power. Again, this would be more accurately assessed using cross-validation, but here is a quick assessment:

```
> y <- as.integer(bank$y) - 1
```

```
> yhat <- round(glout$fitted.values)
> mean(y)
[1] 0.11524
> mean(yhat == y)
[1] 0.9051095
```

So, without covariate information, we would also guess y to be 'no', and would be correct about 1 - 0.11524 = 0.88476 of the time. With the covariates we would do slightly better, 0.9051095. Let's compare the latter figure with the model yielded by **stepAIC()**:

```
> yhatstep <- round(mnout$fitted.values)
> mean(yhatstep == y)
[1] 0.9046671
```

Good, almost identical.

As mentioned, the data package for this bank marketing data contains two versions of the data, first the one we used above, **bank.csv**, and a much larger one, **bank-full.csv**. After going through the same computations as above for this large set (not shown), it turns out that only 2 of the 42 predictors are removed, compared to 15 for the smaller data set. This makes sense; as discussed before, the larger n is, the more predictors the model can handle well.

9.7.5.2   Example: Vertebrae Data

As a multiclass example, let's again use the vertebrae data from Section 5.5.2:

```
> mnout <- multinom(V7 ~ ., data=vert)
. . .
> stepout <- stepAIC(mnout)
. . .
> summary(stepout)
Call:
multinom(formula = V7 ~ V1 + V2 + V5 + V6, data = vert)

Coefficients:
   (Intercept)         V1          V2          V5
1    -20.85643  0.1827638  -0.2634111  0.13592470
2    -21.37424  0.2202886  -0.2183829  0.07716224
                V6
```

```
1  -0.000133527
2   0.311163955
```

```
Std. Errors:
   (Intercept)           V1            V2            V5
1    4.249310  0.03310576  0.04888643  0.02873804
2    5.239924  0.05140014  0.08326775  0.03095909
            V6
1  0.03862044
2  0.05794419
...
```

So, the predictors V3 and V4 were dropped.

9.7.6   Nonparametric Settings

So, how does all this relate to nonparametric regression methods (including classification, of course), say k-nearest neighbor? There are actually two questions here:

- Is dimension reduction even an issue in the nonparametric case?

- If so, how might this goal be accomplished?

9.7.6.1   Is Dimension Reduction Important in the Nonparametric Setting?

The answer to this question is indeed yes. To see why, let's look again at the vertebrae data, estimating $\mu(t)$ for t equal to the first observation in the data set, which for convenience we will call the *prediction point*.

To find the nearest neighbors of the prediction point, we'll use the function **get.knnx()** from the **FNN** package on CRAN [88] (which is also used in our **regtools** package [97]). The call form is

```
get.knnx(dataframe, tvalue)
```

where **dataframe** is our data frame of predictor values, and **tvalue** is our prediction point. This call returns an R list with components **nn.index** and **nn.dist**.

p=1	p=2	p=3
245	1	1
1	254	236
23	308	175
109	23	308
201	17	119
153	201	120
260	231	4
197	236	240
254	119	286
17	144	22

Table 9.1: Indices of Nearest Neighbors

To gauge the effects of different values of the number of predictors p, we'll consider $p = 1, 2, 3$, so that for instance $p = 2$ means that we do prediction based on the variables $V1$ and $V2$. Here is the code:

```
test <- vert[1,,drop=FALSE]
get.knnx(vert[,1],test[,1])$nn.index
get.knnx(vert[,1:2],test[,1:2])$nn.index
get.knnx(vert[,1:3],test[,1:3])$nn.index
```

The results are shown in Table 9.1. The indices of the 10 closest neighbors to the prediction point are shown, one per row, for each value of p.

Suppose that, unknown to the analyst, the regression function $\mu(t_1, t_2, ...)$ depends only on t_i, i.e., only the first predictor has impact on the response variable Y. Then (again, unknown to the analyst) the nearest-neighbor finding process need only consider the first predictor, **V1**. In that case, it turns out that observation number 245 is the closest. Yet that observation doesn't make the nearest-10 list at all for the cases $p = 2$ and $p = 3$. And though there is some commonality among the three columns of the table, it is clear that generally the nearest neighbors of a point for $p = 1$ will differ from those for the other two values of p.

What are the implications of this? Recall the bias-variance tradeoff issue in under/overfitting (Section 9.1). The more distant an observation from the prediction point, the more the bias. So, making a "mistake" in choosing the nearest neighbors will generally give us more-distant neighbors, and increase prediction error.

In other words, yes, dimension reduction is just as much an issue in the nonparametric setting as in the parametric one.

9.7.7   The LASSO

One of the reasons for the popularity of the LASSO is that it does automatic variable selection. We will take a closer look at LASSO methods in this section.

9.7.7.1   Why the LASSO Often Performs Subsetting

First, similar to the ridge case, minimizing (8.18) is equivalent to minimizing[4]

$$q(b) = \sum_{i=1}^{n}(Y_i - \widetilde{X}_i b)^2 \tag{9.21}$$

subject to the constraint

$$||b||_1 \leq \lambda \tag{9.22}$$

This motivates Figure 9.1.

The figure is for the case of $p = 2$ predictors (for simplicity, we assume there is no constant term β_0). Writing $b = (b_1, b_2)'$, then the horizontal and vertical axes are for b_1 and b_2, as shown. The corners of the diamond are at $(\lambda, 0)$, $(0, \lambda)$ and so on. Due to the constraint (9.22), our LASSO estimator $\widehat{\beta}_l$ must be somewhere within the diamond.

What about the ellipses? They are *contours* of q: For a given value of q, say c, then the locus of points b for which $q(b) = c$ takes the form of an ellipse. Each value of c gives us a different ellipse; two of them, out of infinitely many, are shown in the figure, with the smaller one corresponding to a smaller value of c.

But remember, we are trying to minimize q, so we want c to be as small as possible, i.e., we want the countour curve to be small — but our constraint

---

[4]The computational details of the minimization process are beyond the scope of this book.

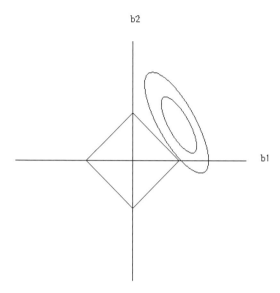

Figure 9.1: Subsetting nature of the LASSO

requires that the curve must include at least one point within the diamond. In our figure here, this implies that we must choose c so that the ellipse is barely touching the diamond, as the larger ellipse does.

Now, here is the key point: The point at which the ellipse barely touches the diamond will typically be one of the four corners of the diamond. And at each of those corners, either b_1 or b_2 is 0 — i.e., $\widehat{\beta}_l$ has selected a *subset* of the predictors, in this case a subset of size 1.

The same geometric argument works in higher dimensions, and this is then the appeal of the LASSO for many analysts:

> The LASSO often does automatic subset selection. The analyst need only use the predictors $X^{(i)}$ for which $\widehat{\beta}_i \neq 0$.

We say that the LASSO tends to produce a *sparse* estimator of β. Needless to say, though this is indeed an appealing property, there is no guarantee that this produces a "good" set of predictors.

Suppose in the figure, the inner ellipse corresponds to the ordinary estimator $b = \widehat{\beta}_{OLS}$, i.e.,

$$c = q(\widehat{\beta}_{OLS}) \qquad (9.23)$$

In order to satisfy the LASSO constraint, we needed to accept a larger value of q, corresponding to the outer ellipse, and thus a smaller $\widehat{\beta}$. This illustrates the shrinkage nature of the LASSO.

On the other hand, the ellipse corresponding to OLS might already dip into the diamond. In this case,

$$\widehat{\beta}_l = \widehat{\beta}_{OLS} \qquad (9.24)$$

So, it is not guaranteed that the LASSO will choose a sparse $\widehat{\beta}$. As was noted earlier for shrinkage estimators in general, for fixed p, the larger n is, the less need for shrinkage, and the above situation may occur.

There is of course the matter of choosing the value of λ. Our old friend, cross-validation, is an obvious approach to this, and others have been proposed as well. The **lars** package includes a function **cv.lars()** to do k-fold cross-validation.

9.7.7.2   Example: Bodyfat Data

Let's continue the example of Section 9.7.4. Let's see what **lars** finds here.

```
> library(lars)
> larsout <- lars(as.matrix(bodyfat[,-1]),bodyfat[,1])
> larsout
Call:
lars(x = as.matrix(bodyfat[, -1]), y = bodyfat[, 1])
R-squared: 0.749
Sequence of LASSO moves:
      abdomen height age wrist neck forearm hip
Var         6      3   1    13    4      12   7
Step        1      2   3     4    5       6   7
      weight biceps thigh ankle chest knee
Var        2     11     8    10     5    9
Step       8      9    10    11    12   13
```

So, at Step 1, the abdomen predictor was brought in, then height at Step 2, and so on. Now look further:

```
> summary(larsout)
...
      Df      Rss        Cp
0      1  15079.0   698.131
1      2   5423.4    93.012
2      3   5230.7    82.893
3      4   4914.9    65.038
4      5   4333.6    30.484
5      6   4313.5    31.225
6      7   4101.8    19.910
7      8   4090.5    21.202
8      9   4006.5    17.919
9     10   3980.0    18.252
10    11   3859.5    12.679
11    12   3793.0    10.495
12    13   3786.0    12.057
13    14   3785.1    14.000
```

Based on the C_p value, we might stop after Step 11, right after the ankle variable is brought in. The resulting model would consist of predictors abdomen, height, age, wrist, neck, forearm, hip, weight, biceps, thigh and ankle.

By contrast, if one takes the traditional approach and selects the variables on the basis of p-values, as discussed in Section 9.5, only 4 predictors would be chosen (see output in Section 9.7.4), rather than 9 as above.

We can also determine what λ values were used:

```
> larsout$lambda
 [1]  99.9203960  18.1246879  15.5110550  10.7746865
 [5]   4.8247693   4.5923026   2.6282871   2.5472757
 [9]   1.9518718   1.7731184   1.0385786   0.3162681
[13]   0.1051796
```

9.8   Post-Selection Inference

Stepwise predictor selection is an *adaptive* technique. This refers to any statistical method that works in stages, with the outcome of any stage determining what action is taken at the next stage. The problem with this is that a proper statistical analysis would be based on the *conditional* distribution in the later stage, given the earlier stage, rather than the unconditional distribution. If for instance we wish to form a confidence interval for β_5 after removing the predictor $X^{(2)}$, the distribution of $\widehat{\beta}_5$ is no longer given by the material in Section 2.8. If the latter is used, then inferences will be incorrect. Following is an intuitive derivation of that property.

For concreteness, let's consider the following simple example. We have $p = 2$ predictors, and use a linear model without an intercept term,

$$E(Y|X^{(1)} = t_1, X^{(2)} = t_2) = \beta_1 t_1 + \beta_2 t_2 \qquad (9.25)$$

Consider a classical variable selection approach, in which the variable to delete is chosen by p-values (Section 9.7.1). The latter are calculated using (2.56), which is based on the standard unconditional distribution assumptions.

Under those assumptions, the pair $(\widehat{\beta}_1, \widehat{\beta}_2)'$ has a bivariate normal distribution with mean vector $(\beta_1, \beta_2)'$ (Section 2.8.4). For simplicity, suppose in actuality $\beta_1 = \beta_2 = 0$. Also, suppose the standard errors of the $\widehat{\beta}_i$ are about 1.0.

In the forward stepwise process, we first test the hypothesis $H_0 : \beta_1 = 0$; using the standard 0.05 significance level, we will reject the hypothesis if and only if

$$|\widehat{\beta}_1| > 1.96 \qquad (9.26)$$

If that hypothesis is rejected, then we retain $X^{(1)}$ as a predictor, and move on to $X^{(2)}$, retaining it if and only if

$$|\widehat{\beta}_2| > 1.96 \qquad (9.27)$$

But (9.27) is based on the unconditional distribution of $\widehat{\beta}_2$. Here, under $H_0 : \beta_2 = 0$,

$$P(|\widehat{\beta}_2| > 1.96) = 0.05 \qquad (9.28)$$

in the setting $\beta_2 = 0$. But the proper probability to use would be

$$P(|\widehat{\beta}_2| > 1.96 \mid |\widehat{\beta}_1| > 1.96) \qquad (9.29)$$

and this could be quite different from 0.05. Indeed, it could be near 1.0! Here is why:

Recall the analysis in Section 8.2.1, especially (8.8). In that setting, $\widehat{\beta}_1$ and $\widehat{\beta}_2$ are negatively/positively correlated if $X^{(1)}$ and $X^{(2)}$ are positively/negatively correlated, i.e., $c > 0$ versus $c < 0$ in (8.8). Moreover, if $|c|$ is near 1.0 there, the correlation is very near -1.0 and 1.0, respectively.

So, when c is near -1.0, for example, $\widehat{\beta}_1$ and $\widehat{\beta}_2$ will be highly positively correlated. In fact, given that they have the same mean and variance, the two quantities will be very close to identical, with high probability (Exercise 11, Chapter 2). In other words, (9.29) will be near 1.0, not the 0.05 value we desire.

The ramification of this is that any calculation of confidence intervals or p-values made on the final chosen model after stepwise selection cannot be taken too much at face value, and indeed could be way off the mark. Similarly, quantities such as the adjusted R-squared value may not be reliable after variable selection.

Some research has been done to develop adaptive methods for such settings, termed *post-selection inference*, but they tend to be hypothesis-testing oriented and difficult to convert to confidence intervals, as well as having restrictive assumptions. See for instance [59] [17]. There is no "free lunch" here.

9.9 Direct Methods for Dimension Reduction

Given this chapter's title, "Dimension Reduction," we should discuss direct methods for that goal.

Typically the methods are applied to X, the vector of predictor variables, excluding Y. The oldest method is the classical statistical technique of *principal components analysis* (PCA), but a number of others have been devised in recent years. Here we will cover PCA and another method now very popular in classification contexts, *nonnegative matrix factorization*, as well as the *parallel coordinates* graphical approach we touched on in Section 6.7.3.

9.9.1 Informal Nature

Black cat, white cat, it doesn't matter as long as it catches mice — former

Chinese leader Deng Xiaoping

Though statisticians tend to view things through the lens of exact models, proper computation of standard errors and so on, much usage of regression models is much less formal, especially in the machine learning community. Research in the area is often presented as intuitive *ad hoc* models that seem to do well on some data sets. Since these models would typically be extremely difficult to analyze mathematically, empirical study is all we have.

Nevertheless, some methods have been found to work well over the years, and must be part of the analyst's toolkit, without worrying too much about assumptions. For example, PCA was originally intended for use in multivariate normal distributions, but one might apply it to dummy variables.

9.9.2   Role in Regression Analysis

The basic approach for using these methods in regression is as follows:

- Center and scale the data.

- Find new predictors, typically linear combinations of the original predictor variables.

- Discard some of the new variables, on the grounds that they are largely redundant.

- Then fit whatever regression model is desired — linear/nonlinear parametric, or even nonparametric — on this new lower-dimensional predictor space, i.e., using the remaining new variables.

9.9.3   PCA

We'll start with Principal Components Analysis. This is a time-honored statistical method, closely related to Singular Value Decomposition (SVD); see for example [84] [44]. PCA was discussed in Section 8.7. Recall from there that we create new variables, called *principal components*, which are linear combinations of the original predictor variables. They are determined through a matrix diagonalization process (Section A.6) applied to the sample covariance matrix S of the original predictors.

We then use these new variables as our predictors, discarding any having a small variance, or equivalently, corresponding to a small eigenvalue.

9.9.3.1   Issues

Though PCA regression would seem to be a natural solution to the dimension reduction problem, of course, this approach is no panacea:

- The sample covariance matrix is subject to sampling error, and thus the ordering of the principal components with respect to their variances may be incorrect, resulting in some components being discarded when they should be retained and *vice versa*. The coefficients in the linear combinations may be similarly unstable. And the uncorrelated nature of the principal components only holds for the sample, and substantial correlations may exist at the population level.

- Even though one of the new predictors has a small variance, it still could have a strong correlation with Y. Indeed, the problem is that the entire dimension reduction process using PCA is oblivious to Y.

- As usual, we are faced with choosing the value of a tuning parameter, the number of principal components to retain, k. A rule of thumb used by many is to choose k so that the variances of the retained variances sum to, say, 90% of the total. Of course, one might choose k via cross-validation on the regression analysis.

9.9.3.2   Example: Bodyfat Data

Let's apply PCA to the bodyfat data:

```
> library(mfp)
Loading required package: survival
Loading required package: splines
> data(bodyfat)
> bodyfat <- bodyfat[,-c(1,3:4)]
> prc <- prcomp(bodyfat[,-1], center=TRUE, scale=TRUE)
> str(prc)
List of 5
 $ sdev : num [1:13] 2.836 1.164 1.001 0.817 0.774 ...
 $ rotation: num [1:13, 1:13] 0.00985 0.34454
        0.10114 0.30559 0.31614 ...
 ...
```

R's **prcomp()** function does PCA analysis, returning an object of class **'prcomp'**. Whenever working with a new R class, it's always useful to get a quick overview, via the **str()** ("structure") function. The full results are not shown here, but they show that **prc** includes various components, notably **sdev**, the standard deviations of the principal components, and **rotation**, the matrix of principal components themselves. Let's look a little more closely at the latter.

For instance, **rotation[1,]** will be the coefficients, called *loadings*, in the linear combination of X that comprises the first principal component (which is the first row of M):

```
> prc$rotation[,1]
          age        weight       height         neck
0.009847707  0.344543793  0.101142423  0.305593448
        chest       abdomen          hip        thigh
0.316137873  0.311798252  0.325857835  0.310088798
         knee         ankle       biceps      forearm
0.308297081  0.230527382  0.299337590  0.249740971
        wrist
0.279127655
```

In other words, the first principal component is

$$0.009847707 \times \text{ age } + 0.344543793 \times \text{ weight } + \ldots + 0.279127655 \times \text{ wrist} \tag{9.30}$$

Keep in mind that we have centered and scaled the original predictors, so that for instance age has been so transformed. For example,

```
> prc$scale[1]
     age
12.60204
> var(bodyfat[,-1]$age)
[1] 158.8114
> var(bodyfat[,-1]$age/prc$scale[1])
[1] 1
```

We could transform (9.30) to the original scaling by multiplying the coefficients by **prc$scale**. On the other hand, **prc$x** contains the new version of our data, expressed in terms of the principal components. Our original data, **bodyfat[,1]**, contained 252 observations on 13 variables, and **prc$x** is a matrix of those same dimensions:

```
> head(prc$x)
          PC1         PC2          PC3          PC4
1  -2.2196368  1.236612  -1.49623685  -0.2846021
```

```
2  -0.8861482  2.009718  -0.03778771  -0.2279745
3  -2.3610815  1.219974  -2.19356148   1.9142027
4  -0.1158392  1.548643  -0.31048859  -0.5128123
5   0.3215947  1.687231  -1.21348165   1.4070365
6   3.0692594  2.364896  -0.12285297   0.3177992
            PC5          PC6          PC7          PC8
1   0.1630507  -0.2355776  -0.2260290   0.26286935
...
```

Whoever was the first person in our dataset has a value of -2.2196368 for the principal component and so on.

Now, which principal components should we retain?

```
> summary(prc)
Importance of components:
                          PC1     PC2     PC3
Standard deviation      2.8355  1.1641  1.0012
Proportion of Variance  0.6185  0.1042  0.0771
Cumulative Proportion   0.6185  0.7227  0.7998
                          PC4      PC5      PC6
Standard deviation      0.81700  0.77383  0.56014
Proportion of Variance  0.05135  0.04606  0.02413
Cumulative Proportion   0.85116  0.89722  0.92136
                          PC7      PC8      PC9
Standard deviation      0.53495  0.51079  0.42776
Proportion of Variance  0.02201  0.02007  0.01408
Cumulative Proportion   0.94337  0.96344  0.97751
                          PC10     PC11     PC12
Standard deviation      0.36627  0.27855  0.23866
Proportion of Variance  0.01032  0.00597  0.00438
Cumulative Proportion   0.98783  0.99380  0.99818
                          PC13
Standard deviation      0.15364
Proportion of Variance  0.00182
Cumulative Proportion   1.00000
```

Applying the 90% rule of thumb (again, cross-validation might be better), we would use only the first 6 principal components. Let's take that as our example. Note that this means that our new predictor dataset is **prc$x[,1:6]**. We can now use that new data in **lm()**:

```
> lmout <- lm(bodyfat[,1] ~ prc$x[,1:6])
> summary(lmout)
```

. . .
Coefficients:

	Estimate	Std. Error	t value
(Intercept)	18.9385	0.3018	62.751
prc$x[, 1:6]PC1	1.6991	0.1066	15.932
prc$x[, 1:6]PC2	-2.6145	0.2598	-10.064
prc$x[, 1:6]PC3	-1.5999	0.3021	-5.297
prc$x[, 1:6]PC4	0.5104	0.3701	1.379
prc$x[, 1:6]PC5	1.3987	0.3908	3.579
prc$x[, 1:6]PC6	2.0243	0.5399	3.750

	$Pr(>|t|)$	
(Intercept)	< 2e-16	***
prc$x[, 1:6]PC1	< 2e-16	***
prc$x[, 1:6]PC2	< 2e-16	***
prc$x[, 1:6]PC3	2.62e-07	***
prc$x[, 1:6]PC4	0.169140	
prc$x[, 1:6]PC5	0.000415	***
prc$x[, 1:6]PC6	0.000221	***

. . .
Multiple R-squared: 0.6271, Adj. R-squared: 0.6179
. . .

The adjusted R-squared value, about 0.62, is considerably less than what we obtained from stepwise regression earlier, or for that matter, than what the full model gave us. This suggests that we have used too few principal components. (Note that if we had used all of them, we would get the full model back again, albeit with transformed variables.)

As mentioned, we could choose the number of principal components via cross-validation: We would break the data into training and test sets, then apply lm() to the training set p times, once with just one component, then with two and so on. We would then see how well each of these fits predicts in the test set.

However, in doing so, we would lose one of the advantages of the PCA approach, which is that it does predictor selection independent of the Y values. Our selection process with PCA does not suffer from the problems cited in Section 9.8.

On the other hand, in situations with very large values of p, say in the hundreds or even more, PCA provides a handy way to cut things down to size. On that scale, it also may pay to use a sparse version of PCA [138].

9.9.3.3   Example: Instructor Evaluations

This dataset, another from the UCI repository,[5] involves student evaluations of instructors in Turkey. It consists of four questions about the student and so on, and student ratings of the instructors on 28 different aspects, such as "The quizzes, assignments, projects and exams contributed to helping the learning." The student gives a rating of 1 to 5 on each question.

We might be interested in regressing Question 9, which measures student enthusiasm for the instructor against the other variables, including the **difficulty** variable ("Level of difficulty of the course as perceived by the student"). We might ask, for instance, how much of a factor is that variable, when other variables, such as "quizzes helpful for learning," are adjusted for.

It would be nice to reduce those 28 rating variables to just a few. Let's try PCA:

```
> turk <−
      read.csv('turkiye−student−evaluation_generic.csv')
> tpca <− prcomp(turk[,−(1:2)])
> summary(tpca)
Importance of components:
                            PC1       PC2       PC3
Standard deviation       6.1372   1.70133   1.40887
Proportion of Variance   0.7535   0.05791   0.03971
Cumulative Proportion    0.7535   0.81143   0.85114
                            PC4       PC5       PC6
Standard deviation      1.05886   0.81337   0.75777
Proportion of Variance  0.02243   0.01323   0.01149
Cumulative Proportion   0.87357   0.88681   0.89830
 . . .
```

So, the first principal component already has about 75% of the total variation of the data, rather remarkable since there are 32 variables. Moreover, the 28 ratings all have about the same coefficients:

```
> tpca$rotation[,1]
        instr           class        nb.repeat
> tpca$rotation[,1]
   nb.repeat       attendance       difficulty
  0.003571047    −0.048347081    −0.019218696
           Q1              Q2               Q3
```

---

[5] *https://archive.ics.uci.edu/ml/datasets/Turkiye+Student+Evaluation.*

−0.178343832	−0.186652683	−0.181963142
Q4	Q5	Q6
−0.183842387	−0.189891414	−0.186746480
Q7	Q8	Q9
−0.187475846	−0.186424737	−0.182169214
Q10	Q11	Q12
−0.192004876	−0.186461239	−0.185863361
Q13	Q14	Q15
−0.192045861	−0.191005635	−0.190046305
Q16	Q17	Q18
−0.195944361	−0.180852697	−0.193325354
Q19	Q20	Q21
−0.192520512	−0.193006276	−0.190947008
Q22	Q23	Q24
−0.190710860	−0.194580328	−0.192843769
Q25	Q26	Q27
−0.188742048	−0.190603260	−0.189426706
Q28		
−0.188508546		

In other words, the school administrators seem to be wasting the students' time with all those questions! One could capture most of the content of the data with just one of the questions — any of them. To be fair, the second component shows somewhat more variation about the student questions.

```
> tpca$rotation [ ,2]
      nb.repeat      attendance      difficulty
 −0.0007927081    −0.7309762335    −0.6074205968
           Q1             Q2             Q3
  0.1124617340     0.0666808104     0.0229400512
           Q4             Q5             Q6
  0.0745878231     0.0661087383     0.0695127306
           Q7             Q8             Q9
  0.0843229105     0.0869128408     0.0303247621
 . . .
```

Also interesting is that the **difficulty** variable is basically missing from the first component, but is there with a large coefficient in the second one.

9.9.4 Nonnegative Matrix Factorization (NMF)

Nonnegative matrix factorization (NMF) is a popular tool in many applications, such as image recognition and text classification.

9.9.4.1 Overview

Given an $u \times v$ matrix A with nonnegative elements, we wish to find nonnegative, rank-k matrices[6] W ($u \times k$) and H ($k \times v$) such that

$$A \approx WH \qquad (9.31)$$

The larger the rank, the better our approximation in (9.31). But we typically hope that a good approximation can be achieved with

$$k \ll \text{rank}(A) \qquad (9.32)$$

The rank here is analogous to the choice of number of components in PCA. However, there may be other criteria for choosing k.

One of the popular methods for computing NMF actually uses regression analysis in a creative manner. See Section 9.12.1.

9.9.4.2 Interpretation

The rows and columns of A above will correspond to some quantities of interest, and as will be seen here, the rows of H are viewed as "typical" patterns of the ways the column variables interact. To make this concrete, consider the text classification example in Section 4.3.6.

In that example, the matrix A had 4601 rows, corresponding to the 4601 e-mail messages in our data. There were 48 columns of word frequency data, corresponding to the 48 words tabulated.

The matrix H will then have dimensions $k \times 48$. Its rows can then be thought of as k synthetic "typical" e-mail messages. And row i of W will give us the coefficients in the linear combination of those synthetic messages that approximates message i in our data.

---

[6]Recall that the rank of a matrix is the maximal number of linearly independent rows or columns.

How are these synthetic messages then used to do machine classification of new messages, to determine spam/no spam status? One can then use these variables as inputs to a logit model or whatever method we wish.

Take for instance our e-mail example above. Say we choose $k = 3$. We now have $p = 3$ in, say, our logit model. The matrix W, of size 4061×3, will take the place of A as our matrix of predictor values.

Note that the factorization (9.31) is not unique. For any invertible $k \times k$ matrix R with a nonnegative inverse, then setting $\tilde{W} = WR^{-1}$ and $\tilde{H} = RH$ will produce the same product. But this should not be a problem (other than possible computational convergence issues); any set of synthetic typical messages in the above sense should work, provided we have a good value of k.

9.9.4.3   Sum-of-Parts Property

Up there in the sky.
Don't you see him?
No, not the moon.
The Man in the Moon. — William Joyce

"Sum-of-Parts" is a hoped-for-property, not guaranteed to arise, and possibly in the eye of the beholder. Due to the nonnegativity of the matrices involved, it is hoped that the matrix H will be sparse. In that case, the locations of the nonzero elements may be directly meaningful to the given application [86]. In facial recognition, for instance, we may find that rows of H correspond to the forehead, eyes, nose and so on.

As noted above, the NMF factorization is not unique. Thus if a nicely interpretable factorization is found, changing the factorization using a matrix R as above will likely destroy that interpretability. Thus interpretation should be done with care, with review by domain experts.

9.9.4.4   Example: Spam Detection

Let's again use the spam dataset from Section 4.3.6.

```
> library(ElemStatLearn)
> data(spam)
> spam$spam <- as.integer(spam$spam == 'spam')
> library(NMF)
> spam48 <- spam[,1:48]
```

```
> rowsums <- apply(spam48,1,sum)
> spam48a <- spam48[rowsums > 0,]
> nmfout <- nmf(spam48a,10)
> h <- nmfout@fit@H
```

In our first try at using NMF on this dataset (not shown), it turned out that some rows of **spam48** consisted of all 0s, preventing the computation from being done. Thus we removed the offending rows.

The choice of 10 for the rank was rather abitrary. It needs to be less than or equal to the minimum of the numbers of rows and columns in the input matrix, in this case 4601 and 48, preferably much less. Again, we might choose the rank via cross-validation.

It turns out that this data set (as is typical in the text classification case) does yield the sum-of-parts property. Here is the first row of H:

```
> h[1,]
          A.1           A.2           A.3           A.4
2.220446e-16  2.220446e-16  2.220446e-16  2.220446e-16
          A.5           A.6           A.7           A.8
2.220446e-16  2.220446e-16  2.220446e-16  2.220446e-16
          A.9          A.10          A.11          A.12
2.220446e-16  2.220446e-16  2.220446e-16  2.220446e-16
         A.13          A.14          A.15          A.16
2.220446e-16  2.220446e-16  2.220446e-16  2.220446e-16
         A.17          A.18          A.19          A.20
2.220446e-16  2.220446e-16  2.220446e-16  2.220446e-16
         A.21          A.22          A.23          A.24
2.220446e-16  2.220446e-16  2.220446e-16  2.220446e-16
         A.25          A.26          A.27          A.28
2.220446e-16  2.220446e-16  2.220446e-16  2.220446e-16
         A.29          A.30          A.31          A.32
2.220446e-16  2.220446e-16  2.220446e-16  2.220446e-16
         A.33          A.34          A.35          A.36
2.220446e-16  2.220446e-16  2.220446e-16  2.220446e-16
         A.37          A.38          A.39          A.40
6.523588e-03  2.220446e-16  4.309196e-03  2.220446e-16
         A.41          A.42          A.43          A.44
2.393124e-03  2.220446e-16  2.307614e-03  2.220446e-16
         A.45          A.46          A.47          A.48
1.650840e-02  9.855154e-03  2.220446e-16  2.220446e-16
```

The entries 2.220446e-16 are basically 0s, so the nonzero entries are for words A.37, '1999', A.39, 'pm', A.41, 'cs', a.43, 'original', A.45, 're' (i.e., *regarding*) and A.46, 'edu'. Since the dataset consists of messages received by the person who compiled the data, a Silicon Valley engineer, this row seems to correspond to the messages about meetings, possibly with acdemia. The second row (not shown) has just two nonzero entries, A.47, 'table', and A.48, 'conference'. Some rows, such as row 4, seem to be flagging spam messages, containing words like 'credit.'

As in the PCA case, we could use these 10 variables for prediction, instead of the original 48. (We would still use the other variables, e.g., related to long all-capital-letter words.)

Note the specific form of these new variables. We see from **h[1,]** above, for instance, that the first new variable is

$$0.0065 * A.37 + 0.0043 * A39 + 0.0024 * A.41 +$$
$$0.0023 * A.43 + 0.0165 \ A.45 + 0.0099 * A.46$$

9.9.5   Use of freqparcoord for Dimension Reduction

The **freqparcoord** package provides us with another tool for dimension reduction.

9.9.5.1   Example: Student Evaluations of Instructors

Consider the Turkish instructor evaluations again. Let's run **freqparcoord**:

```
> library(freqparcoord)
> turk <-
    read.csv('turkiye-student-evaluation_generic.csv')
> freqparcoord(jitter(as.matrix(turk[,-(1:4)])),m=5)
```

The R **jitter()** function adds a small amount of random noise to data. We needed it here for technical reasons; **freqparcoord()** cannot operate if the data includes too many tied values.[7]

The result, shown in Figure 9.2, is rather striking. We seem to be working with just 3 kinds of instructors: those who consistently get high ratings to

---

[7]The function uses k-NN density estimation, Section 5.10.1.1. With lots of ties and moderate k, the (multidimensional analog of the) quantity h will be 0, causing a divide-by-0 problem.

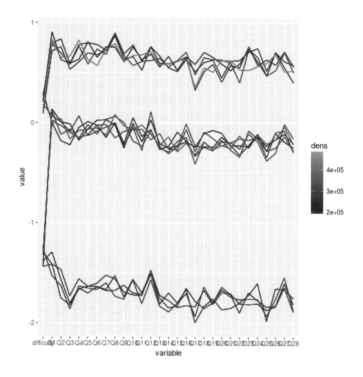

Figure 9.2: Turkish student evaluations

the various questions; those who consistently get medium-high ratings; and those who consistently get low ratings. (In viewing the vertical axis, recall that the data are centered and scaled.) This suggests removing many of the questions from our analysis, so that we can better estimate the effect of the **difficulty** variable.

Of course, given what we learned about this data through PCA above, the results here are not entirely unexpected. But **freqparcoord** is giving us further insight, showing three groups.

9.9.5.2 Dimension Reduction for Dummy/R Factor Variables

As discussed in Section 1.16, in many cases, in order to obtain a good fit in a parametric model, we need to include interaction terms. This is true for any kinds of predictors, including dummy variables (or R factors). To include second-order interactions, for example, we would form new predictors from products of pairs of the original dummies; for three-way interactions, we would add in products of triples and so on.

Such product terms can be numerous. In the bank marketing example in Section 9.7.5.1, there are 12 kinds of jobs, 3 types of marital status, 4 levels of education etc. For three-way interactions, that would already be $11 \times 2 \times 3 = 66$ new terms. In other words, the value of p would increase by 66 just from those added terms, and there would be many more.

If, say, job type and education were independent random variables, we would not need pairs of that type. But of course there is some dependency there, and elsewhere among the dummies for this data set. Thus there may be a real need for dimension reduction.

This need can be met via *log-linear models* [35], which are used to analyze tabular data. Though this topic is beyond the scope of this book, the basic idea is to write the log of the cell probabilities in the given table as a sum of main effects, first-order interactions and so on. Readers who are familiar with analysis of variance (ANOVA) will find the structure to be similar. Computations can be done with for instance, the built-in **loglin()**.

9.10   The Verdict

So, what predictor selection method should one use? Though many empirical studies have been conducted on real and simulated data, there simply is no good answer to this question. The issue of dimension reduction is an unsolved problem.

The classical version of stepwise regression described at the beginning of Section 9.7.1, in which a hypothesis test is run to decide whether to include a predictor, has been the subject of a great deal of research and even more criticism. Sophisticated techniques have been developed (see for example [110] and [112]), but they have stringent assumptions, such as normality, homoscedasticity and exact validity of the linear model. Again, this is an unsettled question.

Nevertheless, a wide variety of methods have been developed for approaching this problem, allowing analysts to experiment and find ones they believe work reasonably well.

We will return to this problem in Chapter 12 to discuss the case $p \gg n$, where it is especially important.

9.11 Further Reading

Alan Miller's book [110] is a *tour de force* on the topic of predictor variable selection.

There is much interesting material on the use of PCA, NMF and so on in classification problems in [44].

As noted earlier, [65] is a comprehensive treatment of the LASSO and related estimators. The fundamental paper presenting the theory behind the **lars** package, notably the relation of LASSO and stepwise regression to another technique called *least-angle regression* is [42]. See [28] for a theoretical treatment.

9.12 Computational Complements

9.12.1 Computation for NMF

The matrices W and H are calculated iteratively, with one of the major methods being regression. (There are other methods, such as a multiplicative update method; see [90].) Here is how:

We make initial guesses for W and H, say with random numbers. Now consider an odd-numbered iteration. Suppose just for a moment that we know the exact value of W, with H unknown. Then for each j we could "predict" column j of A from the columns of W. The coefficient vector returned by **lm()** will become column j of H. (One must specify a model without an intercept term, which is signaled via a -1 in the predictor list; see Section 2.4.5.) We do this for $j = 1, 2, ..., v$.

In even-numbered iterations, suppose we know H but not W. We could take transposes,

$$A' = H'W' \tag{9.33}$$

and then just interchange the roles of W and H above. Here a call to **lm()** gives us a row of W, and we do this for all rows.

Figure 9.3: Mt. Rushmore

R's **NMF** package [54] for NMF computation is quite versatile, with many, many options. In its simplest form, though, it is quite easy to use. For a matrix **a** and desired rank **k**, we simply run

> nout <− nmf(a,k)

The factors are then in **nout@fit@W** and **nout@fit@H**.

Let's illustrate it in an image context, using the image in Figure 9.3. Though typically NMF is used for image classification, with input data consisting of many images,[8] here we have only one image, and we'll use NMF to compress it, not do classification. We first obtain A:

> **library**(pixmap)
> mtr <−
    read.pnm('MtRush.pgm')   # *see this book's website*
> a <− mtr@grey

Now, perform NMF, find the approximation to A, and display it, as seen in Figure 9.4:

---
[8]Each image is stored linearly in one column of the matrix A.

Figure 9.4: Mt. Rushmore, compressed image

```
> aout <- nmf(a,50)
> w <- aout@fit@W
> h <- aout@fit@H
> approxa <- w %*% h
# brightness values must be in [0,1]
> approxa <- pmin(approxa,1)
> mtrnew <- mtr
> mtrnew@grey <- approxa
> plot(mtrnew)
```

This is understandably blurry. The original matrix has dimension 194×259, and thus presumably has rank 194.[9] We've approximated the matrix by

---

[9]This is confirmed by running the function **rankMatrix()** in the **Matrix** package [10].

one of rank only 50, with a 75% storage savings. This is not important for one small p;cture, but possibly worthwhile if we have many large ones. The approximation is not bad in that light, and may be good enough for image recognition or other applications.

Indeed, in many if not most applications of NMF, we need to worry about overfitting. As you will see later, overfitting in this context amounts to using too high a value for our rank, something to be avoided.

9.13   Mathematical Complements

9.13.1   MSEs for the Simple Example

Here are the details of the MSE computations in Section 9.1.1. From Section 1.19.2 we know that

$$MSE = \text{variance } + (\text{bias})^2 \tag{9.34}$$

In the case of (9.6), we know that each $\widehat{\mu}_1$ is an unbiased estimator of μ_i, with variance σ^2/n. So (9.6) has the value $2\sigma^2/n$.

Things are a little more complicated for (9.7). First, the variance term is

$$Var[\tfrac{1}{2}(\overline{X} + \overline{Y})] = \tfrac{1}{4}2Var(\overline{X}) = \tfrac{1}{2}\sigma^2/n \tag{9.35}$$

so that the variance portion of (9.7), σ^2/n, is smaller than that of (9.7). This of course is the goal of using the $\check{\mu}_i$. But that improvement is offset by the nonzero bias. How large is it?

$$\text{bias} \ = E(\check{\mu}_i) - \mu_i = E[\tfrac{1}{2}(\overline{X} + \overline{Y})] - \mu_i = \tfrac{1}{2}(\mu_1 + \mu_2) - \mu_i \tag{9.36}$$

so that

$$\text{bias}^2 = \tfrac{1}{4}(\mu_1 - \mu_2)^2 \tag{9.37}$$

In other words, the total value of (9.7) is

$$\sigma^2/n + \frac{1}{2}(\mu_1 - \mu_2)^2 \tag{9.38}$$

9.14 Exercises: Data, Code and Math Problems

Data problems:

1. Apply **lars()** to the bodyfat data, as in Section 9.7.7.2, this time setting **type = 'stepwise'**, and compare the results.

2. Apply principal component regression to the letters recognition data in Section 5.5.4.

3. Take the NMF factorization in Section 9.9.4.4, and use logit to predict, as in Section 4.3.6. Try various values of the rank k. Compare to the logit full model in Section 4.3.6.

4. In our analysis of evaluations of Turkish instructors in Section 9.9.5, we found that there were three main groups of students; in one group the students gave instructors consistently high evaluations, and so on. The **lme4** package [9] also contains a data set of this kind, **InstEval**. Explore that data, to see whether a similar pattern emerges.

Mini-CRAN and other computational problems:

5. Edit R's **summary.lm** so that in addition to the information printed out by the ordinary version, it also reports C_p.

6. Edit R's **summary.lars** so that in addition to the information printed out by the ordinary version, it prints out R^2.

7. Write a function **stepAR2()** that works similarly to **stepAIC()**, except that this new function uses adjusted R^2 as its criterion for adding or deleting a predictor. The call form will be

```
stepAR2(lmobj, dir='fwd', nsteps=ncol(lmobj$model)-1)
```

where: **lmobj** is an object of class **'lm'**; **dir** is **'fwd'** or **'back'**; and **nsteps** is a positive integer.

Predictors will be added/deleted one at a time, over the course of **nsteps** models, according to which one maximizes adjusted R^2 (even if that value is lower than the present one).

The return value will be an S3 object of type **'stepr2'**. with sole component a data frame of $\widehat{\beta}_i$ values (0s meaning the predictor is not currently in the prediction equation), one row per model. There will also be an R^2 column. Write a **summary()** function for this class, that shows the actions taken at each step of the process.

8. In principal components regression analysis, we transform the original p predictor variables $X^{(i)}$ to p new ones that are linear combinations of the originals, then choose $q \leq p$ of the new ones to use as our actual predictors. Let W^j, $j = 1, ..., q$ denote those new predictors. We have

$$W^j = \sum_{i=1}^{p} w_{ji} X^{(i)} \tag{9.39}$$

Our regression model is now

$$E(Y \mid W) = \gamma_0 + \gamma_1 W^{(1)} + ... + \gamma_q W^{(q)} \tag{9.40}$$

But after fitting that model, we could transform back to X, substituting with (9.39). In this way, we could obtain an estimated regression function of Y on the original X, which might have useful interpretability.

Write an R function with call form

xformback (prc)

that takes the argument **prc**, which is output from **prcomp()**, and returns the vector of estimated regression coefficients with respect to X derived as above.

9. Consider the example in Section 9.8. Write an R function that will evaluate (9.29) for the value of the argument **c**. Call the function for various values in (-1,1) and plot the results. Assume that the distribution of Y given X is normal. Calculate the bivariate normal probabilities using **pmvnorm()** in the **mvtnorm()** package [70].

10. With stepwise variable selection procedures, a central question is of course what *stopping rule* to use, i.e., what policy to use. One approach, say for the forward direction setting, has been to add p artificial noise variables to the predictor set, and then stop the stepwise process the first time an artificial variable is entered by the stepwise procedure [111]. Candes *et al* [30] suggest that the artificial variables, called *knockoffs*, have a distribution matching the distribution of the (real) predictor vector X in certain senses, but we will assume i.i.d. knockoffs here.

(a) Write an R function with call form

    kofilter (dframe , yname , direction)

   that does the following: It first adds knockoffs, then calls **lm()** on the data frame **dframe** with **yname** being the name of the response variable. It then applies **stepAIC()** in the specified direction, stopping according to the above prescription. either stopping the first time the procedure attempts to enter a knockoff variable (forward direction) or the first time all knockoffs are removed (backward direction).

(b) Apply this function to the bodyfat data, comparing the results to the ones obtained in this chapter.

Math problems:

11. In this problem we will extend the analysis of Section 9.1.1.

Suppose we have a categorical variable X with k categories (rather than just 2 categories as in Section 9.1.1). Let μ_i denote the mean of Y in the subpopulation defined by $X = i$, and suppose $\mu_i - \mu_{i-1} = d$, $i = 2, ..., k$. Derive a variance-bias tradeoff analysis like that of (9.1.1).

12. Consider the following simple regression model with $p = 2$:

$$\mu(t_1, t_2) = \beta_1 t_1 + \beta_2 t_2, \quad Var(Y|X = (t_1, t_2)) = \sigma^2 \qquad (9.41)$$

Assume we have an orthogonal design (Section 2.3). Derive a relation similar to (9.11). As in Section 2.8.4, consider the conditional distribution of Y given X in MSE computation.

13. For any k-dimensional vector $u = (u_1, ..., u_k)'$, its l_q norm is defined to be

$$\|u\|_q = (\sum_{i=1}^{k} u_i^q)^{1/q} \qquad (9.42)$$

The familiar cases are $q = 2$, yielding the standard Euclidean norm in (A.1), and $q = 1$, the latter playing a central role in the LASSO, Equation (8.18). But it is defined for general $q > 0$.

In fact, an important special case is $q = \infty$. If one lets $q \to \infty$ in (9.42), it can be shown that one gets

$$||u||_\infty = \max_i |u_i| \tag{9.43}$$

Now, look at Figure 9.1. Since this was for the LASSO, it was based on the use of the l_1 norm in (8.18). Discuss how the figure would change for other l_q norms, in particular the cases $q = 2$, $q = \infty$ and $0 < q$. In particular, argue that the latter case would also lead to a variable selection process.

14. Fill in the gaps in the intuitive derivation in Section 9.8 to make it a careful proof. You may find Exercise 11 of Chapter 2 helpful.

Chapter 10

Partition-Based Methods

These methods partition the X or *feature* space,[1] space into rectangular subregions, sub-subregions and so on. To predict a new case, we determine which region its X value belongs to, and then predict from the Y values in that region.

This approach is similar to k-NN methods, in that we are in essence finding neighbors of the new case, but with a quite different way to find those neighbors. The method was invented by statisticians and is popular in that community, but are widely used in the machine learning community.

Here is a sneak preview, using the Letters Recognition data from Section 5.8.3. The code

```
> lr <- LetterRecognition  # leave lettr as factor!
> library(rpart)
> library(rpart.plot)
> rplr <- rpart(lettr ~ .,data=lr,method='class')
> prp(rplr)
```

produces the "flow chart" in Figure 10.1. For instance, suppose we have a certain letter image for which **x2ybr** is less than 2.5, but for which **y2bar** is greater than or equal to 3.5. Then we predict this to be the letter 'L'.

Given the tree-like structure in Figure 10.1, it is not surprising that it is referred to as a "tree." This kind of approach is very appealing. It is easy to

---

[1]The predictors are called *features* in the machine learning literature, inherited from the electrical engineering community. In a classification problem, the names of the classes are called *labels*, and a point with unknown class, to be predicted, is termed *unlabeled*.

implement and even easier to explain to nonspecialists. Computationally, it can handle large numbers of predictors, works well with dummy variables and so on.

In the next section we will discuss the methodology behind this, known as CART. A number of questions arise, such as how far to take the partitioning process.

We then move on to the refinement, *ensembles* of trees. The idea behind forming an ensemble of trees is to generate many different trees from the same data, and combine the results. The combining is done by averaging in the regression case, and by "voting" (as in the AVA approach) in the classification setting.

The idea of random forests is to address the "discreteness" of CART. If an observation lies near the boundary of some subregion, this produces some instability: If this data point had been just slightly different, the entire portion of the tree below it could change. We address this by looking at random subsets of the data, producing the ensembles used for averaging or voting.

The approach of *boosted trees* again generates an ensemble of trees, using *boosting*, a technique we discussed briefly in Section 6.12.3. At any given step in the process, we assign weights to each of our n data points. At step i, we update the weights that were used in generating tree $i - 1$, with greater weight placed on data points that were not predictd well. We then generate tree i from these new weights.

10.1 CART

Classification and Regression Trees (CART) were developed by Breiman, Friedman, Olshen and Stone [24], building on preliminary ideas from various researchers (see [71] for a history). The basic concept is simple: Build a "flow chart" as in the letter recognition data above.

In the leaf nodes, the estimated $\widehat{\mu}(t)$ is recorded if we have a regression problem. For classification problems, the most likely class for that node is recorded.

We will use the **rpart** package [132], as well as the **rpart.plot** package for plotting [109]. (For larger graphs, use **prp()** from the same package.)

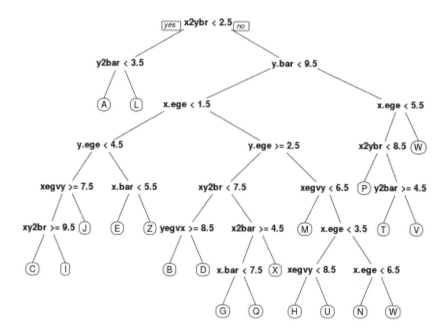

Figure 10.1: Letter recognition flow chart (see color insert)

10.2   Example: Vertebral Column Data

Let's try CART on the dataset in Section 5.5:

```
> vert <- read.table('column_3C.dat',header=FALSE)
> library(rpart)
> library(rpart.plot)
> rpvert <- rpart(V7 ~ .,data=vert,method='class')
```

Note that we needed to inform **rpart()** that this was a classification problem, flagged by **method='class'**.

Now plot the result:

```
> rpart.plot(rpvert)
```

The result is shown in Figure 10.2. We see that the first split is based on the predictor **V6**. If that variable is greater than or equal to 16 (actually 16.08), we would guess $V7 = 2$. Otherwise, we check whether **V4** is less than 28; if so, our prediction is $V7 = 0$, and so on.

Figure 10.3 shows the partioning of the predictor space we would have if we were to just use the variables **V6** and **V4**. We have divided the space into three rectangles, corresponding to the conditions in Figure 10.2. If we were to add **V5** to our predictor set, the rectangle labeled "V6 < 16, V4 >= 28" would be further split, according to the condition V5 < 117, and so on.

Figure 10.3 also shows that CART is similar to a *locally adaptive* k-NN procedure. In the k-NN context, that term means that the number of nearest neighbors used can differ at different locations in the X space. Here, each rectangle plays the role of a neighborhood.

Our estimated class probabilities for any point in a certain rectangle are the proportions of observations in that rectangle in the various classes, just as they would be in a neighborhood with k-NN. For instance, consider the tall, narrow rectangle in the

```
> vgt16 <- vert[vert$V6 > 16.08,]
> mean(vgt16$V7 == 'SL')
[1] 0.9797297
```

shown as 0.98 in the rightmost leaf, i.e., terminal node.

Now take a look inside the returned object, which gives us more detailed information:

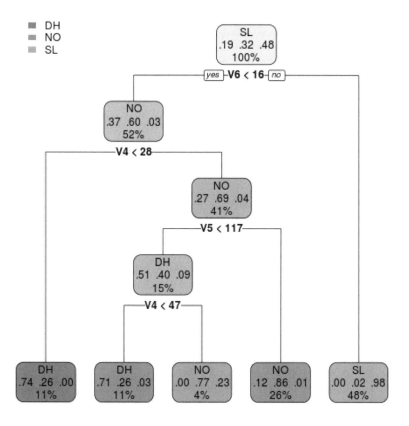

Figure 10.2: Flow chart for vertebral column data (see color insert)

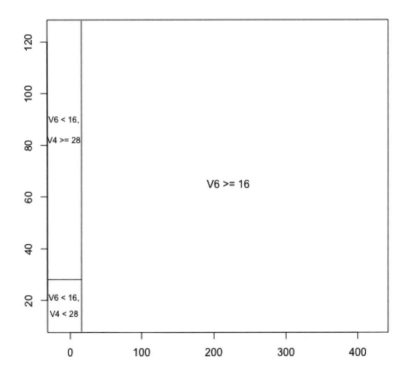

Figure 10.3: Partitioning for vertebral column data

```
> rpvert
n= 310

node), split, n, loss, yval, (yprob)
       * denotes terminal node

 1) root 310 160 SL (0.19354839 0.32258065 0.48387097)
   2) V6< 16.08 162    65 NO
              (0.37037037 0.59876543 0.03086420)
    4) V4< 28.135 35      9 DH
              (0.74285714 0.25714286 0.00000000) *
    5) V4>=28.135 127    39 NO
              (0.26771654 0.69291339 0.03937008)
     10) V5< 117.36 47    23 DH
              (0.51063830 0.40425532 0.08510638)
      20) V4< 46.645 34    10 DH
              (0.70588235 0.26470588 0.02941176) *
      21) V4>=46.645 13     3 NO
              (0.00000000 0.76923077 0.23076923) *
     11) V5>=117.36 80    11 NO
              (0.12500000 0.86250000 0.01250000) *
   3) V6>=16.08 148     3 SL
              (0.00000000 0.02027027 0.97972973) *
```

The line labeled "2)," for instance, tells us that its decision rule is $V6 <$ 16.08; that there are 65 such cases; that we would guess $V7 = NO$ if we were to stop there; and so on. Leaf nodes are designated with asterisks.

Prediction is done in the usual way. For instance, let's re-predict the first case:

```
 z <- vert[1,]
> predict(rpvert,z)
          DH          NO           SL
1  0.7058824  0.2647059  0.02941176
> predict(rpvert,z,type='class')
 1
DH
Levels: DH NO SL
```

The conditional probabilities that **V7** is DH, NO or SL for this case are 0.71 etc., and we would guess $V7 = SL$.

So, how well did we predict (on the original set, no cross-validation)?

```
> rpypred <- predict(rpvert,type='class')
> mean(rpypred == vert$V7)
[1]  0.883871
```

10.3   Technical Details

The devil is in the details — old saying

CART has been the subject of much research over the years, reflected in a number of algorithms and options in the implementation software. The details matter a lot.

The full details of CART are complex and beyond the scope of this book. (And we have used default values for the various optional arguments in **rpart()**.) See [64] [92] for further information on general techniques, and [131] for the details on **rpart** in particular.

However, we briefly discuss implementation in this section.

10.3.1   Split Criterion

At each step, CART must decide (a) whether to make further splits below the given node, (b) if so, on which predictor it should make its next split, and (c) the split itself, i.e., the cutoff value to use for this predictor. How are these decisions made?

One method is roughly as follows. For a candidate split, we test whether the mean Y values on the two sides of the split are statistically significantly different. If so, we take the split with the smallest p-value; if not, we don't split.

10.3.2   Predictor Reuse and Statistical Consistency

Details of some theoretical properties are in [36]. However, the issue of statistical consistency (Section 2.7.3) should be discussed here. As presented above, CART is not consistent. After all, with p predictors, the greatest number of rectangles we can get is 2^p, no matter how large the sample size n is. Therefore the rectangles cannot get smaller and smaller as n grows, which they do in k-NN. Thus consistency is impossible.

However, this is remedied by allowing the same predictor to enter into the process multiple times. In Figure 10.3, for instance, **V6** might come in again at the node involving **V6** and **V6** at the lower left of the figure.

Here is a rough argument as to why statistical consistency can then be achieved. Consider the case of $p = 1$, a single predictor, and think of what happens at the first splitting decision. Suppose we take a simple split criterion based on significance testing, as mentioned above. Then, as long as the regression function is nonconstant, we will indeed find a split for sufficiently large n. The same is true for the second split, and so on, so we are indeed obtaining smaller and smaller rectangles as in k-NN.

10.4 Tuning Parameters

Almost any nonparametric regression method has one or more tuning parameters. For CART, the obvious parameter is the size of the tree. Too large a tree, for example, would mean too few observations in each rectangle, causing a poor estimate, just like having too few neighbors with k-NN. Statistically, this dearth of data corresponds to a high sampling variance, so once again we see the variance-bias tradeoff at work. The **rpart()** function has various ways to tune this, such as setting the argument **minsplit**, which specifies a lower bound on how many points can fall into a leaf node.

Note that the same predictor may be used in more than one node, a key issue that we will return to later. Also, a reminder: Though tuning parameters can be chosen by cross-validation, keep in mind that cross-validation itself can be subject to major problems (Sections 9.3.2 and 1.13).

10.5 Random Forests

CART attracted much attention after the publication of [24], but the authors and other researchers continued to refine it. While engaged in a consulting project, one of the original CART authors discovered a rather troubling aspect, finding that the process was rather unstable [130]. If the value of a variable in a single data point were to be perturbed slightly, the cutoff point at that node could change, with that change possibly propagating down the tree. In statistical terms, this means the variances of key quantities might be large. The solution was to randomize, as follows.

10.5.1   Bagging

To set up this notion, let's first discuss *bagging*, where "bag" stands for "bootstrap aggregation."

The bootstrap [39] is a *resampling* procedure, a general statistical term (not just for regression contexts) whose name alludes to the fact that one randomly chooses samples from our sample! This may seem odd, but the motivation is easy to understand. Say we have some population parameter θ that we estimate by $\widehat{\theta}$, but we have no formula for the standard error of the latter. We can generate many subsamples (typically with replacement), calculate $\widehat{\theta}$ on each one, and compute the standard deviation of those values; this becomes our standard error for $\widehat{\theta}$.

We could then apply this idea to CART. We resample many times, constructing a tree each time. We then average or vote, depending on whether we are doing regression or classification. The resampling acts to "smooth out the rough edges," solving the problem of an observation coming near a boundary.

The method of random forests then tweaks the idea of bagging. Rather than just resampling the data, we also take random subsets of our predictors, and finding splits using them.

Thus many trees are generated, a "forest," after which averaging or voting is done as mentioned above. The first work in this direction was that of Ho [67], and the seminal paper was by Breiman [22].

A popular package for this method is **randomForest** [89].

10.5.2   Example: Vertebrae Data

Let's apply random forests to the vertebrae data. Continuing with the data frame **vert** in Section 10.2, we have

```
> library (randomForest)
> rfvert <- randomForest (V7 ~ ., data=vert)
```

Prediction follows the usual format. Let's look at five random rows:

```
> predict (rfvert, vert [sample (1:310,5),])
267 205 247  64 211
 NO  SL  NO  SL  NO
Levels: DH NO SL
```

So, cases 267, 205 and so on are predicted to have **V7** equal to NO, SL etc.

Let's check our prediction accuracy:

```
> rfypred <- predict(rfvert)
> mean(rfypred == vert$V7)
[1]  0.8451613
```

This is actually a bit less than what we got from CART, though the difference is probably commensurate with sampling error.

The **randomForest()** function has many, many options, far more than we can discuss here. The same is true for the return value, of class 'random-Forest' of course. One we might discuss is the **votes** component:

```
> rfvert$votes
             DH             NO             SL
1    0.539682540   0.343915344   0.116402116
2    0.746268657   0.248756219   0.004975124
3    0.422222222   0.483333333   0.094444444
4    0.451977401   0.254237288   0.293785311
5    0.597883598   0.333333333   0.068783069
6    0.350877193   0.649122807   0.000000000
. . .
```

So, of the 500 trees (this value is queriable in **rfvert$ntree**), observation 1 was predicted to be DH by 0.54 of the trees, while 0.34 of them predicted NO and so on. These are also the conditional probabilities of the various classes.

10.5.3   Example: Letter Recognition

Let's first apply CART:

```
> rplr <- rpart(lettr ~ ., data=lr, method='class')
> rpypred <- predict(rplr, type='class')
> mean(rpypred == lr$lettr)
[1]  0.4799
```

Oh, no! This number, about 48% is far below what we obtained with our logit model in Section 5.5.4, even without adding quadratic terms. But recall that the reason for adding those terms is that we suspected that the regression functions are not monotonic in some of the predictors. That could well be the case here. But if so, perhaps random forests can remedy that problem. Let's check:

```
> rflr <- randomForest(lettr ~ .,data=lr)
> rfypred <- predict(rflr)
> mean(rfypred == lr$lettr)
[1] 0.96875
```

Ah, very nice. So use of random Forests led to a dramatic improvement over CART in this case.

10.6 Other Implementations of CART

A number of authors have written software implementing CART (and other partitioning-based methods). One alternative package to **rpart** worth trying is **partykit** [72], with its main single-tree function being **ctree()**. It uses p-value-based splitting and stopping rules. You may find that it produces better accuracy than **rpart**, possibly at the expense of slower run time.

Here is how it does on the letter recognition data:

```
> library(partykit)
Loading required package: grid
> library(mlbench)
> data(LetterRecognition)
> lr <- LetterRecognition
> ctout <- ctree(lettr ~ .,data=lr)
> ctpred <- predict(ctout,lr)
> mean(ctpred == lr$lettr)
[1] 0.8552
```

This is much better than what we got from **rpart**, though still somewhat below random forests. (The **partykit** version, **cforest()** is still experimental as of this writing.)

Given that, let's take another look at **rpart**. By inspecting the tree generated earlier, we see that no predictor was split more than once. This could be a real problem with non-monotonic data, and may be caused by premature stopping of the tree-building process.

With that in mind, let's try a smaller value of the **cp** argument, which is a cutoff value for split/no split, relative to the ratio of the before-and-after split criterion (default is 0.01),

```
> rplr <- rpart(lettr ~ .,data=lr,method='class',
```

```
    cp=0.00005)
> rpypred <- predict(rplr,type='class')
> mean(rpypred == lr$lettr)
[1]  0.8806
```

Great improvement! It may be the case that this is generally true for non-monotonic data.

10.7 Exercises: Data, Code and Math Problems

Data problems:

1. Fill in the remainder of Figure 10.3.

2. Not surprising in view of the 'R' in CART, the latter can indeed be used for regression problems, not just classification. In **rpart()**, this is specified via **method = 'anova'**. In this problem, you will apply this to the bodyfat data (Section 9.7.4).

Fit a CART model, and compare to our previous results by calculating R^2 in both cases, using the code in Problem 4.

3. Download the New York City taxi data (or possibly just get one of the files), *http://www.andresmh.com/nyctaxitrips/* Predict trip time from other variables of your choice, using CART.

Mini-CRAN and other computational problems:

4. As noted in Section 2.9.2, R^2, originally defined for the classic linear regression model, can be applied to general regression methodology, as it is the squared correlation between Y and predicted-Y values. Write an R function with call form

```
rpartr2(rpartout,newdata=NULL,type='class')
```

that returns the R^2 value in a CART setting. Here **rpartout** is an object of class **'rpart'**, output of a call to **rpart()**; **newdata** is a data frame on which prediction is to be done, with that being taken as the original data set if this argument is NULL; and **type** is as in **predict.rpart()**.

5. It would be interesting, in problems with many predictors, to see which ones are chosen by **rpart()**. Write an R function with call form splitsvars(rpartout)

that returns a vector of the (unique) predictor variable names involved in the splits reported in **rpartout**, an object of class **'rpart'**.

6. In classification problems, it would be useful to identify the hard-to-classify cases. We might do this, for instance, by finding the cases in which the two largest probabilities in the **votes** component of an object of class **'randomForest'** are very close to each other. Write a function with call form

hardcases(rfobj, tol)

that reports the indices of such cases, with the argument **tol** being the percentage difference between the two largest probabilities.

Chapter 11

Semi-Linear Methods

Back in Chapter 1, we presented the contrasting approaches to estimating a regression function:

- Nonparametric methods, such as k-Nearest Neighbors, have the advantage of being model-free, and thus generally applicable.

- Parametric models, if reasonably close to reality, are more powerful, generally yielding more accurate estimates of $\mu(t)$.

In this chapter, we try to "have our cake and eat it too." We present methods that are still model-free, but make use of linearity in various ways. For that reason, they have the potential of being more accurate than the unenhanced model-free methods presented earlier. We'll call these methods *semi-linear*.

Here is an overview of the techniques we'll cover, each using this theme of a "semi-linear" approach::

- **k-NN with Local Linearization:** Instead of estimating $\mu(t)$ using the average Y value in a neighborhood of a given point t, fit a linear regression model to the neighborhood and predict Y at t using that model. The idea here is that near t, $\mu(t_1, ..., t_p)$ should be monotonic in each t_j, indeed approximately linear in those variables,[1] and this approach takes advantage of that.[2]

---

[1]Think of the tangent plane at the regression surface at t.

[2]The reader may notice some similarity to the **loess()** function in R, which in turn is based on work of Cleveland [37].

- **Support Vector Machines:** Consider for example the two-class
  classification problem, and think of the boundary between the two
  classes in the predictor space — on one side of the boundary we guess
  $Y = 1$ while on the other side we guess $Y = 0$. We noted in Section
  4.3.7 that under the logit model, this boundary is linear, i.e., a line
  in two dimensions, a plane in three dimensions, and a hyperplane in
  higher dimensions. SVMs drop the assumption of logit form for $\mu(t)$
  but retain the assumption that the boundary is linear.

  In addition, recall that if we believe the effect of a particular predictor X on Y is nonlinear, we might add an X^2 term to our parametric model, and possibly interaction terms (Section 1.16). SVM does
  something like this too.

  After the hyperplanes are found, OVA or AVA is used (Section 5.5);
  [33], for instance, uses AVA.

- **Neural networks:** This too is based on modeling the class-separating
  hypersurfaces as linear, and again input variables are put through
  nonlinear transformations. The difference is that we go through a series of stages in which new variables are formed from old ones in this
  manner. We first form new variables as linear functions of the original inputs, and transform those new variables. We then find linear
  functions of those transformed variables, and pass *those* new variables
  through a transformation, and so on. Each step takes the variables
  in one *layer*, and forms the next layer from them. The final layer is
  used for prediction.

The latter two methods above are mainly used in classification settings,
though they can be adapted to regression. The basic idea behind all of
these methods is:

> By making use of monotonicity and so on, these methods "should"
> work better than ordinary k-NN or random forests.

But as the use of quotation marks here implies, one should not count on
these things.

11.1 k-NN with Linear Smoothing

Consider the code below, which simulates a setting with

$$E(Y \mid X = t) = t \tag{11.1}$$

```
n <- 1000
z <- matrix(runif(2*n),ncol=2)
x <- z[,1]
y <- z %*% c(1,0.5)
xd <- preprocessx(x,100)
kout <- knnest(y,xd,100)
plot(x,kout$regest)
```

The plot is shown in Figure 11.1. Most of it looks like a 45-degree line, as it should, but near 0 and 1 the curve is flat. Here is why:

Think of a point very close to 1. Most or all of its neighbors will be to its left, so their $\mu(t)$ values will be lower than for our given point. So, averaging them, as straight k-NN does, will produce a downward bias. Similarly, a point near 0 will experience an upward bias, hence the flattening of $\hat{\mu}(t)$ for values of t near 0 and 1.

11.1.1 Extrapolation Via lm()

One solution to this is to fit a linear regression to the points in a neighborhood, rather than just taking the average. This should reduce the bias near 0 and 1, and may produce smoother estimates even in interior points.

In **regtools**, this can be specified by setting **nearf = loclin** in the call to **knnest()**:

```
> koutll <- knnest(y,xd,100,nearf=loclin)
> plot(x,koutll$regest)
```

The result, shown in Figure 11.2, looks much better.

Of course, one could even try a quadratic approximation scheme. This is explored in Exercise 3 at the end of this chapter.

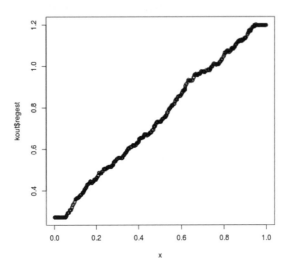

Figure 11.1: k-NN bias on the edges of the data

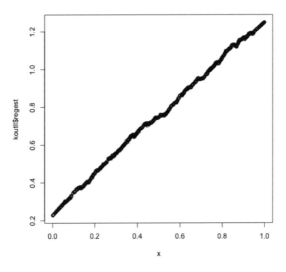

Figure 11.2: Reducing k-NN bias on the edges of the data

11.1.2 Multicollinearity Issues

One issue to keep in mind here is multicollinearity. Recall the comment near the end of Section 8.11.1.3, which in essence said that multicollinearity is exacerbated in small samples. This is relevant in our context here: If our dataset as a whole suffers from multicollinearity, then the problem would be accentuated in the neighborhoods, since "n" is small for them. In such cases, **lm()** may compute some of its coefficients as NA values, or possibly even refuse to do the computation.

Remedies are as before: We can simply use fewer predictors, say choosing them via PCA, or might even try a ridge approach.

11.1.3 Example: Bodyfat Data

Let's apply these ideas to the bodyfat data of Section 9.7.4. Recognizing the possible multicollinearity issues as above, we will only use a few of the predictors, basically the first few found by **stepAIC()** in Section 9.7.4:

```
> bf <- bodyfat[,c(1,2,3,5,7)]
> xd <- preprocessx(bf[,-1],10)
> kout <- knnest(bf[,1],xd,10)
> mean(abs(bf[,1] - kout$regest))
[1]  3.468016
> koutll <- knnest(bf[,1],xd,10,nearf=loclin)
> mean(abs(bf[,1] - koutll$regest))
[1]  2.453403
```

That's quite an improvement. Of course, with n being only 252, part of the difference may be due to sampling error, but the result does make sense. This being data on humans, there are likely some individuals who are on the fringes of the data, say people who are exceptionally thin. Use of the local linear method may help predict such people more accurately.

11.1.4 Tuning Parameter

As before, the tuning parameter here is k, the number of nearest neighbors.

11.2 Linear Approximation of Class Boundaries

In Section 4.3.7, it was noted that if a logistic model holds, then class boundaries — say, on one side, we predict Class A and on the other side we predict B — are linear: With one predictor, the boundary is a single point value, with two it is a line, with three it is a plane, and with more than three predictors it is a hyperplane. (As noted earlier, for the multiclass case, one can apply OVA or AVA to the methods below. We will stay with the two-class case here.)

The remaining two methods to be described in this chapter, Support Vector Machines (SVMs) and Neural Networks (NNs), seek to estimate class boundaries, using linearity assumptions in indirect ways. (Both methods can also be used for general regression purposes, but we will not pursue that aspect much here.) In this sense they differ from the other methods in this book, which involve estimation of conditional class probabilities, $P(Y = i \mid X = t)$.[3] SVMs and NNs are much more "machine learning-ish" than k-NN or random forests, and it is no coincidence that the latter two were developed in the statistics community while the former are from ML. This section is devoted to those ML methods.

A word on notation: In ML, a class membership variable Y is coded as +1 or -1, rather than 1 or 0 as in statistics.

Note: SVMs and NNs are highly complex methods, with many variations and tuning parameters. This chapter can only scratch the surface. For further details, see for instance [64] [83] [129]. Also, though cross-validation can be used to choose the values of the tuning parameters, it must be once again pointed out that cross-validation itself has problems (Section 9.3.2).

11.2.1 SVMs

SVM methodology [38] is used in a wide variety of applications, with a rich theoretical foundation having been developed. We'll introduce the subject here.

We will assume the two-class classification problem, which can then be applied to multiclass settings via OVA or AVA. As noted above, the main motivation of SVM stems from the idea of a linear boundary between classes,

---

[3]NNs can estimate those probabilities, as will be seen below, but tend to be thought of as class predictors rather than probability estimators.

as with for instance the logistic model. But the linear-boundary nature of the logit stems from a model on the regression function,

$$\mu(t) = P(Y = 1 \mid X = t) \tag{11.2}$$

whereas SVM makes NO assumption on the nature of that function; SVM just models the interclass boundary.

How, then, might we estimate the parameters in that linear boundary?[4] Lacking much probabilistic structure, we turn to geometry.[5]

11.2.1.1   Geometric Motivation

Let's generate some artificial data, using this code:

```
library(mvtnorm)
cv <- rbind(c(1,-0.1),c(-0.1,1))
set.seed(99999)
m <- 3.5
means <- rbind(c(0,0),c(m,m))
n <- 7
pts0 <- rmvnorm(n,means[1,],cv)
pts1 <- rmvnorm(n,means[2,],cv)
plot(x=NULL,y=NULL,xlim=c(-2,m+3),ylim=c(-2,m+3),
    xlab='X1',ylab='X2')
points(pts0,pch='x')
points(pts1,pch='o')
```

Our two predictor variables are X1 and X2, so a hyperplane is simply a line. Let's refer to the population line as ℓ and our estimate as $\widehat{\ell}$. Again, the question is, how should we choose $\widehat{\ell}$?

As you can see in Figure 11.3. there is quite a separation between the x and o data, meant to be classes 1 and 0 in this simulation, so much so that there are many lines we might take as our $\widehat{\ell}$ (all of which would give a "perfect" fit, a matter that will come into play later). SVM makes that choice in a manner that should warm the hearts of geometers.

---

[4]SVM was invented in the machine learning community. As mentioned in this book's Preface, that community typically doesn't think in terms of samples from a population. We take the statistical view here.

[5]This idea is certainly not limited to the ML community. After all, the origin of least-squares regression was geometric. A modern example is *concave density estimation*, which fits concave curves without other assumptions [124].

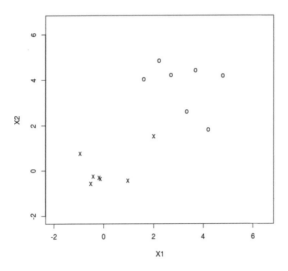

Figure 11.3: Artificial data

To explain, we need to define a term, *convex hull*. A geometric set C is called *convex* if the line segment connecting any two points in the set itself lies in the set. More formally,

$$x, y \in C \implies tx + (1-t)y \in C \text{ for all } 0 \le t \le 1 \qquad (11.3)$$

The convex hull of a collection of points is the smallest set that contains the given points. The convex hulls of the x and o points are shown in Figure 11.4. A common way to describe a convex hull is to imagine a string wrapped tightly around the points, as can be seen in the picture.

I used R's **chull** function to compute those convex hulls:

```
> ch0 <- chull(pts0)
> lines(pts0[c(ch0,ch0[1]),])
# ch0[1] needed for final segment
> ch1 <- chull(pts1)
> lines(pts1[c(ch1,ch1[1]),])
```

With the convex hulls added to the picture, it is clearer what our choices are to choose the line $\widehat{\ell}$, but there are still infinitely many choices. We choose one as follows.

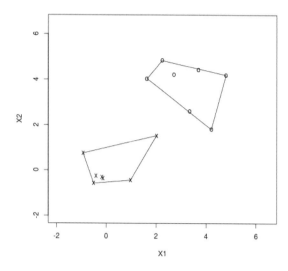

Figure 11.4: Convex hulls

First, we find the points closest to each other in the two sets. If such a point is not a vertex of its set, it will at least be on one of the edges of the set, and we record the vertices at the ends of that edge. In our case here, that gives us vertices labeled SV01, SV11 and SV12 in Figure 11.5. These are called the *support vectors*.

We then drop a line segment between the support vectors, in this case a perpendicular line from SV01 to the line connecting SV11 and SV12. Finally, we take $\widehat{\ell}$ to be the line that perpendicularly bisects this dropped line segment. The result is shown in Figure 11.6. The line $\widehat{\ell}$ extends between the top left and the bottom right.

11.2.1.2 Reduced convex hulls

Of course, the most salient "toy" aspect of the above example is that, as mentioned, the data has been constructed to be *separable*, meaning that a line can be drawn that fully separates the points of the two classes. In reality, though, our data will not be separable. There will be overlap between the two point clouds, making the convex hulls overlap, and the above formulation fails.

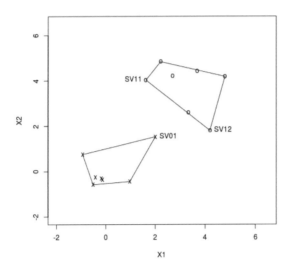

Figure 11.5: Support vectors

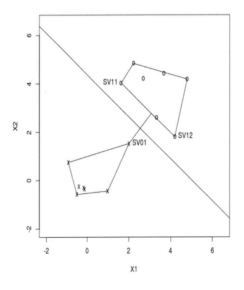

Figure 11.6: Support vectors

One obvious way to solve the problem of overlapping convex hulls is to make them smaller, so that some points are not included. This is done via *reduced* convex hulls (RCHs), defined as follows.

For a set of points $W = \{p_1, ..., p_m\}$, the convex hull of W is defined algebraically as all *convex combinations* of the p_i

$$a_1 p_1 + ... + a_m p_m \qquad (11.4)$$

with

$$a_1 + ... + a_m = 1, \quad a_i \geq 0 \text{ for all } i \qquad (11.5)$$

In the reduced case, we set a *cost c* and impose the additional constraint

$$a_i \leq c \text{ for all } i \qquad (11.6)$$

For any vertex i in the original convex hull, we would need $\alpha_i = 1$ so any c value smaller than 1 would eliminate p_i in the reduced convex hull. This results in smaller convex hulls, and for sufficiently small c, they will be nonoverlapping. The above method can then be used to find the separating hyperplane.

There is an equivalent algebraic *dual* formulation of SVM, outlined in Section 11.4.2 in the Mathematical Complements section at the end of this chapter. There our parameter c above is replaced by a related value, C, termed the *cost*.

11.2.1.3 Tuning Parameter

The smaller we set c, the fewer points there are of the form (11.4) that satisfy (11.6). In other words, the smaller c, the smaller our RCH. The quantity c is then our tuning parameter.

Let's consider the variance-bias tradeoff (Section 1.11) in this setting. First note that in the end, $\widehat{\ell}$ depends only on the support vectors. This makes the solution $\widehat{\ell}$ sensitive to perturbations in the support vectors. In Figure 11.6, suppose the point SV12 were to be moved straight downward a bit. This would force $\widehat{\ell}$ to move downward as well, i.e., have a more negative slope.

However, suppose the two convex hulls in the picture were of the same shape, orientation with respect to each other and so on, but much further apart than in the picture. The same amount of downward motion of SV12 would have a smaller impact on $\widehat{\ell}$ in this case.

The point is that perturbations of, say, SV12 correspond to sampling variation, i.e., from one sample to another. The above thought experiment shows that sampling variation corresponds to variation in $\widehat{\ell}$. We knew that from general statistical principles, of course, but the implication is this:

> A small value of c increases the distance (called the *margin*) between the two RCHs, and thus reduces the variance of (the coefficients in) $\widehat{\ell}$.

On the other hand, if we set c too small, we are introducing bias in $\widehat{\ell}$, as it is ignoring the vital region near the boundary, basing our estimation only on points at the edges of our combined data. In other words, we have a variance-bias tradeoff, with smaller c giving us smaller variation but larger bias, and the opposite for larger c.

But it's not quite so simple as that. We could argue that with very small c, our $\widehat{\ell}$ will be based on two RCHs that each contain a very small number of points. That should *increase* variance.

So, we have competing intuitive arguments. Which one is correct? In view of this conflict, it is not surprising that [133] found that increasing bias does not necessarily decrease variance, and *vice versa*. Thus there is no clear answer. Of course, we can still use cross-validation to choose c, but perhaps less confidently than in other situations.

11.2.1.4   Nonlinear Boundaries

In our discussions of parametric models, we have frequently explored forming squared and interaction terms from our predictor variables (Section 1.16). It would be natural, then, to do so with SVM, and that is indeed what is commonly done, though with a computation-saving extra aspect known as the *kernel trick*.

The details are presented in Section 11.4.3 in the Mathematical Complements section at the end of this chapter. The bottom line, though, is that one can model nonlinearity of ℓ by transforming our X values. The assumption is that in the transformed space, ℓ becomes linear.

The type of transformation is called the *kernel*, and most SVM software packages offer the user a choice of kernels.

11.2.1.5 Statistical Consistency

Under the proper conditions, if we use techniques such as k-NN or random forests, the estimates of our regression function $\mu(t)$ converge to the true values as the sample size n goes to infinity. This will not quite be the case with SVM. Here is the intuition:

First, as noted, we are not really estimating a regression function (though regression versions of SVM do exist). We are estimating a class boundary. But more important, we are assuming that that boundary has a linear form (possibly after a kernel transform). If that assumption is true, then with proper choice of c (or the cost C in Section 11.4.2), one could show statistical consistency. The situation is similar to that of ordinary linear models.

Of course, one can try using kernels of higher and higher order of complexity, say polynomial kernels of higher and higher degree. Theoretically this would give statistically consistent estimation for any interclass boundary surface. But as noted in Section 1.16.4 for the linear-model case, this is not practical.

11.2.1.6 Example: Letter Recognition Data

Here we will apply SVM to the Letter Recognition data analyzed in Sections 5.5.4 and 10.5.3, using the **svm()** function from the popular **e1071** pacakge. As explained in the Preface to this book, we use the default values. notably for the choice of kernel (radial basis) and cost **C** (1). With $n = 20000$ and only 16 predictors, we will not bother with cross validation.

```
> library (mlbench)
> data(LetterRecognition)
> lr <- LetterRecognition
> library (e1071)
> eout <- svm(lettr ~ .,data=lr)
> svmpred <- predict(eout,data=lr)
> mean(svmpred == lr$lettr)
[1] 0.9624
```

This is similar to what we previously obtained with random forests, and possibly slightly better than the result with k-NN.

11.2.2   Neural Networks

The term *neural network* (NN) alludes to a machine learning method that
is inspired by the biology of human thought. In a two-class classification
problem, for instance, the predictor variables serve as inputs to a "neuron,"
with output 1 or 0, with 1 meaning that the neuron "fires" and we decide
Class 1. NNs of several *hidden layers*, in which the outputs of one layer of
neurons are fed into the next layer and so on, until the process reaches the
final output layer, were also given biological interpretation.

The method was later generalized, using *activation functions* with outputs
more general than just 1 and 0, and allowing backward feedback from later
layers to earlier ones. This led development of the field somewhat away from
the biological motivation, and some questioned the biological intepretation
anyway, but NNs have a strong appeal for many in the machine learning
community. Indeed, well-publicized large projects using *deep learning* have
revitalized interest in NNs.

11.2.2.1   Example: Vertebrae Data

Let us once again consider the vertebrae data. We'll use the **neuralnet**
package, available from CRAN.

```
> library(neuralnet)
> vert <- read.table('column_3C.dat',header=FALSE)
> library(dummies)
> ys <- dummy(vert$V7)
> vert <- cbind(vert[,1:6],ys)
> names(vert)[7:9] <- c('DH','NO','SL')
> set.seed(9999)
> nnout <- neuralnet(DH+NO+SL ~ V1+V2+V3+V4+V5+V6,
    data=vert,hidden=3,linear.output=FALSE)
> plot(nnout)
```

Note that we needed to create dummy variables for each of the three classes.
Also, **neuralnet()**'s computations involve some randomness, so for the sake
of reproducibility, we've called **set.seed()**.

As usual in this book, we are using the default values for the many possible
arguments, including using the logistic function for activation.

$$g(t) = \frac{1}{1 + e^{-t}} \tag{11.7}$$

However, we have set a nondefault value of **hidden = 3**. That argument is a vector through which the user states the number of hidden layers and the number of neurons in each layer. By setting the value 3 here, we are specifying three nodes in just one hidden layer, to keep the example simple. We also had to specify **linear.output = FALSE**, to indicate that this is a classification problem rather than regression.

The plot is shown in Figure 11.7, which will serve as our introduction to NNs, as follows. The circles represent neurons. Values coming in from the left of a neuron are fed into the activation function. Lines coming out of the right of a neuron represent inputs going in to the next layer on the right.

The leftmost column of circles represents transforming our original input variables via the activation function; the center column of three circles represents our single hidden layer; and the rightmost column produces the final estimated regression function values, in this case the values of P(V7 | V1,...,V6).

The several lines coming into a neuron from the left are labeled with numbers representing weights. The final value fed into that neuron is the weighted sum of the "ancestor" neurons on the left. That weighted sum is input to the activation function, with the value of that function then being the output value at the neuron.

Weights are computed on the outputs of all circles except the rightmost. For instance, it turns out that the weight of logit(V1) as input to the second circle in the center column is 0.64057.

For any circle except those in the leftmost column, a weighted sum of outputs from the previous column is input. For example, the input to the first circle, center column is $1.0841 \cdot 1 + 0.84311 \ V1 + 0.49439 \ V2 + ...$ The weighted sum is then fed into the logit function, with the final output being the estimated conditional probability.

So, how do we predict with this? Instead of having a generic function for **predict**, here the function is **compute()**. Let's use it to go back and re-predict our data:

```
> predinfo <- compute(nnout, vert[,1:6])
> ypred <- predinfo$net.result
> ypred[56,]
[1] 0.37218683338  0.60285701068  0.02462869721
```

So, for the 56th observation, our estimated probabilities for DH, NO and SL are about 0.37, 0.60 and 0.02.

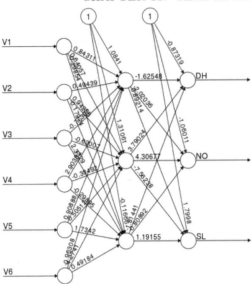

Error: 43.000304   Steps: 1292

Figure 11.7: Verbebrae NN, 1 hidden layer (see color insert)

How about our overall accuracy?

```
> mean(apply(ypred,1,which.max) ==
    apply(vert[,7:9],1,which.max))
[1] 0.783870967
```

This is not quite what we attained with a parametric logistic model in Section 5.5.2.[6] For a more careful analysis, we should use cross validation as we did there.

11.2.2.2   Tuning Parameters and Other Technical Details

Clearly, the number of hidden layers, as well as the number of neurons per layer, are tuning parameters. But there are more of them, involving things such as to what degree iterative feedback (*back propagation*) is used.

The weights are typically calculated via least-squares minimization (in one

---

[6]Of course, the logit model there should not be confused with our use of logit as our activation function here.

variant or another), possibly with a LASSO-type regularization parameter. These and other details are beyond the scope of this book.

11.2.2.3 Dimension Reduction

Say we have ℓ layers and d nodes per layer. The number of weights is then potentially ℓd^2, which could be tremendous in large problems. The danger of overfitting is thus quite grave, even though NNs are typically used with very large n.

The solution is to somehow bar certain connections among the neurons in one layer to those of the next. In other words, we impose *structural 0s* into certain types of weights. This can be application-specific. For instance, *convolutional neural networks* [114] are geared to image classification problems, as follows.

Images are naturally thought of as two-dimensional arrays, but are stored as one-dimensional arrays, say in *column-major* form: First the top column is stored, then the second column and so on. This destroys the two-dimensional locality of an image — points near a pixel are likely to have similar values to it — but we can restore that locality by requiring the weights in our NN to favor neural connections involving neighboring pixels.

11.2.2.4 Why Does It Work (If It Does)?

Hopefully the reader is not the type who is satisfied with "black box" techniques, so this section presents an outline of why all this may actually work!

As with SVM, a geometric view can be very helpful. Toward that end, think of the case $p = 2$, and suppose we use an "ideal" neuron activation functiuon $a(s)$ equal to 1 for $s > 0$ and 0 otherwise. We can interpret any set of weights coming in to a neuron in the first hidden layer as being represented by a line in the (t_1, t_2) plane,

$$w_1 t_1 + w_2 t_2 - c \qquad (11.8)$$

(We can pick up the c term by, for instance, allowing a 1 input to the network, in addition to the predictors $X^{(1)}$ and $X^{(2)}$.) The neuron receiving these inputs fires if and only if we are on one particular side of the line. We have as many lines as there are neurons in the first hidden layer.

Let's do a simple "What if?" analysis. Consider a toy example with $p = 2$ and two classes, in which our regression function is

$$P[\text{class 1} \mid (X^{(1)}, X^{(2)}) = (t_1, t_2)] = \mu(t_1, t_2) =$$

$$\begin{cases} 1, & \text{if } 0.4 < t_1 < 0.6 \text{ and } 0.4 < t_2 < 0.6 \\ 0, & \text{otherwise} \end{cases} \tag{11.9}$$

So, ideally we should predict class 1 if $(X^{(1)}, X^{(2)})$ falls in this little square in the center of the space, and predict class 0 otherwise.

If we knew the above — again, this is a "What if?" analysis — we could in principle predict perfectly with a three-layer network, as follows. (This will be an outline; the reader is invited to fill in the details.)

We would have three inputs, $X^{(1)}$, $X^{(2)}$ and 1, to the left layer, and we would have two neurons in the rightmost layer, for our two classes. Following up on our geometric view above, note that for instance, for the constraint

$$0.4 < t_1 \tag{11.10}$$

the weights for $X^{(1)}$, $X^{(2)}$ and 1 would be 1, 0 and -0.4. From (11.9), we see we need eight such lines, thus eight neurons in the middle layer.

The weights coming out of the second layer into the top neuron of the rightmost layer would all be 1/8, so that that neuron would fire if and only if *all* of the eight constraints in (11.9) are satisfied. The lower neuron would do the opposite.

The word *all* above is key. It basically says that we have formed an "AND" network. But what if in (11.9) there were two square regions in which $\mu(t) = 1$, rather than just one? Then we sould need to effect an "OR" operation. We could do this by having two AND nets, with a second hidden layer playing the role of OR.

Finally, note that any general $\mu(t)$ could be approximated by having many little squares like this, with the value of $\mu(t)$ now being more general than just 0 and 1. We could still use a four-layer AND-OR network, with some modification to account for $\mu(t)$ now having values between 0 and 1.

Now, let's come down to Earth a bit. The above assumes that $\mu(t)$ is known, which generally is not the case; it must be estimated from our data, typically with p (and n) very large. It is hoped that, since the weights are computed by least-squares fits to our data, plus iterative techniques such as back propagation — adjusting our earlier iterates by feedback from the prediction accuracy of the final layer — eventually "it all comes out in the wash." Hopefully in the end we obtain a network that works well. **Of course, there are no guarantees.**

Note too that this shows that in principle we need only two hidden layers (actually even one, by modifying this analysis), no matter how many predictor variables we have. However, in practice, we may find it easier to use more.

You can see that the above intuition could be the basis for a proof of statistical consistency. Some researchers have developed complicated technical conditions under which NNs can be shown to yield statistical consistency [69]. which says that, given enough neurons, NNs can approximate any smooth regression function. Showing statistical consistency then becomes a matter of determining how fast the number of neurons can grow with n. The Stone-Weierstrass Theorem (Section 1.16.4), which states that we approximate any continuous regression function by polynomials, is used in some of this theory.

11.3 The Verdict

The reader, having come this far in the book, is now armed with a number of techniques, linear/nonlinear and parametric/nonparametric. Which is best? There is no good answer to this, and though many research papers or books will say something like "Method A is better in such-and-such settings, while Method B is better in some other situations, etc.", the reader is advised to retain a healthy skepticism.

SVMs and NNs were developed in the machine learning community, and have attracted much attention in the press. These methods, especially NNs, have generated some highly impressive example applications [85], but they have also generated controversy [15]. There has been concern that the science fiction-like names of the methods are overinterpreted as implying that these methods somehow have special powers. As remarked in [64]:

> There has been a great deal of hype surrounding neural networks, making them seem magical and mysterious. As we make clear in this section, they are just nonlinear statistical models...

Again, **none of these methods, or any other, is guaranteed to work**.
There are always challenges that the analyst must face. But these are good
tools available to try. Furthemore, keep in mind that SVM and NN, which
make use of linear functions, may be more powerful, since in many applied
problems there $\mu(t)$ is monotonic in the t_i.

11.4 Mathematical Complements

11.4.1 Edge Bias in Nonparametric Regression

It is important to keep in mind that k-NN and other nonparametric methods
are subject to bias near the boundaries of our data. It will be easier to
explain this in the density estimation context (Section 5.10.1.1). We'll use
(5.30), which for convenience is duplicated here:

$$\widehat{f}(t) = \frac{\#(t-h, t+h)}{2hn} \tag{11.11}$$

Let R denote the variable whose density is of interest. Suppose the true
population density is $f_R(t) = 4t^3$ for t in (0,1), 0 elsewhere. The quantity
in the numerator has a binomial distribution with n trials and probability
of success per trial

$$p = P(t-h < R < t+h) = \int_{t-h}^{t+h} 4u^3 \, du = (t+h)^4 - (t-h)^4 = 8t^3h + 8th^3 \tag{11.12}$$

By the binomial property, the numerator of (11.11) has expected value np,
and thus

$$E[\widehat{f_R}(t)] = \frac{np}{2nh} = 4t^3 + 4th^2 \tag{11.13}$$

Subtracting $f_R(t)$, we have

$$\text{bias}[\widehat{f_R}(t)] = 4th^2 \tag{11.14}$$

So, the smaller we set h, the smaller the bias, consistent with intuition. But
note too the source of the bias: Since the density is increasing, we are likely

to have more neighbors on the right side of t than the left, thus biasing our density estimate upward.

How about the variance? Again using the binomial property, the variance of the numerator of (11.11) is $np(1-p)$, so that

$$Var[[\widehat{f_R}(t)] = \frac{np(1-p)}{(2nh)^2} = \frac{np}{2nh} \cdot \frac{1-p}{2nh} = (4t^3 + 4th^2) \cdot \frac{1-p}{2nh} \quad (11.15)$$

This matches intuition too: On the one hand, for fixed h, the larger n is, the smaller the variance of our estimator — i.e., larger samples are better, as expected. On the other hand, the smaller we set h, the larger the variance, because with small h there just won't be many R_i falling into our interval $(t - h, t + h)$.

So, you can really see the bias-variance tradeoff here, in terms of what value we choose for h.

The nonparametric regression case is similar. For $p = 1$, the numerator of (11.11) now becomes the sum of all Y_i for which X_i is in $(t - h, t + h)$. The expected value of the numerator is now

$$E(Y | t - h < X < t + h) \quad (11.16)$$

The relevant density is then the conditional density of X, given X is in $(t - h, t + h)$. We can then compute the bias and variance as above (Exercise 7).

11.4.2 Dual Formulation for SVM

It can be shown that the vector w of coefficients of $\widehat{\ell}$ can be computed in the separable case as follows.

Find w and a scalar b that minimize

$$\frac{1}{2} ||w||^2 \quad (11.17)$$

subject to

$$Y_i(w'X_i + b) \geq 1, \quad i = 1, ..., n \tag{11.18}$$

The intuition is this: Look at Figure 11.6. The separating line is mathematically

$$w't + b = 0 \tag{11.19}$$

with the value b chosen so that this is the case. Thus we are on one side of the line if

$$w't + b > 0 \tag{11.20}$$

and are on the other side if

$$w't + b < 0 \tag{11.21}$$

There will be two *supporting hyperplanes* on the edges of the two convex hulls. In Figure 11.6, one of those hyperplanes is the line through SV11 and SV12, and the other is the parallel line through SV01. Recall that the *margin* is the distance between the two supporting hyperplanes. We want to maximize the distance between them, that is, separate the classes as much as possible. Simple geometric calculation shows that the margin is equal to $2/||w||^2$. We want to maximize the margin, thus minimize (11.17).

Recall that $Y_i = \pm 1$. So, in the class having all $Y_i = 1$, we will want our prediction to be at least 1, i.e.,

$$w'X_i + b \geq c, \quad i = 1, ..., n \tag{11.22}$$

with equality for the support vectors, while in the other class we will want

$$w'X_i + b \leq -c, \quad i = 1, ..., n \tag{11.23}$$

The scaling of w is defined so that $c = 1$.

Again since $Y_i = \pm 1$, (11.18) neatly captures both cases.

In the general, nonseparable case, for cost C (not numerically equal to c above), the problem becomes:

Find w, a scalar b, and nonnegative quantities ξ_i that minimize

$$\frac{1}{2}||w||^2 + C \sum \xi_i \tag{11.24}$$

subject to

$$Y_i(w'X_i + b) \geq 1 - \xi_i, \quad i = 1, ..., n \tag{11.25}$$

There is an equivalent formulation (in mathematical terminology, the *dual* problem):[7]

Find quantities α_i that minimize

$$\frac{1}{2}\sum_{i=1}^{n}\sum_{j=1}^{n} Y_iY_j\alpha_i\alpha_j X_i'X_j - \sum_{i=1}^{n}\alpha_i \tag{11.26}$$

such that

$$\sum_{i=1}^{n} Y_i\alpha_i = 0 \tag{11.27}$$

and $0 \leq \alpha_i \leq C$. Then

$$w = \sum_{i=1}^{n} \alpha_i Y_i X_i \tag{11.28}$$

and

$$b = Y_j - X_j'w \tag{11.29}$$

---

[7]For readers with background in *Lagrange multipliers*, that is the technique used here. The variables α_i are the Lagrange variables.

for any j having $\alpha_j > 0$, i.e., any support vector.

The guess for Y for a new observation with predictor vector X is

$$\text{sign}\left(\sum_{i=1}^{n} \alpha_i Y_i X_i' X + b\right) \tag{11.30}$$

11.4.3   The Kernel Trick

Take the case $p = 2$, for instance. We would replace $X = (X_1, X_2)'$ by $h(X)$, where

$$h(t_1, t_2) = (t_1, t_1^2, t_2, t_2^2, t_1 t_2)' \tag{11.31}$$

as our new predictor vector. Then instead of minimizing

$$\frac{1}{2} \sum_{i=1}^{m} \sum_{j=1}^{m} Y_i Y_j \alpha_i \alpha_j X_i' X_j - \sum_{i=1}^{n} \alpha_i \tag{11.32}$$

we would minimize

$$\frac{1}{2} \sum_{i=1}^{m} \sum_{j=1}^{m} Y_i Y_j \alpha_i \alpha_j h(X_i)' h(X_j) - \sum_{i=1}^{n} \alpha_i \tag{11.33}$$

In SVM, one can reduce computational cost by changing this a bit, minimizing

$$\frac{1}{2} \sum_{i=1}^{m} \sum_{j=1}^{m} Y_i Y_j \alpha_i \alpha_j K(X_i, X_j) - \sum_{i=1}^{n} \alpha_i \tag{11.34}$$

where the function $K()$ is a *kernel*, meaning that it must satisfy certain mathematical properties, basically that it is an inner product in some space. A few such functions have been found to be useful and are incorporated into SVM software packages. One of them is

$$K(u, v) = (1 + u'v)^2 \tag{11.35}$$

This expression is quadratic in the u_i and v_i, so it achieves our goal of a second-degree polynomial transformation as posed above. And, we could take the third power in (11.35) to get a cubic transformation and so on. Another kernel often found in software packages for SVM is the *radial basis kernel*,

$$K(w, x) = e^{-\gamma ||w - x||^2} \qquad (11.36)$$

11.5 Further Reading

Much theoretical (though nonstatistical) work has been done on SVMs. See for instance [38] [36].

For a statistical view of neural networks, see [120].

11.6 Exercises: Data, Code and Math Problems

Data problems:

1. In the discussion of Figure 6.1, it was noted that ordinary k-NN is biased at the edges of the data, and that this might be remedied by use of local-linear smoothing, a topic treated here in this chapter, Section 11.1.

Re-run Figure 6.1 using local-linear smoothing, and comment on what changes, if any, emerge.

2. Write code to generate **nreps** samples, say 10000, in the example in Section 11.1, and compute the bias for ordinary vs. local-linear k-NN at **r** equally-spaced points in (0,1). In your experiment, vary **k**, **n** and **r**.

Mini-CRAN and other computational problems:

3. Write a function with call form

```
locquad ( predpt , nearxy , prodterms=FALSE)
```

analogous to **regtools'** **loclin()**, except that it is locally quadratic instead of locally linear. The function will add squared terms for the predictors, and if **prodterms** is TRUE, then the interactions will be added as well.

Math problems:

4. Given points $p_1, ..., p_m$ in k-dimensional space $\mathcal{R}^k$, let CH_c denote the reduced convex hull for cost c. For any $c \geq 1$, this becomes the original convex hull of the points. Show that CH_c is indeed a convex set.

5. Following up on the discussion in Section 11.2.2.4, show by construction of the actual weights that any second-degree polynomial function can be reproduced exactly.

6. Show that the kernel in (11.35) is indeed an inner product, i.e.,

$$K(u, v) = q(u)'q(v) \tag{11.37}$$

for

$$q((s, t)') = (as^2, bt^2, cst, ds, et, 1)' \tag{11.38}$$

with some suitable a, b, c.

7. Find the bias and variance in the regression example in Section 11.4.1, taking in that example $\mu(s) = s^2$.

Chapter 12

Regression and Classification in Big Data

Gone are the old days of textbook examples consisting of $n = 25$ observation. Automatic data collection by large companies has led to data sets on the terabyte scale or more. This has enabled applications of previously unimaginable power, some noble such as geographical prediction of epidemics and crime incidents, others that create disturbing invasions of privacy. For good or bad, though, Big Data is here to stay.

All this draws renewed attention on the predictive methods covered in this book, but with a new set of problems in the case of Big Data:

- Computational: Typically any of the methods in this book will have *very long run times*, hours, days or even more in some cases. Equally important, *the data may not fit into available memory*, and thus become impossible to run.

- Statistical: In many applications we have $p >> n$, many more predictors than observations. This is common in genomics problems, for instance. Since we need $p < n$ to avoid indeterminacy in a linear model, and need $p << n$ to avoid overfitting, this presents a real challenge.

This chapter is devoted to addressing such issues. As in [102], we will distinguish here betwen Big-n and Big-p.

12.1   Solving the Big-n Problem

Consider the case of n large, p moderate. Here there is a simple solution to the computational problem, which I call Software Alchemy. (See [103] for the various authors who have independently developed this idea, and for details on the method.)

12.1.1   Software Alchemy

Let's illustrate this notion with **lm()**. The procedure is as follows:

- Break the rows of the data matrix into chunks, say r of them.

- Apply the statistical procedure, say **lm()**, to each chunk.

- Average the r $\widehat{\beta}$ vectors to obtain the overall $\widehat{\beta}$.

Since each chunk is much smaller than the full data set, the run time for each chunk is smaller as well. And since the chunks are run in parallel, a substantial speedup can be attained. An analysis of the types of applications in which speedups are possible is presented in the Mathematical Complements section at the end of this chapter.

It is easy to prove that this does work. One does not obtain the same value of $\widehat{\beta}$ that would be computed from the entire data set, but the result is *just as good* — the chunked and the full estimator have the same asymptotic distribution.

It is easy to apply this to nonparametric models. With k-NN for instance, we would compute $\widehat{\mu}(t)$ on each chunk, and average the resulting values. For nonparametric classification methods that normally return just a predicted class rather than computing $\widehat{\mu}(t)$, say CART, we can "vote" among the r predicted classes, taking our guessed class to be the one that gets the most votes, as with AVA. Similarly, with dimension reduction via PCA or NMF, we cannot average the bases, but we can average the $\widehat{\mu}(t)$ values or vote among the predicted classes.

Software Alchemy is implemented in the **partools** package [96].[1] The main work is done in the function **cabase()**. There are various wrappers for that function, such as **calm()**, which applies Software Alchemy to **lm()**.[2]

---

[1] Version 1.1.5 or higher is needed below.
[2] The prefix 'ca' stands for "chunk averaging."

Some details on the innards of **partools** are given in the Computational Complements section at the end this chapter.

12.1.2 Example: Flight Delay Data

Yet another famous dataset is that of airline flight data [6]. There are records for all U.S. flights between 1987 and 2008, with the focus on arrival and departure delay, ArrDelay and DepDelay.

In order to facilitate efforts by readers to replicate our analysis here, we will focus on just one year, 2008, which is already "big enough," over 7 million records.

```
> library(partools)
> cls <- makeCluster(16)
> setclsinfo(cls)
> y2008 <- read.csv('y2008',header=TRUE)
> mnthnames <-
  c('Jan','Feb','Mar','Apr','May','Jun',
    'Jul','Aug','Sep','Oct','Nov','Dec')
> mnth <- mnthnames[y2008$Month]
> daynames <-
    c('Sun','Mon','Tue','Wed','Thu','Fri','Sat')
> day <- daynames[y2008$DayOfWeek]
> y2008$Month <- as.factor(mnth)
> y2008$DayOfWeek <- as.factor(day)
> system.time(calmout <- calm(cls,'ArrDelay ~
    DepDelay+Distance+TaxiOut+UniqueCarrier+Month+
    DayOfWeek,data=y2008'))
   user    system  elapsed
 40.788    3.748   50.040
> system.time(lmout <- lm(ArrDelay ~
    DepDelay+Distance+TaxiOut+UniqueCarrier+Month+
    DayOfWeek,data=y2008))
   user    system  elapsed
 74.720    2.508   77.376
> distribsplit(cls,'y2008',scramble=TRUE)
```

Here we regress arrival delay against departure delay, distance and so on (UniqueCarrier is the airline, e.g., DL for Delta). Note that we needed to convert the month and day variables from numeric to factor, to avoid having, say, March, count 3 times as much as January, which would be meaningless.

We set up a (virtual) cluster of 16 nodes, which in this case run on 16 cores (the machine had 32, counting hyperthreading); we could run on 16 nodes of a physical cluster and so on. Each node is running its own invocation of R, in communication with the invocation running on the parent node.

We split the dataset **y2008** across the 16 nodes, so that each node had about 1/16th of the data. With much less data to work with, this potentially enables speedier computation, since the nodes run in parallel. Though we cannot expect to attain a 16-fold speedup, for reasons given in the Computational Complements section of this chapter, we did get a speedup of about 60%.[3]

Note the argument **scramble = TRUE** in the call to **distribsplit()**. Many datasets are ordered on one or more of the variables, in this case ordered by date. A straightforward splitting of the data would mean that some nodes may receive data for only one month, making it impossible to compute a **Month** coefficient in the regression. (Also, Software Alchemy assumes i.i.d. data.) The random permutation solves that problem.

So, how does the Software Alchemy output compare with that of the analysis on the full data?

```
> calmout$tht
```

(Intercept)	DepDelay	Distance
−16.230510414	0.991774602	−0.002382778
TaxiOut	UniqueCarrierAA	UniqueCarrierAQ
0.826768548	4.516806436	8.953951702
UniqueCarrierAS	UniqueCarrierB6	UniqueCarrierCO
4.030485591	−0.367642225	0.194950670
UniqueCarrierDL	UniqueCarrierEV	UniqueCarrierF9
1.581873640	2.056932472	6.766798981
UniqueCarrierFL	UniqueCarrierHA	UniqueCarrierMQ
5.670985961	10.692474278	3.469432309
UniqueCarrierNW	UniqueCarrierOH	UniqueCarrierOO
3.449247970	−0.072744312	3.842304091
UniqueCarrierUA	UniqueCarrierUS	UniqueCarrierWN
1.681950224	0.441686744	4.140866177
UniqueCarrierXE	UniqueCarrierYV	MonthAug
2.791769787	2.928448428	−1.400718871
MonthDec	MonthFeb	MonthJan
−0.038449264	−0.017799483	−0.404931144
MonthJul	MonthJun	MonthMar

---

[3]There was overhead in splitting the data. As noted in Section 12.4.1, the real advantage of **partools** is accrued when many operations are done on the distributed data after a split.

−0.951371174	0.275108880	−0.107499646
MonthMay	MonthNov	MonthOct
−0.019260538	−1.767687711	−1.510315535
MonthSep	DayOfWeekMon	DayOfWeekSat
−1.534679083	0.581085330	0.320872459
DayOfWeekSun	DayOfWeekThu	DayOfWeekTue
0.289664672	0.673834948	0.356366605
DayOfWeekWed		
0.581757662		

```
> coef(lmout)
```

(Intercept)	DepDelay	Distance
−16.230600841	0.991777979	−0.002382658
TaxiOut	UniqueCarrierAA	UniqueCarrierAQ
0.826780859	4.516897691	8.957282637
UniqueCarrierAS	UniqueCarrierB6	UniqueCarrierCO
4.030860520	−0.368104197	0.195027499
UniqueCarrierDL	UniqueCarrierEV	UniqueCarrierF9
1.581651876	2.057249536	6.766480218
UniqueCarrierFL	UniqueCarrierHA	UniqueCarrierMQ
5.671174474	10.691920438	3.469761618
UniqueCarrierNW	UniqueCarrierOH	UniqueCarrierOO
3.448967869	−0.072846826	3.842374665
UniqueCarrierUA	UniqueCarrierUS	UniqueCarrierWN
1.681752229	0.441697940	4.141035514
UniqueCarrierXE	UniqueCarrierYV	MonthAug
2.791406582	2.928769069	−1.400844315
MonthDec	MonthFeb	MonthJan
−0.038074477	−0.017720576	−0.405148865
MonthJul	MonthJun	MonthMar
−0.951546924	0.274937723	−0.107709304
MonthMay	MonthNov	MonthOct
−0.019137396	−1.767617225	−1.510315831
MonthSep	DayOfWeekMon	DayOfWeekSat
−1.534714190	0.581024609	0.320610432
DayOfWeekSun	DayOfWeekThu	DayOfWeekTue
0.289584205	0.673469157	0.356256017
DayOfWeekWed		
0.581504814		

The results are essentially identical. Note very carefully again, though, that *the chunked estimator has the same asymptotic variance as the original one,* so any discrepancies that may occur between them should not be interpreted as meaning that one is better than the other. In other words, we saved some computation time here with results of equal quality.

Let's try k-NN on this data. Continuing from our computation above, we have:

```
> y2008mini <- y2008[,c(15,16,19,21)]
# can't have NAs
> y2008mini <- y2008mini[complete.cases(y2008mini),]
> distribsplit(cls,'y2008mini',scramble=TRUE)
> system.time(caknn(cls,'y2008mini[,1]',50,
    xname='y2008mini[,-1]'))
   user   system elapsed
 35.552    3.792 113.514
> library(regtools)
> system.time(xd <- preprocessx(y2008mini[,-1],50))
   user   system elapsed
303.516    4.400 308.068
> system.time(kout <- knnest(y2008mini[,1],xd,50))
   user   system elapsed
701.832   37.720 740.036
```

The speedup here in the fitting stage is large, 113.514 seconds vs. 308.068 + 740.036, almost 10-fold.

Note that **caknn()** has the call form

```
caknn(cls, yname, k, xname = '')
```

The optional argument indicates whether **preprocessx()** has already been called at the cluster nodes. A nonblank value indicates that this is not the case, and contains the expression for the desired matrix/data frame of X values.

12.1.3   More on the Insufficient Memory Issue

As noted earlier, use of parallel computation in large problems is not just a matter of reducing run time, but is also in many cases a solution to lack of sufficient memory. Indeed, the problem may be too large for R's address space.

Software Alchemy can solve that problem, by breaking the task into chunks, each of which can fit into memory. This would be feasible if the task is run on a cluster of machines, rather than one multicore one. In the multicore case, we could run the chunks one at a time.

A subtle problem may lurk here, though. Software Alchemy relies on the Central Limit Theorem, in multivariate form, and the convergence rate depends on the dimension [19]. So, if p/n is not small, Software Alchemy may not be appropriate. How, then, can we do computation on the full data set while circumventing the memory problem?

For a linear or generalized linear model, a solution is available in packages such as **biglm** [94], and the package **biganalytics** [46], which provides a wrapper to **biglm**. (The **biganalytics** package may also provide a solution to memory space problems.)

The **partools** package includes some file-based functions, such as **filesplit()**, counterpart to **distribsplit()**, another way to avoid memory restrictions.

12.1.4 Deceivingly "Big" n

Even with millions of observations, n may not be large enough to estimate what we want. Consider the airline data, for instance. Suppose there is some small, obsolescent airport that is rarely used, so rarely that there are only, say, 5 records in the dataset involving this airport. Then clearly we have insufficient data to be able to say anything about this airport. If for example we have as a predictor a dummy variable for this airport and use a linear model, the resulting estimated coefficient will have a large standard error, despite the huge n.

12.1.5 The Independence Assumption in Big-n Data

Many instances of Big Data arise with multiple observations on the same unit, say the same person. This raises issues with assumptions of statistical independence, the core of most statistical methods.

However, with Big Data, people are generally not interested in statistical inference, since the standard errors will typically be tiny (though not in the situations discussed in the last section). Since in parametric models the assumption of independent observations enters in mainly in deriving standard errors, lack of independence is typically not an issue.

12.2   Addressing Big-p

We now turn to the case of large p, including but not limited to $p >> n$. Note the subdued term in the title here, *addressing* rather than *solving* as in Section 12.1. Fully satisfactory methods have not been developed for this situation.

12.2.1   How Many Is Too Many?

We know that overfitting can be a problem. In terms of our number of observation n and number of variable p, we risk overfitting if p/n is not small. But what is "small"? — not an easy question at all. This question is the subject of dozens of methods developed over the decades, some of which will be discussed in this chapter. None of them completely settles the issue, but the point here is that that issue cannot be addressed without understanding the dynamics underlying it.

From earlier discussions in this book, the reader is familiar with the notion of examining what happens as the number of data points n goes to infinity, as for instance in the Central Limit Theorem, whose implication was that the estimated coefficient vector $\widehat{\beta}$ in a linear model has an approximately multivariate normal distribution, important for forming confidence intervals and finding p-values.

But in the context of variable selection, we have an additional aspect to consider, as seen in our example in Section (9.1.1). The latter suggests that the larger n is, the more predictors p our model can handle well. For this reason, researchers in this field typically not only model n as going to infinity, but also have p do so. A question then arises as to whether p should go to infinity more slowly than n does.

To nonmathematicians, this concept may seem odd, maybe even surreal. After all, in practice one just has a certain sample size and a certain number of predictors to choose from, so why care about those quantities going to infinity? This is a valid concern, but the researchers' findings are useful in a "cultural" way, in the sense that the results give us a general feeling about the interplay between p and n. So, in this section we will take a brief look at the nature of the results researchers have obtained.

12.2.1.1   Toy Model

As a warmup, consider a very simple example, in the fixed-X setting with an orthogonal design (Section 2.3).

Again, in the fixed-X case, the X values are chosen beforehand by the experimenter. We will assume that she has done so in a manner in which none of the values is very big, which in our simple model here we will take to mean that

$$\sum_{i=1}^{n} (X_i^{(j)})^2 = n, \quad j = 1, ..., p \tag{12.1}$$

We will assume a linear model with no intercept term,

$$\mu(t) = \sum_{j=1}^{n} \beta_j t_j \tag{12.2}$$

In our setting here of an orthogonal design, the factor $A'A$ in (2.28) will have 0s for its off-diagonal elements, with the jth diagonal entry being

$$\sum_{i=1}^{n} (X_i^{(j)})^2 \tag{12.3}$$

which by our assumption has the value n. Then (2.28) has the easy closed-form solution

$$\widehat{\beta}_j = \frac{\sum_{i=1}^{n} X_i^{(j)} Y_i}{n}, \quad j = 1, ..., p \tag{12.4}$$

We'll also assume homoscedasticity, which the reader will recall means

$$Var(Y \mid X = (t_1, ..., t_p)) = \sigma^2 \tag{12.5}$$

regardless of the values of the t_j. (All calculations here will be conditional on the X values, since they are fixed constants chosen by the experimenter, but we will not explicitly write the $\mid$ symbol.) Thus we have

$$Var(X_i^{(j)} Y_i) = (X_i^{(j)})^2 \sigma^2 \tag{12.6}$$

and thus from (12.4), we see that

$$Var(\widehat{\beta}_j) = \frac{1}{n}\sigma^2, \quad j = 1, ..., p \tag{12.7}$$

Now finally, say we wish to estimate $\mu(1, ..., 1)$, i.e., enable prediction for the case where all the predictors have the value 1. Our estimate will be

$$\sum_{j=1}^{p} \widehat{\beta}_j \cdot 1 \tag{12.8}$$

and its variance will then be

$$Var[\widehat{\mu}(1, , ..., 1)] = \frac{p\sigma^2}{n} \tag{12.9}$$

Now, what does this have to do with the question, "How many predictors is too many?" The point is that since we want our estimator to be more and more accurate as the sample size grows, we need (12.9) to go to 0 as n goes to infinity. We see that even if at the same time we have more and more predictors p, the variance will go to 0 as long as

$$\frac{p}{n} \to 0 \tag{12.10}$$

Alas, this still doesn't fully answer the question of how many predictors we can afford to use for the specific value of n we have in our particular data set. But it does give us some insight, in that we see that the variance of our estimator is being inflated by a factor of p if we use p predictors. This is a warning not to use too many of them, and the simple model shows that we will have "too many" if p/n is large. This is the line of thought taken by theoretical research in the field.

12.2.1.2   Results from the Research Literature

The orthogonal nature of the design in the example in the last section is not very general. Let's see what theoretical research has yielded.

Stephen Portnoy of the University of Illinois proved some results for general Maximum Likelihood Estimators [115], not just in the regression/classification context, though restricted to distributions in an *exponential family*

(Section 4.2.1). We won't dwell on the technical definition of that term here, but just note again that it includes many of the famous distribution families, such as normal, exponential, gamma and Poisson. Portnoy proved that we obtain consistent estimates as long as

$$\frac{p}{\sqrt{n}} \to 0 \qquad (12.11)$$

This is more conservative than what we found in the last section, but much more broadly applicable.

We might thus take as a rough rule of thumb that we are not overfitting as long as $p < \sqrt{n}$. Strictly speaking, this really doesn't follow from Portnoy's result, but some analysts use this as a rough guideline. As noted earlier, this was a recommendation of the late John Tukey, one of the pioneers of modern statistics.

There is theoretical work on p, n consistency of LASSO estimators, CART and so on, but they are too complex to discuss here. The interested reader is referred to, for instance, [28] and [66].

12.2.1.3 A Much Simpler and More Direct Approach

The discussion in the last couple of subsections at least shows that the key to the overfitting issue is the size of p relative to n. The question is where to draw the dividing line between overfitting and a "safe" value of p. The answer to this question has never been settled, but we do have one very handy measure that is pretty reasonable in the case of linear models: the adjusted R^2 value. If $p < n$, this will provide some rough guidance.

As you will recall (Section 2.9.4), ordinary R^2 is biased upward, due to overfitting (even if only slight). Adjusted R^2 is a much less biased version of R^2, so a substantial discrepancy between the ordinary and adjusted versions of R^2 may be a good indication of overfitting. Note, though, biases in almost any quantity can become severe in *adaptive* methods such as the various stepwise techniques described in Chapter 9.

12.2.1.4 Nonparametric Case

The problems occur with nonparametric methods too.

k-NN:

With p predictor variables, think of breaking the predictor space into little cubes of dimension p, each side of length d. This is similar to a k-NN setting. In estimating $\mu(t)$ for a particular point t, consider the density of X there, $f(t)$. For small d, the density will be approximately constant throughout the cube. Then the probability of an observation falling into the cube is about $d^p f(t)$, where the volume of the cube is d^p. Thus the expected number of observations in the cube is approximately

$$nd^p f(t) \tag{12.12}$$

Now, think of what happens as $n \to \infty$. On one hand, we need the bias of $\widehat{\mu}(t)$ to go to 0, so we need $d \to 0$. But on the other hand, we also need the variance of $\widehat{\mu}(t)$ to go to 0. So we need the number of observations in the cube to go to infinity, which from (12.12) means

$$\log n + p \log d \to \infty \tag{12.13}$$

as $n \to \infty$.

Since d will be smaller than 1, its log is negative, so for fixed d in (12.13), the larger p is, the slower (12.13) will go to infinity. If p is also going to infinity, it will need to do so more slowly than $\log n$. This informal result is consistent with that of [135].

As noted, the implications of the asymptotic analysis for the particular values of n and p in the data we have at hand are unclear. But comparison to the parametric case above, with p growing like $\sqrt{n}$, does suggest that nonparametric methods are less forgiving of a large value of p than are parametric methods.

There is also the related issue of the *Curse of Dimensionality*, to be discussed shortly.

NNs:

Say we have ℓ layers, with d nodes per layer. Then from one layer to the next, we will need d^2 weights, thus ℓd^2 weights in all. Since the weights are calculated using least squares (albeit in complicated ways), we might think of this as a situation with

$$p = \ell d^2 \tag{12.14}$$

Somewhat recklessly applying the Tukey/Portnoy finding, that in the parametric case we need p grow more slowly than $\sqrt{n}$, that would mean that, say, d should grow more slowly than $n^{1/4}$ for fixed ℓ.

12.2.1.5 The Curse of Dimensionality

The term *Curse of Dimensionality* (COD) was introduced by Richard Bellman, back in 1953 [13]. Roughly speaking it says that the difficulty of estimating a statistical model increases exponentially with the dimension of the problem. In our regression/classification context, that would imply that if our number of predictors p is very large, we may need an enormous number of data points n to develop an effective model. This is especially troubling in today's era of Big Data, in which p may be on the order of hundreds, thousands or even more.

One popular interpretation of the COD is that in very high dimensions the data is quite sparse. With k-NN, for instance, one can show that with high values of p, neighboring observations tend to be far from a given one [18], say a point at which prediction is to be done. The neighbors thus tend to be "unrepresentative" of the given point, especially at the edges of the data space. This arguably is a cause of what we found earlier about the need for p to grow slowly when we use nonparametric regression methods.

Perhaps this could be ameliorated using linear smoothing, as in Section 11.1. The general effect should be similar, though.

12.2.2 Example: Currency Data

Here we return to the currency data (Section 6.5). A favorite investigatory approach by researchers of $p \gg n$ situations is to add noise variables to a real data set, and then see whether a given method correctly ignores those variables.

Let's try that here, starting with the LASSO. The dataset has $n = 761$, and we will add 1500 noise variables:

```
> curr <- read.table('EXC.ASC',header=TRUE)
> u <- matrix(rnorm(761*1500),nrow=761)
> curru <- cbind(curr,u)
> library(lars)
> larsout <-
      lars(as.matrix(curru[,-5]),curru[,5],
      normalize=FALSE,use.Gram=FALSE)
> larsout
```

Sequence of LASSO moves:

		Mark	Franc	88	855	1308	1466	310	611	159	59
Var		2		3	92	859	1312	1470	314	615	163	63
Step		1		2	3	4	5	6	7	8	9 10

. . .

This is not a good start! The algorithm has already gone 10 steps, yet has not added in the variables for the Canadian *dollar* and the *pound*, while adding in 8 of the noise variables. Trying the same thing with stepwise regression, calling **lars()** with the argument **type = 'stepwise'**, produces similar results.

Let's try CART as well:

```
> library(rpart)
> rpartout <- rpart(Yen ~ ., data=curru)
> rpartout
n= 761

node), split, n, deviance, yval
      * denotes terminal node

 1) root 761 2052191.00  224.9451
   2) Mark< 2.2752 356   626203.00  185.3631
     4) Franc>=5.32 170   105112.10  147.9460
       8) Mark< 2.20635 148    25880.71  140.4329 *
       9) Mark>=2.20635 22     14676.99  198.4886 *
     5) Franc< 5.32 186    65550.84  219.5616 *
   3) Mark>=2.2752 405   377956.00  259.7381
     6) Canada>=1.08 242    68360.23  237.7370
      12) Canada>-1.38845 14     5393.71  199.2500 *
      13) Canada< 1.38845 228    40955.69  240.1002 *
     7) Canada< 1.08 163    18541.41  292.4025 *
```

Much better. None of the noise variables was selected. One disappointment is that the *pound* was not chosen. Running the analysis with **ctree()** did pick up all four currencies, and again did not choose any noise variables.

12.2.3   Example: Quiz Documents

I give weekly quizzes in my courses. Here I have 143 such documents, in LaTeX, for five courses. We will try to classify documents into their courses, based on frequencies of various words.

There should be a considerable amount of overlap between the courses, making classification something of a challenge. For example, two of the courses, one undergraduate and the other graduate-level, are statistical in nature, thus possibly difficult to distinguish from each other. Among the other three, two are in the computer systems area, thus with common terminology such as *cache*, causing the same problem in distinguishing between them. The fifth is on scripting languages (Python and R).

There are quite a few R packages available for text mining. The **tm** package provides an extensive set of utilities, such as removal of *stop words*, very common words such as *a* and *the* that usually have no predictive power. I used this package (details are presented in Section 12.4.2), and I removed the LaTeX keywords using the **detex** utility.

I ended up with my *term-document matrix* **nmtd**, with $n = 143$ rows and $p = 4670$ columns, the latter figure reflecting the fact that 4670 distinct words were found in the entire document set. So, we certainly have $p >> n$. The element in row i, column j is the number of occurrences of word j in document i. (Various other formulations are possible.)

The distribution of classes was as follows:

```
> table(labels)/sum(table(labels))
labels
      ECS132       ECS145       ECS158       ECS256
0.41958042   0.10489510   0.25874126   0.02797203
      ECS50
0.18881119
```

Once again, we will run our packages naively, just using default values. Here are the results for SVM:

```
> library(tm)
> library(SnowballC)   # supplement to tm
> # tm operations not shown here
> nmtd <- as.matrix(nmtd)
> # 'labels' is the vector of class labels
> dfwhole <-
      as.data.frame(cbind(labels,as.data.frame(nmtd)))
> # cross-validation
> trainidxs <- sample(1:143,72)
> dftrn <- dfwhole[trainidxs,]
> dftst <- dfwhole[-trainidxs,]
> library(e1071)
> svmout <- svm(labels ~ .,data=dftrn)
```

```
> ypredsvm <- predict(svmout, dftst[, -1])
[1]  0.5352113
```

Without the word information, we would always guess the course ECS 132, and would be right about 42% of the time, as seen above. So, a 54% rate does represent substantial improvement.

Let's try CART:

```
> library(rpart)
> rpartout <-
        rpart(labels ~ ., data=dftrn, method='class')
> ypredrpart <-
        predict(rpartout, dftst[, -1], type='class')
> mean(ypredrpart == dftst[, 1])
[1]  0.5774648
```

About 58%, pretty good.

12.2.4   The Verdict

In Chapter 9, it was repeatedly emphasized that there is no good solution to the problem of selecting predictor variables. That statement holds even more strongly for the situation $p >> n$.

In particular, no matter which method we use, the Principle of Rare Events (Section 7.6.1), for p large enough, there is a high probability that some predictor variable with little or no relation to our response variable will be chosen by the method.

It is instructive to consider two conflicting statements by prominent researchers. On the one hand, Hastie *et al* [65, p. 86] remark (citing Chapter 3 of [64]),

> Forward stepwise methods...are hard to beat in terms of finding good, sparse sets of variables.

On the other hand, Gelman [55] writes, in a blog post provocatively titled, "Why We Hate Stepwise Regression,"

> To address the issue more directly: the motivation behind stepwise regression is that you have a lot of potential predictors but

> not enough data to estimate their coefficients in any meaningful way. This sort of problem comes up all the time, for example heres an example from my research, a meta-analysis of the effects of incentives in sample surveys.
>
> The trouble with stepwise regression is that, at any given step, the model is fit using unconstrained least squares. I prefer methods such as factor analysis or lasso that group or constrain the coefficient estimates in some way.

Though Hastie *et al* likely have in mind a more sophisticated use of stepwise regression than Gelman, the latter's s last statement is a little ironic, in view of the leading role that Hastie *et al* have played in the development of the LASSO and related estimation methods.

It must be stated one more time, then, that the analyst needs to keep in mind that all these techniques are merely tools available for possible use. One should also use them in conjunction with cross-validation, though even then one must heed the warning in Section 9.3.2.

These are powerful methods, to be sure, but should be used with care.

12.3 Mathematical Complements

12.3.1 Speedup from Software Alchemy

(Readers who do not have previous background in "big oh" notation should review Section 5.10.2 before continuing.)

Some back-of-the envelope analysis illustrates what kinds of applications can benefit greatly from Software Alchemy (SA), and to what kinds SA might bring only modest benefits.

Consider statistical methods needing time $O(n^c)$. For instance, matrix multiplication, say with both factors having size $n \times n$, has an $O(n^3)$ time complexity.

If we have r processes, say running on r cores of a multicore machine, then SA assigns about n/r data points to each process, each with run time $O((n/r)^c)$. Since the processes run independently *in parallel*, SA would reduce the run time from $O(n^c)$ to $O((n/r)^c) = O(n^c/r^c)$, a speedup of r^c. The larger the exponent c is, the greater the speedup.

Moreover, suppose we run the r chunks sequentially, i.e., one at a time, rather than in parallel. We may do this, for instance, because we cannot fit more than one chunk in memory at a time. Then instead of $O((n/r)^c)$ time, we would have $O(r(n/r)^c)$. That would yield a speedup of $O(r^{c-1})$, so even in this case SA gives us faster runs if $c > 1$.

On the other hand, consider a linear regression computation. In the first phase, we need $O(np^2)$ time to compute $A'A$, and then in the second phase we must compute $(A'A)^{-1}$ or equivalent. A straightforward computation for the latter takes $O(p^3)$ time, or possibly $O(p^2)$ time with QR, depending on what information is desired. Alas, though SA helps in the first phase, it is of no help in that second phase. Thus SA will typically yield limited benefits here.

12.4   Computational Complements

12.4.1   The partools Package

This package consists of a number of functions for parallel computation and data wrangling. Similar to Hadoop and Spark, it is centered on the notion of distributed files and data. In the file case, for instance, this means that a nominal file **x** is actually composed of many files, say **x.001**, **x.002** and so on. But unlike Hadoop and Spark, **partools** does not use MapReduce, which is very constraining, e.g., due to requiring a sort operation whether needed or not.

The **partools** package is itself based on R's **parallel** package. With that package one creates a virtual cluster, running an invocation of R on each node. The virtual cluster could be a real cluster of machines, the cores on a multicore machine, or a combination of both. The original R process, the manager, invokes tasks at the created R processes, the workers, and collects the results.

Here is an example, involving finding the largest row sum of a matrix:

```
> library(parallel)
> cls <- makeCluster(2)
> x <- matrix(runif(100),nrow=10)
> maxsum <- function(m) max(apply(m,1,sum))
# distribute the top and bottom halves of x to the
# nodes, # have them each call maxsum on their halves,
# and collect the results
> tmp <-
```

```
        clusterApply(cls, list(x[1:5,], x[6:10,]), maxsum)
> tmp
[[1]]
[1] 6.052948
[[2]]
[1] 5.906015
> Reduce(max, tmp)
[1] 6.052948
# check
> maxsum(x)
[1] 6.052948
```

Of course, for this to pay off, one must (a) have much larger scale and (b) use the distributed data set repeatedly, in various operations.

For more information on parallel computation, see [101].

12.4.2   Use of the tm Package

Here is the code I used with the **tm** package on the quiz documents:

```
library(tm)
library(SnowballC)
nm <- Corpus(DirSource('MyQuizTexts'))
nm <- tm_map(nm, tolower)
nm <- tm_map(nm, removePunctuation)
nm <- tm_map(nm, removeNumbers)
nm <- tm_map(nm, removeWords, stopwords("english"))
nm <- tm_map(nm, stemDocument, language = "english")
nm <- tm_map(nm, PlainTextDocument)
nmtd <- DocumentTermMatrix(nm)
nmtd <- as.matrix(nmtd)
```

The call to **DirSource()** expects to see a number of documents in the specified directory, where I had my quizzes. There are various other methods in **tm** to input data as well.

The function **tolower()** changes all letters to lower-case, and **removePunctuation() and removeNumbers()** are self-explanatory. As noted earlier, we also wish to remove stop words such as "a" and "the."

Finally, **DocumentTermMatrix()** creates the matrix, though in **tm()**'s form, hence the call to **as.matrix()**.

12.5   Exercises: Data, Code and Math Problems

Data problems:

1. The dataset used in the example in Section 12.2.3 is available in the **regtools** package. Try alternative analyses using the LASSO, say from the package **glmnet** and NMF.

2. Analyze the airline dataset as to the possible difference among the various destination airports, regarding arrival delay.

3. Download the New York City taxi data (or possibly just get one of the files), *http://www.andresmh.com/nyctaxitrips/*. Predict trip time from other variables of your choice, but instead of using an ordinary linear model, fit median regression, using the **quantreg** package (Section 6.9.1). Compare run times of Software Alchemy vs. a direct fit to the full data.

Mini-CRAN and other computational problems:

4. In the example in Section 12.2.3, we performed various text operations such as removal of stop words, but could have gone further. One way to do this would be to restrict the analysis only to the most frequently appearing words, rather than say all 4670 in the example.

Write an R function with call form

```
cullwords(td, howmany=0.80)
```

aimed in this direction. Here **td** is the term-document matrix, which was **nmtd** in our example. The argument **howmany** specifies how many words we want to choose. If it is a value less than 1.0, the meaning is that we wish to extract the top 80% of the words in terms of frequency. If instead it is an integer, say 500, it means extract the 500 most-frequent words. Have the function return the column numbers in **td** of the extracted words.

5. Write an R function, **svplot()**, modeled after **pwplot()** in the **regtools** package (Section 6.13), with call form

```
svplot(svmout, pairs=combn(length(svmout$terms-1), 2))
```

that will plot the support vectors, two predictors at a time.

Appendix A

Matrix Algebra

This book assumes the reader has background in matrix algebra, or is willing to learn the material as part of reading this book. This appendix is intended as a review for the former group, or a quick treatment for the latter group. Even the latter group may find some of the topics new.

It is recommended that both groups of readers skim through this appendix for now, and then consult it for details later as the need arises.

A.1 Terminology and Notation

A **matrix** is a rectangular array of numbers. A **vector** is a matrix with only one row (a **row vector** or only one column (a **column vector**).

The expression, "the (i,j) element of a matrix A," will mean its element in row i, column j, denoted A_{ij}.

If A is a **square** matrix, i.e., one with equal numbers n of rows and columns, then its **diagonal** elements are a_{ii}, i = 1,...,n.

A square matrix is called **upper-triangular** if $a_{ij} = 0$ whenever $i > j$, with a corresponding definition for **lower-triangular** matrices.

The **norm** (or **length**) of an n-element vector X is

$$||X||_2 = \sqrt{\sum_{i=1}^{n} x_i^2} \tag{A.1}$$

451

The norm is thus nonnegative.

This is actually one of many possible norms, called the *Euclidean norm*. Another commonly used norm is the l_1 norm,

$$||X||_1 = \sum_{i=1}^{n} |x|_i \qquad \text{(A.2)}$$

A.2   Matrix Addition and Multiplication

- For two matrices having the same numbers of rows and same numbers of columns, addition is defined elementwise, e.g.,

$$\begin{pmatrix} 1 & 5 \\ 0 & 3 \\ 4 & 8 \end{pmatrix} + \begin{pmatrix} 6 & 2 \\ 0 & 1 \\ 4 & 0 \end{pmatrix} = \begin{pmatrix} 7 & 7 \\ 0 & 4 \\ 8 & 8 \end{pmatrix} \qquad \text{(A.3)}$$

- Multiplication of a matrix by a **scalar**, i.e., a number, is also defined elementwise, e.g.,

$$0.4 \begin{pmatrix} 7 & 7 \\ 0 & 4 \\ 8 & 8 \end{pmatrix} = \begin{pmatrix} 2.8 & 2.8 \\ 0 & 1.6 \\ 3.2 & 3.2 \end{pmatrix} \qquad \text{(A.4)}$$

- The **inner product** or **dot product** of equal-length vectors X and Y is defined to be

$$\sum_{k=1}^{n} x_k y_k \qquad \text{(A.5)}$$

- The product of matrices A and B is defined if the number of rows of B equals the number of columns of A (A and B are said to be **conformable**). In that case, the (i,j) element of the product C is defined to be

$$c_{ij} = \sum_{k=1}^{n} a_{ik} b_{kj} \qquad \text{(A.6)}$$

For instance,

$$\begin{pmatrix} 7 & 6 \\ 0 & 4 \\ 8 & 8 \end{pmatrix} \begin{pmatrix} 1 & 6 \\ 2 & 4 \end{pmatrix} = \begin{pmatrix} 19 & 66 \\ 8 & 16 \\ 24 & 80 \end{pmatrix} \quad (A.7)$$

It is helpful to visualize c_{ij} as the inner product of row i of A and column j of B, e.g., as shown in bold face here:

$$\begin{pmatrix} \mathbf{7} & \mathbf{6} \\ 0 & 4 \\ 8 & 8 \end{pmatrix} \begin{pmatrix} \mathbf{1} & 6 \\ \mathbf{2} & 4 \end{pmatrix} = \begin{pmatrix} 19 & 66 \\ 8 & 16 \\ 24 & 80 \end{pmatrix} \quad (A.8)$$

- Matrix multiplication is associative and distributive, but in general not commutative:

$$A(BC) = (AB)C \quad (A.9)$$

$$A(B + C) = AB + AC \quad (A.10)$$

$$AB \neq BA \quad (A.11)$$

A.3 Matrix Transpose

- The transpose of a matrix A, denoted A' or A^T, is obtained by exchanging the rows and columns of A, e.g.,

$$\begin{pmatrix} 7 & 70 \\ 8 & 16 \\ 8 & 80 \end{pmatrix}' = \begin{pmatrix} 7 & 8 & 8 \\ 70 & 16 & 80 \end{pmatrix} \quad (A.12)$$

- If $A + B$ is defined, then

$$(A + B)' = A' + B' \quad (A.13)$$

- If A and B are conformable, then

$$(AB)' = B'A' \quad (A.14)$$

- A square matrix A is said to be *symmetric* if $A' = A$.

- Note that for a vector x, written as a one-column matrix,

$$x'x = \| x \|_1 \; = \; \sum_{i=1}^{n} |x|_i \qquad (A.15)$$

A.4 Linear Independence

Vectors $X_1,...,X_k$ are said to be **linearly independent** if it is impossible for

$$a_1 X_1 + ... + a_k X_k = 0 \qquad (A.16)$$

unless all the a_i are 0.

By the way, an expression of the form on the left side of (A.16) is called a *linear combination* of the X_i.

A.5 Matrix Inverse

- The **identity** matrix I of size n has 1s in all of its diagonal elements but 0s in all off-diagonal elements. It has the property that $AI = A$ and $IA = A$ whenever those products are defined.

- If A is a square matrix and $AB = I$, then B is said to be the **inverse** of A, denoted A^{-1}. Then $BA = I$ will hold as well.

- A^{-1} exists if and only if its rows (or columns) are linearly independent.

- A^{-1} exists if and only if $det(A) \neq 0$.

- If A and B are square, conformable and invertible, then AB is also invertible, and

$$(AB)^{-1} = B^{-1}A^{-1} \qquad (A.17)$$

- A matrix U is said to be **orthogonal** if its rows each have norm 1 and are orthogonal to each other, i.e., their inner product is 0. U thus has the property that $UU' = I$, i.e., $U^{-1} = U$.

A special case that is used in this book to create simple examples of various phenomena is the inverse of a 2×2 matrix:

$$\begin{pmatrix} a & b \\ c & d \end{pmatrix}^{-1} = \frac{1}{ad - bc} \begin{pmatrix} d & -b \\ -c & a \end{pmatrix} \tag{A.18}$$

providing $ad - because \neq 0$. The reader should verify that the product of the two matrices is indeed the 2×2 identity matrix.

Typically one does not compute matrix inverses directly. A common alternative is the **QR decomposition**: For a matrix A, matrices Q and R are calculated so that $A = QR$, where Q is an orthogonal matrix and R is upper-triangular.

If A is square and invertible, A^{-1} is easily found:

$$A^{-1} = (QR)^{-1} = R^{-1}Q' \tag{A.19}$$

Again, though, in some cases A is part of a more complex system, and the inverse is not explicitly computed.

A.6 Eigenvalues and Eigenvectors

Let A be a square matrix.[1]

- A scalar λ and a nonzero vector X that satisfy

$$AX = \lambda X \tag{A.20}$$

  are called an **eigenvalue** and **eigenvector** of A, respectively.

- If A is symmetric and real, then it is **diagonalizable**, i.e., there exists an orthogonal matrix U such that

$$U'AU = D \tag{A.21}$$

  for a diagonal matrix D. The elements of D are the eigenvalues of A, and the columns of U are the eigenvectors of A (scaled to have length 1).

---

[1]For nonsquare matrices, the discussion here would generalize to the topic of **singular value decomposition**.

A different sufficient condition for A.21 is that the eigenvalues of A are distinct. In this case, U will not necessarily be orthogonal.

By the way, this latter sufficient condition shows that "most" square matrices are diagonalizable, if we treat their entries as continuous random variables. Under such a circumstance, the probability of having repeated eigenvalues would be 0.

A.7   Rank of a Matrix

Definition: The rank of a matrix A is the maximal number of linearly independent columns in A.

Rank has the following properties:

- $\text{rank}(A') = \text{rank}(A)$

- Thus the rank of A is also the maximal number of linearly independent rows in A.

- Let A be $r \times s$. Then

$$\text{rank}(A) \leq \min(r, s) \qquad \text{(A.22)}$$

- If A is $n \times n$, it is invertible if and only if it has *full rank*, i.e. $rank(A) = n$.

- The rank of a square matrix is equal to the number of nonzero eigenvalues.

A.8   Matrices of the Form B'B

Let B be any rectangular matrix, and write $C = B'B$. Then C has the following properties:

- C is symmetric (follows from (A.14)).

- The rank of C is that of B.

- C is *nonnegative definite*: For any conformable vector x,

$$x'Cx \geq 0 \tag{A.23}$$

(follows from writing $x'Cx = (x'B')(Bx)$ and noting that the latter is the squared norm of Bx, thus nonnegative).

A.9 Partitioned Matrices

In certain senses, we can treat submatrices in a matrix almost like scalars, through a process called *partitioning*. This turns out to be very useful.

For example, let

$$A = \begin{pmatrix} 1 & 5 & 12 \\ 0 & 3 & 6 \\ 4 & 8 & 2 \end{pmatrix} \tag{A.24}$$

and

$$B = \begin{pmatrix} 0 & 2 & 5 \\ 0 & 9 & 10 \\ 1 & 1 & 2 \end{pmatrix}, \tag{A.25}$$

so that

$$C = AB = \begin{pmatrix} 12 & 59 & 79 \\ 6 & 33 & 42 \\ 2 & 82 & 104 \end{pmatrix}. \tag{A.26}$$

We could partition A as

$$A = \begin{pmatrix} A_{00} & A_{01} \\ A_{10} & A_{11} \end{pmatrix}, \tag{A.27}$$

where

$$A_{00} = \begin{pmatrix} 1 & 5 \\ 0 & 3 \end{pmatrix}, \tag{A.28}$$

$$A_{01} = \begin{pmatrix} 12 \\ 6 \end{pmatrix}, \tag{A.29}$$

$$A_{10} = \begin{pmatrix} 4 & 8 \end{pmatrix} \tag{A.30}$$

and

$$A_{11} = \begin{pmatrix} 2 \end{pmatrix}. \tag{A.31}$$

Similarly we would partition B and C into blocks of a compatible size to A,

$$B = \begin{pmatrix} B_{00} & B_{01} \\ B_{10} & B_{11} \end{pmatrix} \tag{A.32}$$

and

$$C = \begin{pmatrix} C_{00} & C_{01} \\ C_{10} & C_{11} \end{pmatrix}, \tag{A.33}$$

so that for example

$$B_{10} = \begin{pmatrix} 1 & 1 \end{pmatrix}. \tag{A.34}$$

The key point is that multiplication still works if we pretend that those submatrices are numbers! For example, pretending like that would give the relation

$$C_{00} = A_{00}B_{00} + A_{01}B_{10}, \tag{A.35}$$

which the reader should verify really is correct as matrices, i.e. the computation on the right side really does yield a matrix equal to C_{00}.

A.10   Matrix Derivatives

There is an entire body of formulas for taking derivatives of matrix-valued expressions. One of particular importance to us is for the vector of derivatives

$$\frac{dg(u)}{du} \tag{A.36}$$

for a vector u of length k. This is the *gradient* of $g(u)$, i.e. the vector

$$\left(\frac{\partial g(u)}{\partial u_1}, ..., \frac{\partial g(u)}{\partial u_k}\right)' \tag{A.37}$$

A bit of calculus shows that the gradient can be represented compactly. in some cases, such as

$$\frac{d}{du}(Mu + w) = M' \tag{A.38}$$

for a matrix M and vector w that do not depend on u. The reader should verify this by looking at the individual $\frac{\partial g(u)}{\partial u_i}$.

Another example is the *quadratic form*

$$\frac{d}{du}u'Qu = 2Qu \tag{A.39}$$

for a symmetric matrix Q and a vector u.

And there is a Chain Rule. For example if $u = Mv + w$, then

$$\frac{\partial}{\partial v}u'u = 2M'u \tag{A.40}$$

A.11 Matrix Algebra in R

The R programming language has extensive facilities for matrix algebra, introduced here. Note by the way that R uses column-major order.

A linear algebra vector can be formed as an R vector, or as a one-row or one-column matrix.

```
> # constructing matrices
> a <- rbind(1:3,10:12)
> a
```

```
        [,1]  [,2]  [,3]
[1,]      1     2     3
[2,]     10    11    12
> b <- matrix(1:9, ncol=3)
> b
        [,1]  [,2]  [,3]
[1,]      1     4     7
[2,]      2     5     8
[3,]      3     6     9
# multiplication, etc.
> c <- a %*% b;  c + matrix(c(1,-1,0,0,3,8),nrow=2)
        [,1]  [,2]  [,3]
[1,]     15    32    53
[2,]     67   167   274
> c %*% c(1,5,6)   # note 2 different c's
        [,1]
[1,]    474
> # transpose, inverse
> t(a)  # transpose
        [,1]  [,2]
[1,]      1    10
[2,]      2    11
[3,]      3    12
> u <- matrix(runif(9),nrow=3)
> u
            [,1]         [,2]        [,3]
[1,]  0.08446154  0.86335270  0.6962092
[2,]  0.31174324  0.35352138  0.7310355
[3,]  0.56182226  0.02375487  0.2950227
> uinv <- solve(u)
> uinv
            [,1]        [,2]        [,3]
[1,]   0.5818482  -1.594123   2.576995
[2,]   2.1333965  -2.451237   1.039415
[3,]  -1.2798127   3.233115  -1.601586
> u %*% uinv   # note roundoff error
               [,1]            [,2]             [,3]
[1,]  1.000000e+00  -1.680513e-16  -2.283330e-16
[2,]  6.651580e-17   1.000000e+00   4.412703e-17
[3,]  2.287667e-17  -3.539920e-17   1.000000e+00
> # eigenvalues and eigenvectors
> eigen(u)
$values
```

[1]   1.2456220+0.0000000 i   −0.2563082+0.2329172 i
−0.2563082−0.2329172 i

$vectors
                              [,1]   [,2]
[,3]
[1 ,] −0.6901599+0 i −0.6537478+0.0000000 i
−0.6537478+0.0000000 i
[2 ,] −0.5874584+0 i −0.1989163−0.3827132 i
−0.1989163+0.3827132 i
[3 ,] −0.4225778+0 i   0.5666579+0.2558820 i
0.5666579−0.2558820 i

```
> # diagonal matrices (off−diagonals 0)
> diag(3)
     [ ,1] [ ,2] [ ,3]
[1 ,]   1    0    0
[2 ,]   0    1    0
[3 ,]   0    0    1
> diag((c(5,12,13)))
     [ ,1] [ ,2] [ ,3]
[1 ,]   5    0    0
[2 ,]   0   12    0
[3 ,]   0    0   13
```

We can obtain matrix inverse using **solve()**, e.g.,

```
> m <− rbind(1:2,3:4)
> m
     [ ,1] [ ,2]
[1 ,]   1    2
[2 ,]   3    4
> minv <− solve(m)
> minv
      [ ,1] [ ,2]
[1 ,] −2.0   1.0
[2 ,]  1.5  −0.5
> m %*% minv  # should get I back
     [ ,1]         [ ,2]
[1 ,]   1  1.110223 e−16
[2 ,]   0  1.000000 e+00
```

Note the roundoff error, even with this small matrix. We can try the QR method, provided to us in R via **qr()**. In fact, if we just want the inverse,

qr.solve() will compute (A.19) for us.

We can in principle obtain rank from, for example, the **rank** component from the output of **qr()**. Note however that although rank is clearly defined in theory, the presence of roundoff error in computation may make rank difficult to determine reliably.

A.12   Further Reading

A number of excellent, statistics-oriented books are available on these topics. Among them are [62] [56] [7] on the mathematics, and [134] for usage in R.

Bibliography

[1] AGGARWAL, C. *Recommender Systems: The Textbook.* Springer International Publishing, 2016.

[2] AGRESTI, A. *Categorical Data Analysis.* Wiley Series in Probability and Statistics. Wiley, 2014.

[3] AGRESTI, A. *Foundations of Linear and Generalized Linear Models.* Wiley Series in Probability and Statistics. Wiley, 2015.

[4] ALFONS, A. cvtools: Cross-validation tools for regression models. https://cran.r-project.org/web/packages/cvTools/index.html.

[5] AMBLER, G., ET AL. mfp: Multivariable fractional polynomials. https://cran.r-project.org/web/packages/mpf/index.html.

[6] ASA. Data expo '09. http://stat-computing.org/dataexpo/2009/the-data.html, 2009.

[7] BANERJEE, S., AND ROY, A. *Linear Algebra and Matrix Analysis for Statistics.* Chapman & Hall/CRC Texts in Statistical Science. Taylor & Francis, 2014.

[8] BARTLETT, P. L., AND TRASKIN, M. Adaboost is consistent. *Journal of Machine Learning Research 8* (2007), 2347–2368.

[9] BATES, D., ET AL. lme4: Linear mixed-effects models using 'eigen' and s4. https://cran.r-project.org/web/packages/lme4/index.html.

[10] BATES, D., AND MAECHLER, M. Matrix: Sparse and dense matrix classes and methods. https://cran.r-project.org/web/packages/Matrix/index.html.

[11] BATY, F., ET AL. nlstools: Tools for nonlinear regression analysis. https://cran.r-project.org/web/packages/nlstools/index.html.

[12] BECKER, R. The variance drain and Jensen's inequality. AEPR Working Paper 2012-004, Indiana University, 2010.

[13] BELLMAN, R. *Computational Problems in the Theory of Dynamic Programming*. Rand paper series. Rand Corporation, 1953.

[14] BENDEL, R. B., AND AFIFI, A. A. Comparison of stopping rules in forward stepwise regression. *Journal of the American Statistical Association 72*, 357 (1977), 46–53.

[15] BENNETT, K. P., AND CAMPBELL, C. Support vector machines: hype or hallelujah? *ACM SIGKDD Explorations Newsletter 2*, 2 (2000), 1–13.

[16] BERK, R. *Statistical Learning from a Regression Perspective*. Springer Texts in Statistics. Springer International Publishing, 2016.

[17] BERK, R., BROWN, L., BUJA, A., ZHANG, K., AND ZHAO, L. Valid post-selection inference. *Ann. Statist. 41*, 2 (04 2013), 802–837.

[18] BEYER, K. S., GOLDSTEIN, J., RAMAKRISHNAN, R., AND SHAFT, U. When is "nearest neighbor" meaningful? In *Proceedings of the 7th International Conference on Database Theory* (London, UK, 1999), ICDT '99, Springer-Verlag, pp. 217–235.

[19] BHATTACHARYA, R., AND HOLMES, S. An exposition of Götze's estimation of the rate of convergence in the multivariate central limit theorem. arXiv 1003.4254, 2010.

[20] BICKEL, P. J., HAMMEL, E. A., AND O'CONNELL, J. W. Sex Bias in Graduate Admissions: Data from Berkeley. *Science 187*, 4175 (1975), 398–404.

[21] BRAUN, W., AND MURDOCH, D. *A First Course in Statistical Programming with R*. Cambridge University Press, 2016.

[22] BREIMAN, L. Random forests. *Machine Learning 45*, 1 (2001), 5–32.

[23] BREIMAN, L. Statistical modeling: The two cultures (with comments and a rejoinder by the author). *Statist. Sci. 16*, 3 (08 2001), 199–231.

[24] BREIMAN, L., FRIEDMAN, J., STONE, C., AND OLSHEN, R. *Classification and Regression Trees*. The Wadsworth and Brooks-Cole statistics-probability series. Taylor & Francis, 1984.

[25] BRETZ, F., HOTHORN, T., AND WESTFALL, P. *Multiple Comparisons Using R*. CRC Press, 2016.

[26] BROWN, C. dummies: Create dummy/indicator variables flexibly and efficiently. https://cran.r-project.org/web/packages/dummies/index.html.

[27] BUHLMANN, P. Statistical significance in high-dimensional linear models. *Bernoulli 19*, 4 (09 2013), 1212–1242.

[28] BÜHLMANN, P., AND VAN DE GEER, S. *Statistics for High-Dimensional Data: Methods, Theory and Applications*. Springer Series in Statistics. Springer Berlin Heidelberg, 2011.

[29] BURGETTE, L., ET AL. twang: Toolkit for weighting and analysis of nonequivalent groups. https://cran.r-project.org/web/packages/twang/index.html.

[30] CANDES, E., FAN, Y., JANSON, L., AND LV, J. Panning for gold: Model-free knockoffs for high-dimensional controlled variable selection. arXiv 1610.02351, 2016.

[31] CARD, D. Estimating the Return to Schooling: Progress on Some Persistent Econometric Problems. *Econometrica 69*, 5 (2001), 1127–1160.

[32] CARROLL, R., AND RUPPERT, D. *Transformation and Weighting in Regression*. Chapman & Hall/CRC Monographs on Statistics & Applied Probability. Taylor & Francis, 1988.

[33] CHANG, C.-C., AND LIN, C.-J. Libsvm: A library for support vector machines. *ACM Trans. Intell. Syst. Technol. 2*, 3 (May 2011), 27:1–27:27.

[34] CHEN, S. X., QIN, J., AND TANG, C. Y. Mann-Whitney test with adjustments to pretreatment variables for missing values and observational study. *Journal of the Royal Statistical Society: Series B (Statistical Methodology) 75*, 1 (2013), 81–102.

[35] CHRISTENSEN, R. *Log-linear models and logistic regression*, 2nd ed. Springer texts in statistics. Springer, New York, c1997. Earlier ed. published under title: Log-linear models. 1990.

[36] CLARKE, B., FOKOUE, E., AND ZHANG, H. *Principles and Theory for Data Mining and Machine Learning*. Springer Series in Statistics. Springer New York, 2009.

[37] CLEVELAND, W. S., AND LOADER, C. *Smoothing by Local Regression: Principles and Methods.* Physica-Verlag HD, Heidelberg, 1996, pp. 10–49.

[38] CRISTIANINI, N., AND SHAWE-TAYLOR, J. *An Introduction to Support Vector Machines: And Other Kernel-based Learning Methods.* Cambridge University Press, New York, NY, USA, 2000.

[39] DAVISON, A., AND HINKLEY, D. *Bootstrap Methods and Their Application.* Cambridge Series in Statistical and Probabilistic Mathematics. Cambridge University Press, 1997.

[40] DENG, H., AND WICKHAM, H. Densty estimation in R. http://vita.had.co.nz/papers/density-estimation.pdf, 2011.

[41] DOBSON, A., AND BARNETT, A. *An Introduction to Generalized Linear Models, Third Edition.* Chapman & Hall/CRC Texts in Statistical Science. CRC Press, 2008.

[42] EFRON, B., HASTIE, T., JOHNSTONE, I., AND TIBSHIRANI, R. Least angle regression. *Ann. Statist. 32*, 2 (04 2004), 407–499.

[43] EICKER, F. Limit theorems for regressions with unequal and dependent errors. In *Proceedings of the Fifth Berkeley Symposium on Mathematical Statistics and Probability, Volume 1: Statistics.* University of California Press, Berkeley, Calif., 1967, pp. 59–82.

[44] ELDÉN, L. *Matrix Methods in Data Mining and Pattern Recognition.* Fundamentals of Algorithms. Society for Industrial and Applied MathematicsS, 2007.

[45] ELZHOV, T., ET AL. minpack.lm: R interface to the Levenberg-Marquardt nonlinear least-squares algorithm found in minpack, plus support for bounds. https://cran.r-project.org/web/packages/minpack.lm/index.html.

[46] EMERSON, J., AND KANE, M. biganalytics: Utilities for 'big.matrix' objects from package 'bigmemory'. https://cran.r-project.org/web/packages/biganalytics/index.html.

[47] FARAWAY, J. *Extending the Linear Model with R: Generalized Linear, Mixed Effects and Nonparametric Regression Models.* Chapman & Hall/CRC Texts in Statistical Science. CRC Press, 2016.

[48] FONG, W. M., AND OULIARIS, S. Spectral tests of the martingale hypothesis for exchange rates. *Journal of Applied Econometrics 10*, 3 (1995), 255–71.

[49] FOX, J., ET AL. car: Companion to applied regression. `https://cran.r-project.org/web/packages/car/index.html`.

[50] FOX, J., AND WEISBERG, S. *An R Companion to Applied Regression.* SAGE Publications, 2011.

[51] FRIEDMAN, J. Another approach to polychotomous classification. Tech. rep., Stanford University, 1996.

[52] GALECKI, A., AND BURZYKOWSKI, T. *Linear Mixed-Effects Models Using R: A Step-by-Step Approach.* Springer Texts in Statistics. Springer New York, 2013.

[53] GALILLI, T. Is it harder to advertise to the more educated? `https://www.r-statistics.com/tag/data-set/`, 2010.

[54] GAUJOUX, R., AND SEOIGHE, C. Nmf: Algorithms and framework for nonnegative matrix factorization (nmf). `https://cran.r-project.org/web/packages/NMF/index.html`.

[55] GELMAN, A. Why we hate stepwise regression. `https://www.r-statistics.com/tag/data-set/`, 2010.

[56] GENTLE, J. *Matrix Algebra: Theory, Computations, and Applications in Statistics.* Springer Texts in Statistics. Springer New York, 2010.

[57] GILBERT, P., AND VARADHAN, R. numderiv: Accurate numerical derivatives. `https://cran.r-project.org/web/packages/numDeriv/index.html`.

[58] GILES, D. The adjusted r-squared, again. http://davegiles.blogspot.com/2013/07/the-adjusted-r-squared-again.html, 2013.

[59] GILES, J. A., AND GILES, D. Pre-test estimation and testing in econometrics: Recent developments. *Journal of Economic Surveys 7*, 2 (1993), 145–97.

[60] GOLDENBERG, A., ZHENG, A., FIENBERG, S., AND AIROLDI, E. *A Survey of Statistical Network Models.* Foundations and Trends(r) in Machine Learning. Now Publishers, 2010.

[61] HARRELL, F. *Regression Modeling Strategies: With Applications to Linear Models, Logistic and Ordinal Regression, and Survival Analysis.* Springer Series in Statistics. Springer International Publishing, 2015.

[62] HARVILLE, D. *Matrix Algebra From a Statistician's Perspective.* Springer New York, 2008.

[63] HASTIE, T., AND EFRON, B. lars: Least angle regression, lasso and forward stagewise? https://cran.r-project.org/web/packages/lars/index.html.

[64] HASTIE, T., TIBSHIRANI, R., AND FRIEDMAN, J. *The Elements of Statistical Learning.* Springer Series in Statistics. Springer New York Inc., New York, NY, USA, 2001.

[65] HASTIE, T., TIBSHIRANI, R., AND WAINWRIGHT, M. *Statistical Learning with Sparsity: The Lasso and Generalizations.* Chapman & Hall/CRC Monographs on Statistics & Applied Probability. CRC Press, 2015.

[66] HASTIE, T., TIBSHIRANI, R., AND WAINWRIGHT, M. *Statistical Learning with Sparsity: The Lasso and Generalizations.* Chapman & Hall. Taylor & Francis, 2015.

[67] HO, T. K. The random subspace method for constructing decision forests. *IEEE Transactions on Pattern Analysis and Machine Intelligence 20*, 8 (1998), 832–844.

[68] HOERL, A. E., AND KENNARD, R. W. Ridge regression: Biased estimation for nonorthogonal problems. *Technometrics 42*, 1 (Feb. 2000), 80–86.

[69] HORNIK, K., STINCHCOMBE, M., AND WHITE, H. Multilayer feedforward networks are universal approximators. *Neural Netw. 2*, 5 (July 1989), 359–366.

[70] HOTHORN, T., ET AL. mvtnorm: Multivariate normal and t distributions. https://cran.r-project.org/web/packages/mvtnorm/index.html.

[71] HOTHORN, T., HORNIK, K., AND ZEILEIS, A. Unbiased recursive partitioning: A conditional inference framework. *Journal of Computational and Graphical Statistics 15*, 3 (2006), 651–674.

[72] HOTHORN, T., AND ZEILEIS, A. partykit: A toolkit for recursive partytioning. https://cran.r-project.org/web/packages/partykit/index.html.

[73] HSU, J. *Multiple Comparisons: Theory and Methods.* Taylor & Francis, 1996.

[74] INSELBERG, A. *Parallel Coordinates: Visual Multidimensional Geometry and Its Applications.* Advanced series in agricultural sciences. Springer New York, 2009.

[75] JACKSON, M. *Social and Economic Networks.* Princeton University Press. Princeton University Press, 2010.

[76] JAMES, G., WITTEN, D., HASTIE, T., AND TIBSHIRANI, R. *An Introduction to Statistical Learning: with Applications in R.* Springer Texts in Statistics. Springer New York, 2013.

[77] JIANG, J. *Linear and Generalized Linear Mixed Models and Their Applications.* Springer Series in Statistics. Springer, Dordrecht, 2007.

[78] JORDAN, M. I. Leo Breiman. *Ann. Appl. Stat. 4*, 4 (12 2010), 1642–1643.

[79] KANG, H., ET AL. ivmodel: Statistical inference and sensitivity analysis for instrumental variables model. `https://cran.r-project.org/web/packages/ivmodel/index.html`.

[80] KAYE, D., AND FREEDMAN, D. Reference guide on statistics. `https://www.nap.edu/read/13163/chapter/7`.

[81] KOENKER, R., ET AL. quantreg: Quantile regression. `https://cran.r-project.org/web/packages/quantreg/index.html`.

[82] KRAEMER, N., AND SCHAEFER, J. parcor: Regularized estimation of partial correlation matrices. `https://cran.r-project.org/web/packages/parcor/index.html`.

[83] KUHN, M., AND JOHNSON, K. *Applied Predictive Modeling.* Springer New York, 2013.

[84] LANGE, K. *Numerical Analysis for Statisticians.* Springer, 2001.

[85] LE, Q., RANZATO, M., MONGA, R., DEVIN, M., CHEN, K., CORRADO, G., DEAN, J., AND NG, A. Building high-level features using large scale unsupervised learning. In *International Conference in Machine Learning* (2012).

[86] LEE, D. D., AND SEUNG, H. S. Learning the parts of objects by nonnegative matrix factorization. *Nature 401* (1999), 788–791.

[87] LEISCH, F., AND DIMITRIADOU, E. mlbench: Machine learning benchmark problems. `https://cran.r-project.org/web/packages/mlbench/index.html`.

[88] LI, S., ET AL. Fnn: Fast nearest neighbor search algorithms and applications. https://cran.r-project.org/web/packages/FNN/index.html.

[89] LIAW, A. randomForest: Breiman and Cutler's random forests for classification and regression. https://cran.r-project.org/web/packages/randomForest/index.html.

[90] LIN, C.-J. On the convergence of multiplicative update algorithms for nonnegative matrix factorization. Trans. Neur. Netw. 18, 6 (Nov. 2007), 1589–1596.

[91] LITTLE, R., AND RUBIN, D. Statistical Analysis with Missing Data. Wiley Series in Probability and Statistics. Wiley, 2014.

[92] LOH, W.-Y. Classification and regression trees. Data Mining and Knowledge Discovery 1 (2011), 14–23.

[93] LUKE, D. A Users Guide to Network Analysis in R. Use R! Springer International Publishing, 2015.

[94] LUMLEY, T. biglm: bounded memory linear and generalized linear models. https://cran.r-project.org/web/packages/sandwich/index.html.

[95] LUMLEY, T., AND ZEILEIS, A. sandwich: Robust covariance matrix estimators. https://cran.r-project.org/web/packages/sandwich/index.html.

[96] MATLOFF, N. partools: Tools for the 'parallel' package. https://cran.r-project.org/web/packages/partools/index.html.

[97] MATLOFF, N. regtools: Regression tools. https://cran.r-project.org/web/packages/regtools/index.html.

[98] MATLOFF, N. Estimation of internet file-access/modification rates from indirect data. ACM Trans. Model. Comput. Simul. 15, 3 (July 2005), 233–253.

[99] MATLOFF, N. From Algorithms to Z Scores: Probabilistic and Statistical Modeling in Computer Science. http://heather.cs.ucdavis.edu/probstatbook, 2010.

[100] MATLOFF, N. The Art of R Programming: A Tour of Statistical Software Design. No Starch Press Series. No Starch Press, 2011.

[101] MATLOFF, N. *Parallel Computing for Data Science: With Examples in R, C++ and CUDA*. Chapman & Hall/CRC The R Series. CRC Press, 2015.

[102] MATLOFF, N. Big-n and big-p in big data. In *Handbook of Big Data*, P. Bühlmann, P. Drineas, M. Kane, and M. J. van der Laan, Eds. CRC, 2016.

[103] MATLOFF, N. Software alchemy: Turning complex statistical computations into embrassingly-parallel ones. *Journal of Statistical Software 71*, 1 (2016).

[104] MATLOFF, N., AND XIE, Y. freqparcoord: Novel methods for parallel coordinates. https://cran.r-project.org/web/packages/freqparcoord/index.html.

[105] MATLOFF, N. S. Ergodicity conditions for a dissonant voting model. *Ann. Probab. 5*, 3 (06 1977), 371–386.

[106] MATLOFF, N. S. Use of regression functions for improved estimation of means. *Biometrika 68*, 3 (Dec. 1981), 685–689.

[107] MATLOFF, N. S. Use of covariates in randomized response settings. *Statistics & Probability Letters 2*, 1 (1984), 31–34.

[108] MCCANN, L. *Written Testimony of Laurie A. McCann, AARP Foundation Litigation, EEOC*. 2016.

[109] MILBORROW, S. rpart.plot: Plot 'rpart' models: An enhanced version of 'plot.rpart'. https://cran.r-project.org/web/packages/rpart.plot/index.html.

[110] MILLER, A. *Subset Selection in Regression*. Chapman & Hall/CRC Monographs on Statistics & Applied Probability. CRC Press, 2002.

[111] MILLER, A. J. Selection of subsets of regression variables. *Journal of the Royal Statistical Society. Series A (General)* (1984), 389–425.

[112] MIZON, G. Inferential procedures in nonlinear models: An application in a UK industrial cross section study of factor substitution and returns to scale. *Econometrica 45*, 5 (1977), 1221–42.

[113] MÜLLER, U. U. Estimating linear functionals in nonlinear regression with responses missing at random. *Ann. Statist. 37*, 5A (10 2009), 2245–2277.

[114] MURPHY, K. *Machine Learning: A Probabilistic Perspective*. Adaptive computation and machine learning series. MIT Press, 2012.

[115] PORTNOY, S. Asymptotic behavior of m-estimators of p regression parameters when p^2/n is large. *Ann. Statist. 12*, 4 (12 1984), 1298–1309.

[116] RAO, C., AND TOUTENBURG, H. *Linear Models: Least Squares and Alternatives.* Springer, 1999.

[117] RAO, J., AND MOLINA, I. *Small Area Estimation.* Wiley Series in Survey Methodology. Wiley, 2015.

[118] RAO, R. B., AND FUNG, G. On the dangers of cross-validation. An experimental evaluation. SIAM, pp. 588–596.

[119] RIFKIN, R., AND KLAUTAU, A. In defense of one-vs-all classification. *J. Mach. Learn. Res. 5* (Dec. 2004), 101–141.

[120] RIPLEY, B. *Pattern Recognition and Neural Networks.* Cambridge University Press, 2007.

[121] RIPLEY, B., AND VENABLES, W. nnet: Feed-forward neural networks and multinomial log-linear models. https://cran.r-project.org/web/packages/nnet/index.html.

[122] ROSE, R. *Nonparametric Estimation of Weights in Least-Squares Regression Analysis.* PhD thesis, University of California, Davis, 1978.

[123] RUDIN, W. *Principles of Mathematical Analysis.* McGraw-Hill, 1964.

[124] RUFIBACH, K., AND DUEMBGEN, L. logcondens: Estimate a log-concave probability density from iid observations. https://cran.r-project.org/web/packages/logcondens/index.html.

[125] SARKAR, D. lattice: Trellis graphics for r. https://cran.r-project.org/web/packages/lattice/index.html.

[126] SCHOERKER, B., ET AL. Ggally: Extension to 'ggplot2'. https://cran.r-project.org/web/packages/GGally/index.html.

[127] SCOTT, D. *Multivariate Density Estimation: Theory, Practice, and Visualization.* Wiley Series in Probability and Statistics. Wiley, 2015.

[128] SHAO, J. Linear model selection by cross-validation. *J. of the American Statistical Association 88* (1993), 486–494.

[129] SHAWE-TAYLOR, J., AND CRISTIANINI, N. *Kernel Methods for Pattern Analysis.* Cambridge University Press, 2004.

[130] STONE, C. Private communication, 2016.

[131] THERNEAU, T., ET AL. Introduction to recursive partitioning using the rpart routines. `https://cran.r-project.org/web/packages/rpart/vignettes/longintro.pdf`.

[132] THERNEAU, T., ET AL. rpart: Recursive partitioning and regression trees. `https://cran.r-project.org/web/packages/rpart/index.html`.

[133] VALENTINI, G., DIETTERICH, T. G., AND CRISTIANINI, N. Bias-variance analysis of support vector machines for the development of svm-based ensemble methods. *Journal of Machine Learning Research 5* (2004), 725–775.

[134] VINOD, H. *Hands-on Matrix Algebra Using R: Active and Motivated Learning with Applications.* World Scientific, 2011.

[135] WASSERMAN, L. *All of Nonparametric Statistics.* Springer Texts in Statistics. Springer New York, 2006.

[136] WASSERMAN, L. *All of Statistics: A Concise Course in Statistical Inference.* Springer Texts in Statistics. Springer New York, 2013.

[137] ZEILEIS, A. Object-oriented computation of sandwich estimators. *Journal of Statistical Software 16*, 1 (2006), 1–16.

[138] ZOU, H., HASTIE, T., AND TIBSHIRANI, R. Sparse principal component analysis. *Journal of Computational and Graphical Statistics 15* (2004), 2006.

Index

N

W